Computation of Conduction and Duct Flow Heat Transfer

Computation of Conduction and Duct Flow Heat Transfer

Suhas V. Patankar
Professor of Mechanical Engineering
University of Minnesota

CRC Press
Taylor & Francis Group
Boca Raton   London   New York

CRC Press is an imprint of the
Taylor & Francis Group, an informa business

First published 1991 by Innovative Research, Inc.

Published 2009 by CRC Press
Taylor & Francis Group
6000 Broken Sound Parkway NW, Suite 300
Boca Raton, FL 33487-2742

© 1991 by Taylor & Francis Group, LLC
CRC Press is an imprint of Taylor & Francis Group, an Informa business

First issued in paperback 2019

No claim to original U.S. Government works

ISBN 13: 978-0-367-45059-5 (pbk)
ISBN 13: 978-1-56032-511-6 (hbk)

Visit the Taylor & Francis Web site at
http://www.taylorandfrancis.com

and the CRC Press Web site at
http://www.crcpress.com

The Cover
Front: Temperature distribution in a solid with heat sources
    (The arrows show heat-flux vectors.)
Back: Axial velocity contours for a nonNewtonian flow in a semicircular duct

These color displays were generated using microGRAPHICS,

To the memory of my late father
Vasant D. Patankar
(1912–1984)
and to my mother
Sindhu V. Patankar

Preface

This book has a rather long history. After teaching a graduate-level course on computational heat transfer and fluid flow at the University of Minnesota during 1975–1979, I wrote a textbook for that course (Numerical Heat Transfer and Fluid Flow, Hemisphere, 1980). Subsequently, I also started teaching a two-day short course on the subject at various national meetings of the ASME. Whereas the textbook and the short courses were very well received, there seemed a need for a longer course that went into greater detail and provided a first-hand experience of working with a computer program. In 1981, such a course was designed; it was a five-day course for practicing engineers, originally taught by me along with Dr. Rabi Baliga. By now, I have offered the course over 12 times in a few different versions.

For the five-day course, the computer program to be used had to satisfy two types of requirements. On one hand, it had to be interesting to use, applicable to practical problems, and capable of illustrating many important features of our calculation procedure. On the other hand, the program had to be simple enough so that the course participants could, during the five-day period, understand its structure, appreciate its capabilities and limitations, learn the variable names, and apply it to a number of physical situations. (I did not want them to treat the program as a black box and push a few buttons.) It was, therefore, decided to limit the program to conduction-type problems but illustrate its use for both heat conduction and duct flow. Thus, the computer program CONDUCT was born.

While concentrating on the teaching of the numerical techniques to graduate students and practicing engineers, I also began to look for ways of bringing computational methods into the undergraduate curriculum. Once again, the considerations used in designing the five-day course seemed to apply to an undergraduate course as well. Therefore, a course based on the computer program CONDUCT was an obvious choice. In the summer of 1987, I extracted the material related to the computer program from the lecture notes of the five-day course and expanded it in the form of a preliminary version of this textbook. Since then, the book has been included in the materials of the five-day course and also used as a textbook for a new course on Computational Heat Transfer started in Fall 1987 at the University of Minnesota. The course is taken by senior-level undergraduate students and also by graduate students from various engineering departments.

In teaching this course (and other similar courses at the graduate level), I use a certain format. The course emphasizes physical understanding and the ability to solve physical problems. I begin with the physical and mathematical description of the phenomena of interest. develop the numerical method. and describe the general-purpose computer program. The rest of the course is devoted to the presentation of numerous examples of the adaptation of the program to particular problems. Whereas we begin with very simple examples. we eventually deal with quite sophisticated applications. While this goes on. the students apply the computer program. as a part of their home assignments. to a large number of additional problems. When they are exposed to the solution activity for so many problems. they not only develop the problem-solving skills but also enhance their understanding of the physical processes.

As a climax of the course activity. I require the students to complete two projects. Each project consists of the application of the computer program to a substantial problem that is independently chosen by the student. The project report includes the problem description. its computer implementation. results. and comments on the results. For me. the end of the course brings a special joy when I see the interesting and imaginative work done by the students. Over the past four years. the students have applied the CONDUCT program to a variety of problems including conduction in engine cylinders. heat transfer around buried cables. insulation in building walls. cooling of electronic circuits. flow over array of rods or tubes. moisture migration in stored grain. electric field in EKG applications. and cooling of an anode for an electric arc. In many cases. the work in this course has been the beginning of the thesis work or a journal article. In choosing the examples and problems for this book. I have made liberal use of the ideas that I have picked up from the students.

The computer program CONDUCT has two parts: a large invariant part that is common to all applications and a problem-dependent part that consists of a subroutine to be written by the user for each application. Thus. instead of simply providing the input data. the user writes the specification of the problem in a special subroutine. This makes the program extremely flexible and widely applicable: more importantly. the students get a sense of participation in the whole process of solving a physical problem.

The publication of this book gives me an opportunity to share with the readers my enthusiasm about a number of interesting ideas. The main theme in this book is that the computational activity is to be used for enhancing physical understanding as well as for obtaining quantitative results. The two-part structure of the program is a novel concept. Such a program provides the user with the main machinery of the calculation scheme. while allowing considerable flexibility and participation. This is what makes a program like CONDUCT an effective teaching/learning tool. I am also very fond of a few small innovations that I have used in programming. These include the use of ENTRY statements to create assemblies of many subroutines. the subroutine VALUES designed for the convenience of assigning values to many quantities. and the practice of making two-dimensional arrays equivalent to parts of a three-dimensional array.

The primary purpose of this book is to introduce the science and the art of numerical computation of physical phenomena. In addition, the computer program CONDUCT can be used by educators for another purpose. They can prepare particular adaptations of the program for computer simulation of practical situations. For example, one can set up the program for the flow in a tube with internal fins. Then the students can run the program to learn about the effect of the number of fins, fin height, fin thickness, etc. Here, the emphasis is not on the numerical solution method; learning about the physical behavior is the focus. When coupled with suitable computer graphics, such an activity can be both enlightening and entertaining.

Whereas an early version of this book has been extensively class-tested, I have found it difficult to resist the temptation to introduce new material into the book before it went to press. In the final rush of the preparation of the manuscript, although I have, along with those who helped me, been extremely careful, it is possible that a few errors may have slipped past us. I shall be very grateful to the readers if they would kindly bring to my attention any mistakes that they discover.

In all my professional activities, I have been fortunate to receive generous help and support from my teachers, colleagues, and students. I express my sincere thanks to all of them. In the construction of CONDUCT and this textbook, my thanks are due to many individuals who have provided direct assistance. Rabi Baliga worked as a co-instructor with me in the five-day course in which CONDUCT was introduced. At that time, Chander Prakash, Sumanta Acharya, and Chia-Fu Hsu helped in the construction of an early version of the program. In typing a preliminary form of this book in the summer of 1987, Rod Schmidt and Kanchan Kelkar spent endless hours. The final period of 3–4 weeks before the book went to press was extremely busy. I have no words to express my gratitude to Kailash Karki, Prabhu Sathyamurthy, and John Chai for their untiring and dedicated work in this period on all aspects of typing, word-processing, formatting, proof-reading, and figure drawing, and for their numerous useful suggestions.

Finally, I would like to thank my family. During the hectic days of working on this book, my wife along with my daughters provided a delightful atmosphere of support and understanding.

Suhas V. Patankar
Minneapolis, Minnesota
January 1991

Copyright Notice

Contents

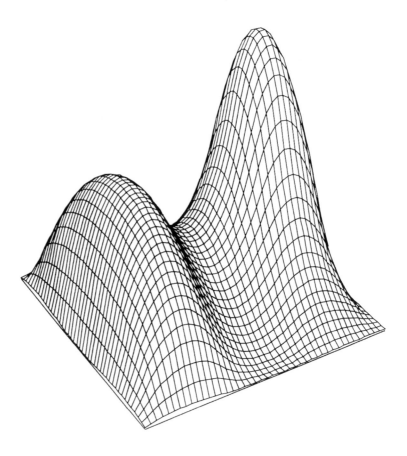

Temperature distribution in a solid with multiple heat sources and sinks

Chapter 1

An Overview

We who live in this computer age are fortunate because we can use computational methods for simulating physical phenomena. In this book, you will be introduced to this exciting activity, which also has great practical utility. An overview of the topics in the book is provided in this chapter.

1.1   Purpose of the Book

Numerical prediction methods play an important role in the analysis and design of practical devices involving heat transfer and fluid flow. The methods, when embodied in suitable computer programs, represent a speedy and economical alternative to experimental measurement. The computational analysis can incorporate the actual details of the geometry, material properties, and boundary conditions and produce complete and detailed information about the fields of temperature, velocity, etc. and about the associated fluxes. In some practical situations, the analysis and design can be wholly done by the use of a computer program. In situations where some experimental testing is desirable, the computational prediction can be used to plan and design the experiments, to significantly reduce the amount of experimentation, and to supplement and enrich the experimental results.

The computer simulation of practical situations involving heat transfer, fluid flow, and related phenomena requires, in general, the solution of a set of nonlinear partial differential equations in three space coordinates and time. Although numerical methods exist for obtaining such a solution, the task of writing and using a sufficiently general computer program for all practical thermofluid processes is quite formidable. Especially to a beginner, such an undertaking may be rather intimidating. A more comfortable entry into the computer simulation activity can be provided by means of a computer program restricted to a *subset* of the general heat transfer and fluid flow processes. The purpose of this book is to illustrate the construction and application of a general-purpose computer program for a particular subset.

For the computer program presented in this book, two main restrictions are adopted. The first restriction is that of two-dimensionality. There are many physical situations that can be satisfactorily approximated as two-dimensional, and in general, the *qualitative* features of most practical problems can be studied in a two-dimensional context. Moreover, the framework for a three-dimensional computer program can be conveniently illustrated via a two-dimensional program. The addition of the third dimension to the program makes it more difficult to use by increasing the complexity of the problem and by requiring much greater amounts of computer time and computer memory. Therefore, a two-dimensional program can be considered a more appropriate learning tool. (By the way, a one-dimensional program will be even easier to construct and run. However, for most practical problems, a one-dimensional representation is often too crude to capture, even qualitatively, many interesting features of multidimensional situations.)

The other restriction is that the computer program is designed for conduction-type phenomena. This means that the program solves for the distribution of any scalar variable that is governed by a differential equation similar to the one used for heat conduction. Thus, the program is primarily designed for heat conduction, although it can be used for many other processes that are analogous.

One important class of problems that can be considered as conduction-type is the fully developed flow and heat transfer in ducts. Beyond the entrance region in a duct, there usually exists a region where the longitudinal velocity and the fluid temperature exhibit a certain regular behavior. In this region, the velocity and temperature fields are governed by equations similar to those used for two-dimensional heat conduction.

Other examples of conduction-type situations include: mass diffusion with chemical reaction, flow through porous materials, lubrication flows, heat and moisture transport in soil, potential flows, and electromagnetic fields. Thus, even within the main restrictions of the computer program presented here, a variety of interesting physical phenomena can be simulated and analyzed.

The primary focus in this book is on the analysis of heat conduction and duct flow heat transfer. For that reason, the computer program is given the name CONDUCT, which refers to heat CONduction and DUCT flow. The purpose of this book is to describe the computer program CONDUCT in terms of its physical, mathematical, and computational details and to illustrate the application of the program to many problems of engineering interest.

Generally, we look at a computer program as a means of producing quantitative numerical results for a practical problem. However, the computational activity serves an additional, and possibly more important, purpose. Through the computer simulation of physical situations, we can develop a better understanding of, and an insight into, a number of complex physical processes. As you will see, throughout the development of the method and the associated computer program in this book, there is an emphasis on the physical understanding.

1.2   Capabilities and Limitations of CONDUCT

CONDUCT is designed for the solution of partial differential equations of the heat conduction type. It can calculate the distribution of scalar quantities such as the temperature in heat conduction, the concentration in mass diffusion, the velocity and the temperature in fully developed duct flows, the potential in potential flows, the pressure in porous-material flows, and so on. As we shall see later, such phenomena are governed by the general differential equation expressed as Eq. (3.6) in Chapter 3. Therefore, CONDUCT can be used for the calculation of any variable that is governed by a differential equation of the form specified by Eq. (3.6). Further, we restrict attention to only two-dimensional situations; i.e., the quantities of interest can have significant variations in only two space coordinates. The program can be used for steady or unsteady situations.

CONDUCT is designed to employ three coordinate systems: (1) Cartesian (x, y), (2) axisymmetric (x, r), and (3) polar (θ, r). These are illustrated in Fig. 1.1. For each coordinate system, the program uses a grid network of lines drawn in the two coordinate directions. As a result, the program is ideally suited for shapes of the calculation domain that conveniently fit into one of these coordinate systems. The program can still be used for domain shapes that exhibit a number of geometrical irregularities; but, if the domain is too irregular, the use of CONDUCT is not convenient. Thus, the use of grids in three standard coordinate systems imposes some limitation on the geometrical shapes that can be conveniently handled. We could have removed the limitation, but this would have been at the expense of making the program much more complicated to construct, understand, and use.

For the polar (θ, r) coordinate system, a specific restriction must be mentioned. The circumferential coordinate θ is not allowed to cover the full circle of 360° unless the values of the dependent variables are known at θ = 0 and $θ = 360^\circ$ or unless these locations represent lines of symmetry. This restriction is not as serious as it appears. For most problems with the full 360 degrees of the circular or annular geometry, there are two or more radial lines of symmetry; so the solution is required only over a sector bounded by two successive symmetry lines. This can be handled by CONDUCT, since the extent of θ will then be less than 360°. For example, consider the fully developed flow in a circular duct with radial internal fins. If the fins are identical in shape and uniformly spaced over the duct perimeter (which is usually the case in practice), it is sufficient to use CONDUCT to analyze the region between two successive fins (or rather, between the centerline of one fin and the location halfway towards the next fin). If the fins

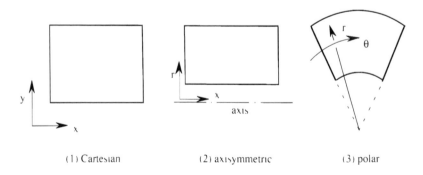

(1) Cartesian   (2) axisymmetric   (3) polar

Fig. 1.1 The three coordinate systems

are nonuniformly spaced with no discernable lines of symmetry. CONDUCT cannot be used for such a problem.

Within these overall limits, CONDUCT can be applied to a wide variety of problems in heat conduction, fully developed duct flow, and analogous phenomena. The properties like the conductivity or viscosity can be nonuniform; they may depend on position (as in a composite material) and on the temperature and other variables. The duct flow can be laminar or turbulent, Newtonian or nonNewtonian. There can be internal heat generation in a conduction problem; the generation may depend on position and/or on temperature. For all problems, a variety of boundary conditions can be present. Once you fully understand the scope and limitations of the program, you will be able to design a large variety of imaginative applications.

1.3   The Structure of CONDUCT

One purpose in introducing you to the program CONDUCT is to illustrate the construction of a general-purpose computer program, which can be used to solve, within its overall limitations, an endless variety of seemingly different physical problems.

CONDUCT is constructed in two parts: the invariant part and the adaptation part. The *invariant* part contains the general calculation scheme that is common to all possible applications within the overall restrictions of the program. It is written without any knowledge or assumption about the particular details of the problem to be solved. Normally, there would be no need for you to make any changes in the invariant part of the program. The *adaptation* part provides the problem specification. It is here that the actual details of the problem such as the geometry, material properties, heat sources, reaction rates, boundary conditions, desired output, etc. are supplied. It then follows that the adaptation part cannot be written "in advance" for the endless variety of practical problems to which the program can be applied. What can be provided is the *framework* for the adaptation part; but its *contents* must be written "on demand" to specify the problem at hand. Thus, a general-purpose program of this type consists of the complete invariant part and a *skeleton* of the adaptation part. The latter is to be completed by the user according to the general instructions provided with the program. Within some overall restrictions, there is considerable freedom in designing the adaptation part. Very complex adaptations can be designed, the possibilities being limited only by the imagination of the user.

This type of program structure may be somewhat unfamiliar to you. Many of you are probably accustomed to programs that are complete and require only the numerical input data. Such programs, even when they are highly complicated, cannot match the virtually unlimited flexibility and the open-ended applicability of a program like CONDUCT, in which you can provide the problem description and design the desired output by *writing* a Fortran subroutine.

Of course, you do not need this flexibility when you use the program for routine computations for a small class of problems. For such applications, you can design the adaptation part such that it works in an automatic manner, simply asking for the values of a few input parameters. Thus, you treat CONDUCT as a toolkit for constructing specialized programs that, although of limited applicability, are easy to use.

It may appear that the use of a general-purpose computer program for a simple problem is rather cumbersome, because the program asks too many questions. It needs to know: the number of dependent variables; variations of viscosity, conductivity, and diffusion coefficients; the distributions of the source and sink terms for all the variables; the details of the boundary conditions for the relevant equations; and so on. However, certain facilities in CONDUCT have been designed to overcome this difficulty; they make the program delightfully easy to apply to simple problems.

This desirable characteristic of the program has been achieved by a judicious use of "default values" for many parameters and variables. In other words, certain quantities are assumed to have the most commonly encountered values, unless they are overwritten by you in the adaptation part. As a result, the adaptation part for a simple problem can be very short. The length and the complexity of the adaptation part increases only in proportion to the complexity of the problem at hand.

Your responsibility of designing the adaptation part is further lightened by the provision of a number of general utilities in the invariant part of the program. These utilities are not an essential part of the program, but are provided for your convenience. Examples of such utilities are: routines for the construction of uniform and nonuniform grids, and a routine for the printout of the grid and the field variables.

Because successful utilization of CONDUCT depends on how correctly you prepare the adaptation part, a complete appreciation of the entire program is essential. This book provides all the necessary details of the invariant part and gives instructions and guidance for the design of the adaptation part. Illustrative adaptations of the program (given in Chapters 8, 10, and 11) serve as examples of the actual implementation of these ideas.

A computer program such as CONDUCT can be endlessly improved and streamlined. In this sense, there is no "perfect" or "best" program. What is presented in this book is the version that I think is adequate for our purposes. But it can certainly be improved. At first, it will be better for you to understand the program as it is and to learn to use it well. Later on, you may wish to design improved versions of the program for your own purposes.

CONDUCT is written in Fortran 77 and can be used on nearly all computers without modification. The results presented in this book have been obtained by using the Microsoft FORTRAN Compiler Version 4.1 on an IBM PC. Depending on the word length used by your computer, your results may differ slightly. This is especially true of certain quantities that should be theoretically zero, but acquire a small value because of computer roundoff. Unless the discrepancies are very large, there is no need to worry about them.

1.4   Outline of the Book

The general concept of numerical methods is described in Chapter 2, where our particular numerical technique is developed in detail with reference to the one-dimensional heat conduction situation. Although limited to one dimension, this chapter explains nearly all important ideas needed for our later work. Therefore, a good understanding of the material in this chapter is highly desirable.

The mathematical formulation for the general physical phenomena of interest is presented in Chapter 3. There we discuss the heat conduction equation and generalize it to represent other analogous processes. The computer program CONDUCT provides a calculation scheme for the solution of this general equation. Please note that Chapter 3 is not intended for providing complete information about the derivation of the heat conduction equation, the formulations for other processes, the dependence of the thermal conductivity on the temperature, and other similar topics. We are primarily concerned with the *form* of the differential equations to be solved. For more complete details of the mathematical representation of heat conduction and other phenomena, you should turn to other textbooks.

Since it is convenient to describe the numerical scheme together with the relevant programming details, an early overview of the entire computer program is presented in Chapter 4. It will enable you to appreciate the references to and the interrelationships of various subprograms. There the two-part structure of CONDUCT is described in detail. At that time, you will find it helpful to glance at the listing of the invariant part of CONDUCT given in Appendix A. In later chapters, you should frequently refer to the listing so that you become completely familiar with the structure and details of the program.

The numerical method developed for a one-dimensional situation in Chapter 2 is extended to general two-dimensional situations in Chapter 5. It is here that you will see some of the details of the design of CONDUCT and learn about numbering schemes, sign conventions, and some Fortran names. Chapter 5 also gives complete details of the algebraic equations, boundary conditions, solution algorithm, source terms, nonlinearity, and so on. This chapter is your main reference for the information on the numerical method incorporated in CONDUCT.

In Chapters 6 and 7. attention is returned to the structure of the computer program. The subprograms appearing in the invariant part of CONDUCT are described in Chapter 6. while the information on the role and design of various ingredients of the adaptation part is provided in Chapter 7. A thorough study of this chapter is essential before you embark on the use of CONDUCT.

The book contains fifteen illustrative applications of CONDUCT. These are presented in Chapters 8. 10. and 11. In the description of each application. we start with the details of the problem selected. discuss the design of the adaptation routine. explain the new Fortran names introduced. give a listing of the actual Fortran routine used. provide the corresponding computer output. and comment on those results. The example problems are chosen not because they represent interesting practical applications: they are selected to provide an all-round experience in the use of CONDUCT. Irrespective of whether you are interested in a particular application. you should study all the adaptations presented since each one of them is designed to illustrate one or more features of the computer program. Your success in using CONDUCT for a variety of problems depends on how well you study the fifteen examples included in the book.

The adaptations to steady and unsteady heat conduction problems are described in Chapter 8. Before we apply the program to duct flows. we review in Chapter 9. the necessary mathematical background for fully developed flow and heat transfer in ducts. This is followed by the descriptions of adaptations to duct flows problems in Chapter 10.

Chapter 11 is devoted to additional adaptations. which include some advanced duct flows. potential flows. and a flow through porous medium. Chapter 12 contains concluding remarks about the method and the computer program. some suggestions on further extensions of CONDUCT. and final reminders on its proper use for a variety of applications.

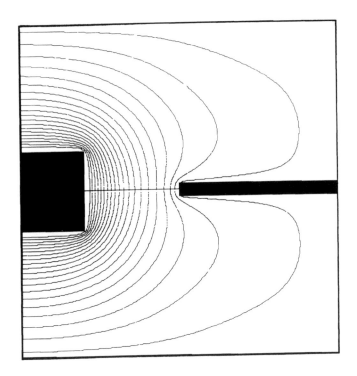

Temperature distribution in a cylindrical region with
embedded solids at different temperatures

Chapter 2

Introduction to Numerical Methods

Before we embark on the formulation of the numerical method for general two-dimensional situations, it will be useful to study the solution of one-dimensional heat conduction. Here the physical picture is simple and the mathematical complication is minimal. Therefore, many ideas can be easily learned in the one-dimensional context. Later, we shall be able to extend the technique quickly to general two-dimensional problems. The information presented in this chapter is thus very important for all our subsequent work.

2.1 Concept of Numerical Solution

For steady one-dimensional heat conduction, let us consider that the temperature T depends only on the coordinate x. The governing differential equation can be written as

$$\frac{d}{dx}\left(k\frac{dT}{dx}\right) + S = 0 \tag{2.1}$$

where k is the thermal conductivity, and S is the source term, i.e., the rate of heat generation (per unit volume) in the material. Initially, we shall treat k and S as constants.

In obtaining a numerical solution of Eq. (2.1), we choose a number of locations along the x direction and seek the values of the temperature there. Let us call these locations "grid points". Figure 2.1 shows a set of uniformly spaced grid points denoted by the sequence numbers i–1, i, i+1, etc. The distance between two consecutive grid points is δx. The task of the numerical method is to obtain the values of temperatures such as T_{i-1}, T_i, T_{i+1}.

An *analytical* solution of Eq. (2.1) consists of an *expression* for T in terms of x. A numerical solution, by contrast, is given in the form of the *numerical values* of T at a finite number of locations (grid points). The discrete values of T_{i-1}, T_i, T_{i+1}, etc. are governed by algebraic equations, which we call discretization equations. These are derived from the differential equation (2.1). How we derive the discretization equations will be discussed in Section 2.2.

When only a small number of grid points are used to discretize the calculation domain, the discretization equations represent an approximation to the differential equation. Then, the resulting numerical solution would normally not coincide with the exact solution of the differential equation. As we increase the number of grid points, the numerical solution becomes more accurate and approaches the exact solution. For many problems, even a modest number of grid points can lead to solutions that are sufficiently accurate for practical purposes. This will be demonstrated in this chapter and throughout the book.

For steady one-dimensional heat conduction, Eq. (2.1) can usually be solved analytically in a straightforward manner. However, for complex multidimensional

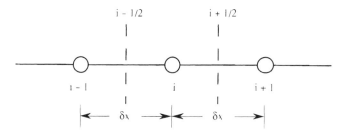

Fig. 2.1 A one-dimensional uniform grid

situations. to obtain an analytical solution for heat conduction is either very difficult or impossible. In these cases. a numerical method (which involves the solution of *algebraic* equations) offers a practical alternative. The strength of a numerical method lies in the replacement of the differential equation by a set of *algebraic* equations. which are easier to solve. We shall now see how such algebraic discretization equations can be derived.

2.2 Derivation of Discretization Equations

There are a number of ways in which a differential equation such as Eq. (2.1) can be converted into its discrete counterpart. namely an algebraic equation containing T_{i-1}. T_i. T_{i+1}. etc. as unknowns. Here. we shall look at a few ways of deriving such a discretization equation and choose one of the ways for our later work.

Approximation of the derivatives. Since the differential equation contains derivatives. we can obtain the discretization equation by replacing the derivatives by a suitable approximation. For example. the second derivative in Eq. (2.1) can be approximated with reference to Fig. 2.1 as

$$\frac{d}{dx}\left(k\frac{dT}{dx}\right) = \frac{1}{\delta x}\left[\left(k\frac{dT}{dx}\right)_{i+1/2} - \left(k\frac{dT}{dx}\right)_{i-1/2}\right] \tag{2.2}$$

Further. the derivative (dT/dx) at locations $i+1/2$ and $i-1/2$ can be written in terms of T_{i-1}. T_i. T_{i+1} as

$$\left(k\frac{dT}{dx}\right)_{i+1/2} = k\frac{T_{i+1} - T_i}{\delta x} \tag{2.3}$$

$$\left(k \frac{dT}{dx}\right)_{i-1/2} = k \frac{T_i - T_{i-1}}{\delta x} \tag{2.4}$$

Substitution of these relationships into Eq. (2.1) leads to

$$\frac{k}{(\delta x)^2} \left[T_{i+1} + T_{i-1} - 2T_i\right] + S = 0 \tag{2.5}$$

where k and S are assumed constant for this initial exploration. Equation (2.5) is an algebraic counterpart of the differential equation (2.1). If we have an equation like (2.5) for every grid point in the domain, we can obtain a numerical solution for the temperature T by solving this set of algebraic equations.

Taylor-series formulation. There is a more formal procedure to arrive at Eq. (2.5). If we expand the temperature T in Taylor series, we can write

$$T_{i+1} = T_i + (\delta x)\left(\frac{dT}{dx}\right)_i + \frac{1}{2}(\delta x)^2 \left(\frac{d^2T}{dx^2}\right)_i + \dots \tag{2.6}$$

Similarly, T_{i-1} can be expressed as

$$T_{i-1} = T_i - (\delta x)\left(\frac{dT}{dx}\right)_i + \frac{1}{2}(\delta x)^2 \left(\frac{d^2T}{dx^2}\right)_i + \dots \tag{2.7}$$

If the higher-order terms in Eqs. (2.6)–(2.7) are neglected, a sum of the two equations gives

$$\left(\frac{d^2T}{dx^2}\right)_i = \frac{T_{i+1} + T_{i-1} - 2T_i}{(\delta x)^2} \tag{2.8}$$

This is the same approximation as we obtained earlier by combining Eqs. (2.2)–(2.4). Thus, we can again obtain the discretization equation (2.5) by substituting Eq. (2.8) into Eq. (2.1).

Control-volume method. The region bounded by the locations i−1/2 and i+1/2 (shown by dashed lines in Fig. 2.1) is a small part of the total one-dimensional domain considered. Such a region is known as a subdomain, a finite volume, or a control volume. It is possible to derive a discretization equation by satisfying the heat balance over a control volume. In other words, we integrate the differential equation (2.1) over the control volume and then express the result as an algebraic equation. Integration of Eq. (2.1) with respect to x between the limits of i–1/2 and i+1/2 leads to

$$\left(k\frac{dT}{dx}\right)_{i+1/2} - \left(k\frac{dT}{dx}\right)_{i-1/2} + S(\delta x) = 0 \tag{2.9}$$

Since $(-k\,dT/dx)$ represents the local heat flux, Eq. (2.9) is clearly a balance between the heat flow rates at the two faces of the control volume and the amount of heat generation within the control volume.

To express the temperature gradient dT/dx in Eq. (2.9) in an algebraic form, we shall make a "profile assumption", that is, assume how T varies with x *between* the grid points. Figure 2.2 shows a simple profile assumption, known as a piecewise-linear profile, in which the temperature T is taken to be linear with x between two neighboring grid points. Using this profile, we can write the temperature gradients at the control-volume faces as

$$\left(\frac{dT}{dx}\right)_{i+1/2} = \frac{T_{i+1} - T_i}{\delta x} \tag{2.10}$$

$$\left(\frac{dT}{dx}\right)_{i-1/2} = \frac{T_i - T_{i-1}}{\delta x} \tag{2.11}$$

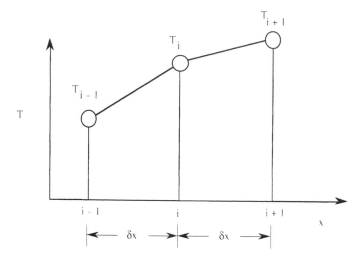

Fig. 2.2 Piecewise-linear profile for temperature

When we substitute these expressions into Eq. (2.9), we get

$$\frac{k}{(\delta x)} \left[T_{i+1} + T_{i-1} - 2T_i \right] + S(\delta x) = 0 \tag{2.12}$$

which is algebraically equivalent to Eq. (2.5) derived earlier. This equation can be rearranged as

$$\left(\frac{2k}{\delta x} \right) T_i = \left(\frac{k}{\delta x} \right) T_{i+1} + \left(\frac{k}{\delta x} \right) T_{i-1} + S(\delta x) \tag{2.13}$$

In this book, we shall adopt this particular *form* of the equation for *all* our discretization equations.

Note that we made no approximation in obtaining the heat balance equation (2.9) for the control volume. In other words, Eq. (2.9) is as exact as Eq. (2.1). The profile assumption, however, is an approximation, which improves as we increase the number of grid points.

Discussion of the control-volume method. In this book, we shall use the control-volume method as our choice for the derivation of discretization equations. The main reason is that the discretization equation obtained by this method has a clear physical meaning; it is not simply a formal mathematical approximation. The differential equations of our interest represent conservation principles. For example, the heat conduction equation is based on energy conservation. Later, we shall encounter momentum conservation in duct flows and mass conservation for flow in porous materials. When the discretization equations are derived by the control-volume method, they represent the conservation of energy, momentum, mass, etc. for each control volume. It then follows that the resulting numerical solution correctly satisfies the conservation of these quantities over the whole calculation domain.

In our simple illustration in this section, different methods for deriving the discretization equation led to the same final equation. This may not always be the case. Here we used a very simple differential equation and chose a particular profile assumption in the control-volume method. For more complex differential equations, or for other profile assumptions, the resulting discretization equations may not be the same when obtained by the Taylor-series formulation, the control-volume method, and other techniques. The solution obtained by the control-volume method will always give a perfect overall balance (of energy, momentum, etc.) for the whole calculation domain; the same cannot be said about the solution given by other methods.

2.3   An Illustrative Example

Let us obtain a numerical solution of Eq. (2.1) for a simple situation. For a domain of length L, let the boundary conditions be

$$x = 0 : \qquad T = T_A \tag{2.14a}$$

$$x = L : \qquad T = T_B \tag{2.14b}$$

Further, let us use the numerical values given by

$$k = 1, \quad S = 2, \quad L = 5, \quad T_A = 0, \quad \text{and} \quad T_B = 15 \tag{2.15}$$

Discretization equations. For this simple illustration, we shall use the uniform grid of $\delta x = 1$ shown in Fig. 2.3. The boundary temperatures T_1 and T_6 have known values given by Eq. (2.15); thus

$$T_1 = 0 \qquad \text{and} \qquad T_6 = 15 \tag{2.16}$$

For the internal grid points 2, 3, 4, and 5, discretization equations of the form (2.13) can now be written. Therefore, with the values given in Eq. (2.15), we obtain

$$2 T_2 = T_3 + 2 \tag{2.17}$$

$$2 T_3 = T_4 + T_2 + 2 \tag{2.18}$$

$$2 T_4 = T_5 + T_3 + 2 \tag{2.19}$$

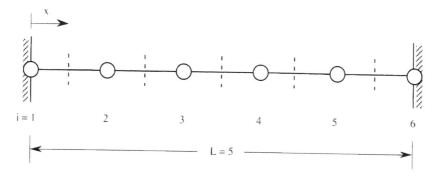

Fig. 2.3   A six-point uniform grid

$$2 T_5 = T_4 + 17 \tag{2.20}$$

Please note that the terms involving the boundary temperatures T_1 and T_6 have been given the known values according to Eq. (2.16).

Solution of the algebraic equations. Now what remains to be done is to solve Eqs. (2.17)–(2.20) for the unknown temperatures. Because these equations have a special form, a particularly easy method of solution can be employed.

To begin, we shall write Eq. (2.17) as

$$T_2 = 0.5 T_3 + 1 \tag{2.21}$$

This can be substituted into Eq. (2.18) as

$$2 T_3 = T_4 + (0.5 T_3 + 1) + 2 \tag{2.22}$$

which gives

$$T_3 = (2/3) T_4 + 2 \tag{2.23}$$

Substitution of this into Eq. (2.19) leads to

$$T_4 = (3/4) T_5 + 3 \tag{2.24}$$

Finally, when this is substituted into Eq. (2.20), we get

$$2 T_5 = [(3/4) T_5 + 3] + 17 \tag{2.25}$$

Now, T_5 being the only unknown in this equation, we can evaluate T_5 as

$$T_5 = 16 \tag{2.26}$$

If we use this value in Eq. (2.24), we can obtain the value of T_4; then Eq. (2.23) will yield T_3, and Eq. (2.21) will give T_2. In this manner, the numerical solution for this problem can be expressed as

$$T_2 = 7, \quad T_3 = 12, \quad T_4 = 15, \quad T_5 = 16 \tag{2.27}$$

The solution method employed here is a very powerful method for the solution of a set of algebraic equations of a certain form. It will be described later in Section 2.4-4 as a general solution algorithm.

Comparison with the exact solution. Now that we have obtained the numerical solution, it will be interesting to examine its accuracy with reference to the exact solution. Equation (2.1) can be solved for constant k and S and for boundary conditions (2.14) to give

$$T = T_A + (T_B - T_A)\left(\frac{x}{L}\right) + \left(\frac{S}{2k}\right)(L - x) x \tag{2.28}$$

When the values given in Eq. (2.15) are substituted we get

$$T = 8 x - x^2 \tag{2.29}$$

This can be used to obtain the exact values of T_2, T_3, T_4, and T_5 for the locations $x = 1, 2, 3,$ and 4 respectively. In this particular case, it so happens that the exact values given by Eq. (2.29) agree perfectly with our numerical solution (2.27).

Such a perfect agreement is rather rare and occurs in only some simple problems. Nevertheless, it is useful to know that occasionally a numerical solution entails no error even when only a few grid points are used. In the problem considered here, the exact temperature distribution is parabolic as given by Eq. (2.29). To replace it by a piecewise-linear profile is clearly an approximation. Yet, the temperature gradients obtained in Eqs. (2.10) and (2.11) from this profile happen to be the correct gradients for the parabolic variation of temperature. It is this fortuitous coincidence in the present problem that makes our numerical solution identical to the exact solution.

More commonly, a numerical solution obtained with only a small number of grid points will exhibit some error with reference to the exact solution. As we increase the number of grid points, the error will diminish. After a certain stage, the error will be so small that any further increase in the number of grid points will not noticeably alter the solution. In this condition, the numerical solution can be regarded as the exact solution for all practical purposes. For many problems for which exact analytical solutions may not be available, we can treat the numerical solution as sufficiently accurate when a further increase in the number of grid points does not alter the solution.

2.4 Steady One-Dimensional Heat Conduction

After an initial exploration of the control-volume method, we shall now develop the method for steady one-dimensional heat conduction. The basic details of the method will be given in this section, while further refinements will be added in Section 2.5. Finally, we shall extend the method to unsteady situations in Section 2.6.

Although the development in this chapter is restricted to *one-dimensional* situations, all the details worked out here will be used in Chapter 5, where the method for two-dimensional problems will be described. In fact, once the technique is properly constructed for one-dimensional cases, very few additional ingredients are needed for two dimensions. Therefore, the information in this chapter serves as the main foundation for our later work.

2.4-1 Discretization Equation

For steady one-dimensional heat conduction, Eq. (2.1) continues to be the governing differential equation. We shall now allow the possibility that the conductivity k and the source term S may not be constant. Let us consider a portion of a one-dimensional grid shown in Fig. 2.4. Unlike in Fig. 2.1, here the grid spacing is not necessarily uniform. The letters W, P, and E denote the grid points, where P is the point under consideration and W and E are its "west" and "east" neighbors. The dashed lines represent the faces of the control volume around P; the lower case letters w and e are used to denote these faces. The exact locations of the control-volume faces will be discussed later in Section 2.5-7; they may not always be located midway between the grid points. The distance between the grid points W and P is given by $(\delta x)_w$ and that between P and E by $(\delta x)_e$. The width or the thickness of the control volume is denoted by Δx.

Let us write Eq. (2.1) as

$$-\frac{dq}{dx} + S = 0 \qquad\qquad (2.30)$$

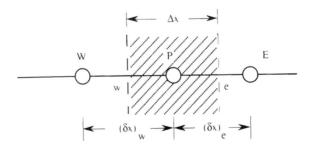

Fig. 2.4 A general one-dimensional grid

where the heat flux q is given by

$$q = -k \frac{dT}{dx} \tag{2.31}$$

If we integrate Eq. (2.30) with respect to x over the control volume. i.e.. between the limits of w and e. we get

$$q_w - q_e + \int_w^e S \, dx = 0 \tag{2.32}$$

Here the first two terms represent the inflow and outflow of heat. while the integral stands for the total heat generation within the control volume.

For the evaluation of q_w and q_e in terms of the grid-point temperatures. we shall use the piecewise-linear profile introduced in Fig. 2.2. It is also shown here in Fig. 2.5 for the general nonuniform grid. Now let us first express q_w and q_e according to Eq. (2.31) and then substitute for the temperature gradient from our profile assumption. The result is

$$q_w = \frac{k_w}{(\delta x)_w} (T_W - T_P) \tag{2.33}$$

$$q_e = \frac{k_e}{(\delta x)_e} (T_P - T_E) \tag{2.34}$$

The subscripts for the conductivity k are intended to account for the variation of

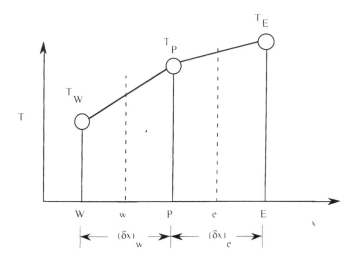

Fig. 2.5 Piecewise-linear profile

conductivity with position. We shall discuss this matter later in Section 2.5-2.

For the calculation of the source-term integral. let us denote the average heat generation rate in the control volume by $\overline{S}$. Then

$$\int_w^e S \, dx = \overline{S} \, \Delta x \tag{2.35}$$

Now, we substitute Eqs. (2.33)–(2.35) into (2.32) and obtain the discretization equation as

$$a_P T_P = a_E T_E + a_W T_W + b \tag{2.36}$$

where

$$a_E = \frac{k_e}{(\delta x)_e} \tag{2.37a}$$

$$a_W = \frac{k_w}{(\delta x)_w} \tag{2.37b}$$

$$a_P = a_E + a_W \tag{2.37c}$$

$$b = \overline{S} \, \Delta x \tag{2.37d}$$

In the general discretization equation (2.36), the coefficient a_P multiplies the temperature T_P; a_E and a_W are the coefficients of the neighbor temperatures; and b represents the constant term. When the discretization equation is written in the form (2.36), the values of the coefficients a_P, a_E, and a_W are always positive. This property is very important, as we shall see later, for proper physical and mathematical behavior of the method.

Since Eq. (2.36) is written for a one-dimensional situation, the grid point P has only two neighbors (one in the positive x direction and the other in the negative x direction). It then follows that there will be four neighbors in a two-dimensional formulation and six neighbors in three dimensions. It is convenient to write Eq. (2.36) in the general form

$$a_P T_P = \sum a_{nb} T_{nb} + b \tag{2.38}$$

where the subscript nb indicates a neighbor and the summation is taken over all neighbors.

This may be a convenient place to comment on the physical significance of the coefficient expressions in Eq. (2.37). For the one-dimensional grid shown in Fig. 2.4. let us assume that the cross-sectional area (normal to the x direction) is unity. Then, the control-volume faces have a unit area and hence Δx represents the volume of the control volume. Since $\overline{S}$ is the heat generation rate *per unit volume*, the constant term b given by Eq. (2.37d) becomes the total heat generation in the control volume (of volume Δx). From Eq. (2.34). $(\delta x)_e/k_e$ will be seen as the *resistance* to heat transfer between the grid points P and E. The expression (2.37a) for a_E is the reciprocal of this. We shall, therefore. interpret the neighbor coefficients a_E and a_W as the *conductances* of the links connecting the neighbor grid points to P. The coefficient a_p is then simply the sum of these conductances.

2.4-2   Treatment of the Source Term

Consider a fin attached to a solid surface at a temperature T_A as shown in Fig. 2.6. The fin exchanges heat with a surrounding fluid at temperature T_∞ with a heat transfer coefficient h. It is common to treat this situation as a one-dimensional heat conduction problem, since the temperature of the fin is nearly uniform over its cross section and mainly varies in the x direction. The heat transfer between the fin and the surrounding fluid is then treated as a source term. The governing differential equation for a fin of uniform cross section is

$$\frac{d}{dx}\left(k\,\frac{dT}{dx}\right)+\frac{hP}{A}\,(T_\infty - T) = 0 \tag{2.39}$$

This equation can be found in many introductory textbooks on heat transfer. In the second term in Eq. (2.39). P and A represent the perimeter and area of the fin cross section. The quantity $(hP/A)(T_\infty - T)$ gives the rate of heat transfer to the fin per unit volume. If we compare Eqs. (2.1) and (2.39). we can identify the source term S as:

$$S = \frac{hP}{A}\,(T_\infty - T) \tag{2.40}$$

In our derivation in Section 2.4-1. we treated the source term S as known. Here we see that S can depend on the unknown temperature T. Under these

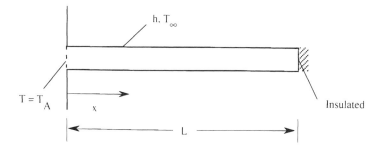

Fig. 2.6 A one-dimensional fin problem

situations, it is important to include this dependence in our formulation.

For the grid shown in Fig. 2.4, T_P can be taken as the representative temperature for the control volume. Then the average source term $\bar{S}$ may depend on T_P. We shall express $\bar{S}$ as a linear function of T_P. Thus

$$\bar{S} = S_C + S_P T_P \tag{2.41}$$

where S_C and S_P are the constants in the linear expression. (We chose a *linear* function here so that our resulting discretization equation would remain linear; this facilitates the solution process. However, as we shall see later in Section 2.5-4, it will still be possible to handle nonlinear S~T relationships in our linear framework.)

If we substitute Eq. (2.41) into (2.35) and complete the derivation of the discretization equation, we shall get slightly different expressions for a_P and b. For completeness, let us write here the final discretization equation and the expressions for all coefficients.

$$a_P T_P = a_E T_E + a_W T_W + b \tag{2.42}$$

where

$$a_E = \frac{k_e}{(\delta x)_e} \tag{2.43a}$$

$$a_W = \frac{k_w}{(\delta x)_w} \tag{2.43b}$$

$$a_P = a_E + a_W - S_P \Delta x \tag{2.43c}$$

$$b = S_C \Delta x \tag{2.43d}$$

When the source term does not depend on the temperature, S_P becomes zero, and S_C equals the known value of $\bar{S}$. Then the expressions in Eq. (2.43) revert to (2.37).

For the one-dimensional fin problem, the source term given by Eq. (2.40) implies

$$S_C = \left(\frac{hP}{A}\right) T_\infty \tag{2.44}$$

$$S_P = -\left(\frac{hP}{A}\right) \tag{2.45}$$

It is important to note that here the value of S_P is negative. Therefore, the term $(-S_P \Delta x)$ in Eq. (2.43c) makes a positive contribution to a_P. Since we must keep a_P positive, it is highly desirable that we always work with a negative S_P. Fortunately, many physical processes, like the fin problem considered here, lead to a naturally negative S_P. How to deal with situations that may indicate a positive S_P will be explained in Section 2.5-4. You can also find a more extensive discussion of this treatment in Patankar (1980).

2.4-3   Boundary Conditions

The solution of a physical problem depends not just on the differential equation but also on boundary conditions. For the domain shown in Fig. 2.7, we have a discretization equation like (2.42) for the control volume around each *internal* grid point. To solve this set of equations, additional information is needed at the boundary points.

If the boundary temperatures are given, then we have enough algebraic equations for the unknown temperatures at the interior grid points. This is why we could solve the simple problem in Section 2.3 without any elaborate treatment of the boundary conditions.

When the boundary temperatures are not known, something else (such as the heat flux at the boundary face) is prescribed as the boundary condition. Then we need to construct an additional equation for the unknown temperature at the respective boundary. In our control-volume method, this equation comes from the heat balance principle applied to the "half" control volume adjacent to the boundary. The one-dimensional domain shown in Fig. 2.7 has two boundaries. Since their treatment is identical, we shall discuss it in terms of the left-hand boundary only. Similar considerations apply to the other boundary.

Figure 2.8 shows an enlarged view of the half control volume near the left boundary. When Eq. (2.30) is integrated over this control volume, we get

$$q_B - q_i + (S_C + S_P T_B) \Delta x = 0 \qquad (2.46)$$

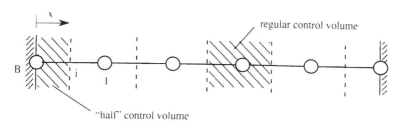

Fig. 2.7 A complete one-dimensional grid including boundary points

where the source term has been expressed as a linear function of T_B. Please note that Δx is the width of the half control volume: this Δx may not be the same as the width Δx of a regular control volume. The heat flux q_1 is given by expressions like (2.34); thus

$$q_1 = \frac{k_1}{(\delta x)_1} (T_B - T_1) \qquad (2.47)$$

The boundary flux q_B comes from the boundary-condition information.

Now, there are two common possibilities. One possibility is that the value of q_B is given. In that case, Eqs. (2.46) and (2.47) can be combined to obtain the discretization equation

$$a_B T_B = a_1 T_1 + b \qquad (2.48)$$

where

$$a_1 = \frac{k_1}{(\delta x)_1} \qquad (2.49a)$$

$$a_B = a_1 - S_P \Delta x \qquad (2.49b)$$

$$b = S_C \Delta x + q_B \qquad (2.49c)$$

Here, the known value of q_B enters the expression for the constant term b. The value of q_B can be zero or nonzero; the zero value is appropriate for an adiabatic boundary or a symmetry boundary.

The other possibility for the boundary condition is that the boundary heat flux

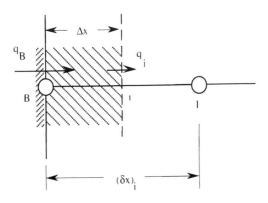

Fig. 2.8 Half control volume near left boundary

q_B is given as a linear function of the boundary temperature T_B. Let us express this as

$$q_B = f_C + f_P T_B \qquad (2.50)$$

where f_C and f_P are the coefficients in the linear expression. Such a boundary condition usually arises when the convective heat transfer coefficient h and the surrounding fluid temperature T_∞ are given at a boundary . Then

$$q_B = h (T_\infty - T_B) \qquad (2.51)$$

which implies

$$f_C = h T_\infty \quad \text{and} \quad f_P = -h \qquad (2.52)$$

It is useful to note the similarity between Eqs. (2.41) and (2.50). Further, just as S_P must have a negative value, we require that f_P should be negative as well. As seen from Eq. (2.52), this condition is easily met in normal convective boundary conditions.

When the boundary heat flux is given by Eq. (2.50), we can obtain the required discretization equation by combining Eqs. (2.46), (2.47), and (2.50). Thus

$$a_B T_B = a_I T_I + b \qquad (2.53)$$

where

$$a_I = \frac{k_I}{(\delta x)_I} \qquad (2.54a)$$

$$a_B = a_I - S_P \Delta x - f_P \qquad (2.54b)$$

$$b = S_C \Delta x + f_C \qquad (2.54c)$$

Since we require that $S_P \le 0$ and $f_P \le 0$, the terms $(-S_P \Delta x)$ and $(-f_P)$ make positive contributions to a_B. The boundary equations (2.48) and (2.53) are similar to our regular discretization equation (2.42) except that, in the boundary equation, T_B has only one neighbor T_I.

The boundary condition in which the value of the heat flux q_B is given can be considered as a special case of Eq. (2.50). If f_C equals the given flux q_B, and f_P is set equal to zero, then Eqs. (2.53)–(2.54) can be used for the given-flux condition as well. What should we do when q_B is a nonlinear function of T_B? Well, this can be handled too; we shall discuss this later in Section 2.5-5.

To summarize, when the boundary temperature is given, we do not construct the half-control-volume equation. When the heat flux q_B is specified, either as a constant or in terms of the temperature T_B, we create an extra equation like Eq. (2.48) or (2.53) to accommodate the additional unknown T_B.

The half-control-volume equation has utility even when the boundary temperature T_B is given. In that situation, *after* we solve for the grid-point temperatures, we can use Eqs. (2.46) and (2.47) to obtain the unknown heat flux q_B as

$$q_B = \frac{k_1}{(\delta x)_1} (T_B - T_1) - (S_C + S_P T_B) \Delta x \qquad (2.55)$$

It will be interesting to use this formula for the simple example we worked out in Section 2.3 and obtain the boundary heat fluxes. Then, we can verify the overall heat balance by checking whether the total heat loss at the boundaries equals the total heat generation within the solid. If you wish to go through the exercise, here are some answers that you should get. The heat *loss* at the left boundary is 8, and the heat loss at the right boundary is 2; thus the total heat loss is 10. This exactly balances SL (= 10), which is the total heat generation in the rod of length L and a unit cross-sectional area. For this simple problem, the numerical values of the boundary heat fluxes are in perfect agreement with the values obtained by differentiating the exact solution for the temperature distribution given by Eq. (2.29).

2.4-4   Solution of Discretization Equations

The discretization equations of the form (2.42) for internal grid points and of the form (2.53) for boundary grid points can be solved as a set of simultaneous linear equations. We have already experienced this solution process for the simple problem in Section 2.3. Here, we shall construct the procedure as a general solution algorithm. Because our discretization equations have a particularly regular and simple form, they can be solved by a very efficient algorithm, known as the TDMA (TriDiagonal-Matrix Algorithm). The name TDMA comes from a certain property of the coefficient matrix of our discretization equations; in that matrix, the nonzero coefficients lie only along three adjacent diagonals of the matrix.

Let us number the grid points as shown in Fig. 2.9. Here 1 and N denote the boundary points and 2, 3, ..., N−1 form the internal points. The discretization equations for the temperatures at these grid points can be written as:

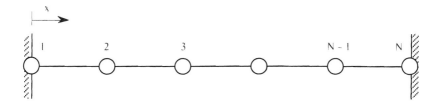

Fig. 2.9 Numbering of grid points

$$a_i T_i = b_i T_{i+1} + c_i T_{i-1} + d_i \qquad (2.56)$$

For i = 2, 3, ..., N–1, this equation represents the regular discretization equation (2.42); thus the coefficients a_i, b_i, c_i, and d_i obviously correspond to a_P, a_E, a_W, and b respectively. For the boundary points 1 and N, we shall have equations of the form (2.53). This means

$$a_1 T_1 = b_1 T_2 + d_1 \qquad (2.57)$$

$$a_N T_N = c_N T_{N-1} + d_N \qquad (2.58)$$

where a_1, b_1, and d_1 are the same as a_B, a_I, and b in Eq. (2.53) and a similar correspondence can be worked out for a_N, b_N, and d_N. Because the boundary temperatures have only one neighbor, Eqs. (2.57) and (2.58) can be regarded as members of Eq. (2.56) if we set

$$c_1 = 0 \quad \text{and} \quad b_N = 0 \qquad (2.59)$$

Strictly speaking, the boundary equations (2.57) and (2.58) come from Eq. (2.53) when the boundary heat flux is specified in some form. What should we do when the boundary temperature is given? We can still continue to use Eqs. (2.57) and (2.58) for the boundary temperatures but set their coefficients so as to include the given value. Thus, when T_1 is given, we can set

$$a_1 = 1, \quad b_1 = 0, \quad \text{and} \quad d_1 = \text{the given value of } T_1 \qquad (2.60)$$

Similar specifications can be worked out for the case of given T_N. This expedient of always treating T_1 and T_N as unknowns (even when their values are given) enables us to employ N equations for the N unknowns for any combination of boundary conditions.

In the tridiagonal-matrix algorithm, we begin by writing Eq. (2.57) as

$$T_1 = P_1 T_2 + Q_1 \tag{2.61}$$

where

$$P_1 = \frac{b_1}{a_1} \quad \text{and} \quad Q_1 = \frac{d_1}{a_1} \tag{2.62}$$

This relation is substituted into Eq. (2.56) for $i = 2$. The result is that T_2 is expressed in terms of T_3. As we continue the successive-substitution process, we are able to express each T_i in terms of T_{i+1}. Thus

$$T_i = P_i T_{i+1} + Q_i \tag{2.63}$$

where P_i and Q_i are new coefficients created in the substitution process. To derive a formula for P_i and Q_i, let us imagine that we are at a stage in the substitution process when T_{i-1} has just been expressed as

$$T_{i-1} = P_{i-1} T_i + Q_{i-1} \tag{2.64}$$

Now, if we substitute this relation into Eq. (2.56), we get

$$a_i T_i = b_i T_{i+1} + c_i (P_{i-1} T_i + Q_{i-1}) + d_i \tag{2.65}$$

which can be written in the form of Eq. (2.63). Then we can identify the expressions for P_i and Q_i as

$$P_i = \frac{b_i}{a_i - c_i P_{i-1}} \tag{2.66a}$$

$$Q_i = \frac{d_i + c_i Q_{i-1}}{a_i - c_i P_{i-1}} \tag{2.66b}$$

Incidentally, for computational efficiency, note that the denominators in the expressions for P_i and Q_i are identical.

Equations (2.66) are recursive relations; this means that the values of P_i and Q_i depend on the values of P_{i-1} and Q_{i-1}. Such a recursive process needs a starting point; it is conveniently provided by Eq. (2.62), which is not recursive.

When we come to the evaluation of P_N and Q_N, we find that, since $b_N = 0$, P_N will be zero. (See Eqs. (2.59) and (2.66a) for details.) As a result, according to Eq. (2.63), Q_N will indeed be the value of T_N. Once the value of T_N is known, we can begin the "back-substitution" process, in which we use Eq. (2.63) for obtaining the values of $T_{N-1}, T_{N-2}, ..., T_3, T_2,$ and T_1.

Summary of the solution algorithm. The TDMA can be performed in the following steps.

1. Calculate P_1 and Q_1 from Eq. (2.62).
2. Obtain P_i and Q_i for i = 2, 3, ..., N by using the recursive relations (2.66).
3. Set $T_N = Q_N$.
4. Substitute into Eq. (2.63) for i = N-1, N-2, ..., 3, 2, 1 to obtain T_{N-1}, T_{N-2}, ..., T_3, T_2, T_1.

It is true that such a convenient direct solution method is available for only one-dimensional problems. However, as we shall see later, the TDMA plays an important role in our solution of the two-dimensional equations as well. This will be described later in Section 5.6.

2.4-5   A Sample Problem

Before we leave Section 2.4, let us apply the technique developed so far to the one-dimensional fin problem outlined in Fig. 2.6. This problem enables us to experience nearly all aspects of the calculation procedure and appreciate the power of the numerical method.

The problem is governed by Eq. (2.39) and the boundary conditions

$$x = 0 : \qquad T = T_A \tag{2.67a}$$

$$x = L : \qquad q_B = 0 \text{ (insulated tip)} \tag{2.67b}$$

Let us use the following numerical values

$$T_A = 200, \ T_\infty = 100, \ k = 50, \ L = 5, \ h = 12, \ P = 6, \ A = 3 \tag{2.68}$$

Here, as in Section 2.3, no units are mentioned for these physical quantities. It is understood that the values are in any set of *consistent* units. We need not state the actual system of units being used. In a practical situation, you will of course find it convenient to use a common system of units and give the actual values of the physical parameters.

Discretization equations. Once again, we shall use the simple uniform grid shown in Fig. 2.3, for which $\delta x = 1$. Here, the regular control volumes have a width $\Delta x = 1$, while the half control volumes near the two boundaries have a width $\Delta x = 0.5$.

For the control volumes around grid points 2, 3, 4, and 5, the discretization equation (2.42) applies. The source term quantities S_C and S_P for this problem are given by Eqs. (2.44) and (2.45). Substituting the numerical values from Eq. (2.68), we get

$$S_C = 2400 \quad \text{and} \quad S_P = -24 \tag{2.69}$$

We can now evaluate the coefficients in Eq. (2.42) according to the expressions (2.43). Thus

$$a_E = 50, \quad a_W = 50, \quad a_P = 124, \quad \text{and} \quad b = 2400 \tag{2.70}$$

for the four regular control volumes.

Since the boundary temperature T_6 is unknown, we need a half control volume equation for the right-hand boundary. This equation has the form of (2.48). The boundary heat flux q_B is given to be zero in the boundary condition (2.67). The numerical values of the coefficients in this half-control-volume equation can be obtained from Eq. (2.49) as

$$a_1 = 50, \quad a_B = 62, \quad \text{and} \quad b = 1200 \tag{2.71}$$

Using Eqs. (2.70)–(2.71), we can write the entire set of discretization equations as

$$T_1 \quad = \quad 200 \tag{2.72a}$$

$$124\ T_2 \quad = \quad 50\ T_3 \ + \ 50\ T_1 \ + \ 2400 \tag{2.72b}$$

$$124\ T_3 \quad = \quad 50\ T_4 \ + \ 50\ T_2 \ + \ 2400 \tag{2.72c}$$

$$124\ T_4 \quad = \quad 50\ T_5 \ + \ 50\ T_3 \ + \ 2400 \tag{2.72d}$$

$$124\ T_5 \quad = \quad 50\ T_6 \ + \ 50\ T_4 \ + \ 2400 \tag{2.72e}$$

$$62\ T_6 \quad = \quad 50\ T_5 \ + \ 1200 \tag{2.72f}$$

Solution of the equations. We can implement the tridiagonal-matrix algorithm described in Section 2.4-4 by filling the following table. The values of the coefficients a_i, b_i, c_i, and d_i come from Eq. (2.72). The other quantities are calculated according to the algorithm.

i	1	2	3	4	5	6
a_i	1.	124.	124.	124.	124.	62.
b_i	0.	50.	50.	50.	50.	0.
c_i	0.	50.	50.	50.	50.	50.
d_i	200.	2400.	2400.	2400.	2400.	1200.
P_i	0.	0.4032	0.4815	0.5004	0.5051	0.
Q_i	200.	100.	71.26	59.68	54.39	106.68
T_i	200.	150.84	126.09	113.86	108.28	106.68

Comparison with the exact solution. For the fin problem, Eq. (2.39) with the boundary conditions (2.67) can be solved analytically. The solution is

$$\frac{T - T_\infty}{T_A - T_\infty} = \frac{\cosh [m(L - x)]}{\cosh (mL)} \tag{2.73}$$

where

$$m^2 = \frac{hP}{kA} \tag{2.74}$$

We can now compare our numerical solution with the values of T obtained at the grid points from Eq. (2.73). This comparison is shown in the following table.

i	1	2	3	4	5	6
T_{num}	200.	150.84	126.09	113.86	108.28	106.68
T_{exact}	200.	150.16	125.38	113.28	107.82	106.25

We can see that the numerical solution agrees very well with the exact solution. This is particularly impressive when you consider that we used only a few grid points. You may wish to explore this further by increasing the number of grid points; you will find that the numerical solution moves even closer to the exact solution.

Boundary heat flux. The heat flux q_A at the root of the fin ($x = 0$) can be calculated from the half-control-volume equation for that boundary, as given by Eq. (2.55). With the numerical values in this problem, we get

$$q_{A,num} = 3657.9 \tag{2.75}$$

The boundary heat flux q_A obtained from the exact solution (2.73) is given by

$$q_A = (hPk/A)^{1/2} (T_A - T_\infty) \tanh (mL) \tag{2.76}$$

With our present data, this evaluates to

$$q_{A,exact} = 3457.3 \tag{2.77}$$

Once again, the agreement between numerical and exact solutions can be seen to be very good.

Finally, you may wish to verify the overall heat balance implied in our numerical solution. Calculate the source term $(S_C + S_P T_P) \Delta x$ for each control volume (including the two half control volumes). The sum of all these terms must equal the heat flux q_A at the root of the fin. In this manner, you can show that, even when you use only a few grid points, our heat balance is perfect (within the roundoff error of your calculator or computer).

An important digression. For the lack of a better place, let us consider here some special characteristics of the role of the temperature in heat conduction. For most part, the temperature acts as a potential, i.e., a temperature *difference* causes a heat flow, but the value of the temperature does not have an influence. In the problem considered here, instead of using $T_A = 200$ and $T_\infty = 100$, we could have used $T_A = 250$ and $T_\infty = 150$, or $T_A = 100$ and $T_\infty = 0$, or $T_A = 2000$ and $T_\infty = 1900$; then our solution for T would have simply shifted by a constant. What controls the problem is the temperature difference $(T_A - T_\infty)$, not the actual values of T_A and T_∞.

It is true that the *value* of the temperature would influence physical properties such as the conductivity and specific heat. But these should be regarded as second-order effects: the main process in heat conduction, and in convection for that matter, is controlled by the temperature *differences*, and not by the actual values of the temperature. When thermal radiation is considered, the radiative heat transfer is proportional to the fourth power of the *absolute* temperature; then the value of the temperature has a specific meaning and cannot be shifted by an arbitrary constant.

With this background, let us discuss a certain mistake that people commonly make. After we have obtained the numerical and the exact values of the temperature, we would like to calculate the percentage error or the relative error incurred in the numerical solution. We tend to define the relative error in the temperature as $(T_{num} - T_{exact})/T_{exact}$. It is a mistake to define the relative error in this manner: for it gives importance to the *value* of T_{exact}, which can shift up or down by an arbitrary constant. The errors would appear much smaller if we used $T_A = 2000$ and $T_\infty = 1900$, instead of $T_A = 200$ and $T_\infty = 100$. Yet, we are solving exactly the same problem. The error will become enormous if the local value of T_{exact} happens to be close to zero. In heat conduction, we must remember that only the temperature *differences* are important, not the *values* of the temperature.

A satisfactory definition of the relative error in temperature would be $(T_{num} - T_{exact})/(T_A - T_\infty)$. The physical problem is controlled by the temperature difference $(T_A - T_\infty)$; therefore, it is reasonable to judge the difference of the numerical solution and the exact solution by reference to the overall temperature difference specified in the problem definition. In general, a convenient definition of the relative error is $(T_{num} - T_{exact})/(T_{max} - T_{min})$, where T_{max} and T_{min} are the maximum and minimum temperatures in the temperatures in the temperature field given by the exact solution.   •

These considerations about the temperature also apply to other quantities such as the pressure, the velocity potential, and the electrostatic potential. The reference values of these quantities can be varied arbitrarily; only their differences are meaningful.

The same considerations do not apply to other physical quantities. If we were to compare the values of the heat flux in Eqs. (2.75) and (2.77), the relative error can be properly defined as $(q_{A,num} - q_{A,exact})/q_{A,exact}$.

2.5   Further Refinements

We have just completed the basic construction of our numerical method for steady one-dimensional heat conduction. We shall now discuss a number of advanced features in the one-dimensional context. This discussion will provide the necessary background for our later work on two-dimensional problems.

2.5-1   Grid Spacing

We have already noted that the grid points in Fig. 2.4 need not be uniformly spaced. Although uniform grid spacing in convenient, there may often be an advantage in using a nonuniform grid. In most problems, there are regions where T varies steeply. In these regions, we can get a better resolution of these regions and improved accuracy if we crowd the grid points there. In regions where there is a very gradual variation of T, the grid points can be spaced further apart. In this manner, a nonuniform grid enables us to deploy a given number of grid points in the most effective manner. To decide the actual distribution of the grid points in a given problem. we can use our physical knowledge of the problem. some insight from analytical solutions. and experience gained from exploratory numerical solutions.

2.5-2 Nonuniform Conductivity

It is quite common to encounter situations where the thermal conductivity k varies with the distance x. This may occur because the solid is made of different materials at different locations or because the conductivity depends on temperature. In most instances. we shall not have a formula for the k~x variation; instead. we shall normally know the discrete values of k at the grid points. Then our task is to obtain the coefficients a_E and a_W in Eq. (2.43) in terms of the conductivities k_W. k_P. k_E at the grid points.

We can work out different averaging techniques for this purpose. The technique that we shall use in this book is based on a simple physical concept and has many advantages. A detailed discussion of this topic can be found in Patankar (1978. 1980). Here only the recommended formulation will be given.

If the conductivity is given only at the grid points. it is reasonable to consider that the conductivity at a grid point remains constant over the control volume surrounding it. In other words. each control volume is filled with a material of uniform conductivity (corresponding to the grid-point conductivity).

Consider the interface e shown in Fig. 2.10. Its distances from the grid points P and E are denoted by $(\delta x)_{e-}$ and $(\delta x)_{e+}$. which in general may not be equal. As we have seen before. the coefficient a_E in Eq. (2.43) represents the *conductance* of the material from P to E. The conductance is the reciprocal of the resistance. The *resistance* from P to E is the sum of the resistances for the segments Pe and eE. Thus.

$$a_E = \frac{k_e}{(\delta x)_e} = \left[\frac{(\delta x)_{e-}}{k_P} + \frac{(\delta x)_{e+}}{k_E} \right]^{-1} \qquad (2.78)$$

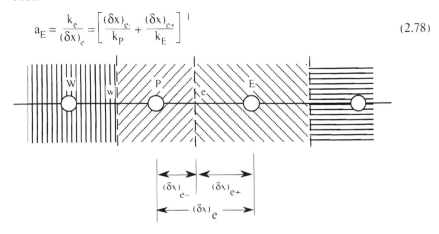

Fig. 2.10 Distances of the face e from grid points

A similar formula can be worked out for a_W.

One advantage of this formula is that we can handle large discontinuities in conductivity at the control-volume faces without having to use a particularly fine grid or other special treatment at the discontinuity. Thus, we can include highly conducting regions and nonconducting regions in the same calculation domain.

It is interesting to examine one more implication of Eq. (2.78). In adopting this formula, we have effectively replaced the piecewise-linear profile in Fig. 2.5 by a two-part profile shown in Fig. 2.11. Now the T~x variation between P and E is not just a single straight line. Because the conductivities k_P and k_E are different, there is a discontinuity of slope (dT/dx) at the control-volume face e. The heat flux q_e can now be expressed as

$$q_e = \frac{k_P}{(\delta x)_{e-}} (T_P - T_e) = \frac{k_E}{(\delta x)_{e+}} (T_e - T_E) \tag{2.79}$$

On one hand, we can eliminate T_e from this expression to get

$$q_e = \left[\frac{(\delta x)_{e-}}{k_P} + \frac{(\delta x)_{e+}}{k_E} \right]^1 (T_P - T_E) \tag{2.80}$$

which directly leads to Eq. (2.78). On the other hand, we can get from Eq. (2.79) a formula for T_e as

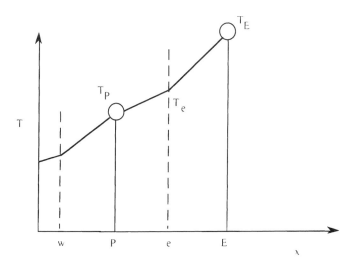

Fig. 2.11 Two-part profile between grid points

$$T_e = \frac{F_P T_P + F_E T_E}{F_P + F_E} \tag{2.81}$$

where

$$F_P = \frac{k_P}{(\delta x)_{e-}} \quad \text{and} \quad F_E = \frac{k_E}{(\delta x)_{e+}} \tag{2.82}$$

In our calculation scheme. we normally obtain the temperatures at the grid points: we do not calculate the temperatures at the control-volume faces. However, whenever there is a need to get the temperature at a control-volume face, Eq. (2.81) should be used.

2.5-3   Nonlinearity

In Section 2.4. we constructed our discretization equations as *linear* algebraic equations and solved them by an algorithm designed for linear equations. However. it is quite common to encounter *nonlinear* situations in heat conduction. For example. the conductivity k may depend on temperature or the heat generation rate S may be a nonlinear function of T: the boundary conditions can also be nonlinear. In these situations. the coefficients a_P. a_E. a_W. and b in the discretization equation depend on the temperature T. Then the discretization equation is not a true linear equation.

To solve nonlinear problems. we use iteration. We start by guessing or estimating the temperature at all grid points. Treating this as the known temperature field. we calculate tentative values of the coefficients in the discretization equations. Then we solve the equations to obtain a new temperature field. Regarding this as a better estimate for temperature. we recalculate the coefficients and solve the equations again. This process is repeated until the solution ceases to change from iteration to iteration. This unchanging solution is called the converged solution. (In practice. the iterations are terminated when the change in the solution is smaller than a prescribed small number. Such a condition is known as a convergence criterion.)

It is possible that the successive iterations sometimes do not converge but lead to a drifting or oscillating solution. This is called divergence. Of course. it is essential to prevent divergence of the iterative process. In Section 2.4-1. we noted that the coefficients in the discretization equation must be positive. One reason for this requirement is that it promotes convergence. Thus. our insistence that S_P in the source-term expression (2.41) must have a negative value is crucial for the success of the iterative process: for the same reason. we require that the boundary condition coefficient f_P in Eq. (2.50) must be less than zero. Further measures for avoiding divergence will be discussed in Section 2.5-6.

The iterative technique allows us, at least in principle, to solve any nonlinear problem. Although we proceed through a series of nominally linear equations, the final converged solution is indeed the correct solution of the nonlinear problem.

2.5-4   Source-Term Linearization

General considerations. In Eq. (2.41), we expressed the average source term in a control volume as a linear function of T_P. When the source term is a nonlinear function of temperature, the quantities S_C and S_P would themselves become functions of temperature and would have to be iteratively updated. There are many ways to "linearize" the source term, i.e., to find the expressions for S_C and S_P. We shall now look at a practice that seems most desirable. However, other ways need not be ruled out.

Let T_P^* represent the current estimate of T_P. This may be the initial guess or the value obtained from the previous iteration. If the source term S is a function of T, we can approximate it as

$$\bar{S} = \bar{S}^* + \left(\frac{d\bar{S}}{dT}\right)^* (T_P - T_P^*) \tag{2.83}$$

where T_P is the unknown value of temperature. If we compare this with our linearized formula

$$\bar{S} = S_C + S_P T_P \tag{2.84}$$

we get

$$S_C = \bar{S}^* - \left(\frac{d\bar{S}}{dT}\right)^* T_P^* \tag{2.85a}$$

$$S_P = \left(\frac{d\bar{S}}{dT}\right)^* \tag{2.85b}$$

We shall use this practice whenever it leads to an S_P that is less than zero. Otherwise, S_P must be set equal to zero and the entire source term expressed through S_C. Some examples will make this clear.

Examples. Let us apply these ideas to some specific cases.

1. $S = 7 - 3\,T$. Here it is obvious that $S_C = 7$ and $S_P = -3$.

2. $S = 6 + 5\,T$. If we set $S_C = 6$ and $S_P = 5$, we get a positive value of S_P, which is undesirable. Therefore, we use $S_C = 6 + 5\,T_P^*$ and $S_P = 0$. Of course, S_C will have to be iteratively updated.

3. $S = 2 - 3\,T^3$. Now we use the prescription given in Eq. (2.85). The result is $S_C = 2 + 6\,T_P^{*3}$ and $S_P = -9\,T_P^{*2}$. Since S_P will always be less than zero, this is a satisfactory linearization.

4. $S = 2 + 3\,T^3$. Here, the use of Eq. (2.85) would lead to a positive value of S_P. Therefore we use $S_C = 2 + 3\,T_P^{*3}$ and $S_P = 0$.

Final remarks. The source-term linearization is a very useful technique for handling temperature-dependent source terms. It allows us to anticipate the change in the source term due to the change in temperature. However, to avoid the possibility of divergence, we do not use formulations that would lead to a positive value of S_P. You have considerable freedom in deciding the expressions for S_C and S_P provided that S_P does not take a positive value and that the total source term given by Eq. (2.84) agrees with the prescribed S~T relationship at convergence ($T_P = T_P^*$). You may wish to test different alternative formulations for S_C and S_P and see the corresponding rate of convergence. In general, you will find that the formulation based on Eq. (2.85) converges most rapidly.

In addition to using the linearized source term for its intended purpose, we sometimes employ it to handle irregular geometry. This nonstandard use of the source term will be described in Section 7.6.

2.5-5   Boundary-Condition Linearization

In our treatment of boundary conditions, we expressed in Eq. (2.50) the boundary flux q_B as a linear function of the boundary temperature T_B. The coefficients f_C and f_P in Eq. (2.50) play the same role as S_C and S_P in Eq. (2.41). Therefore, our considerations for the source-term linearization apply to the linearization of the boundary heat flux as well. When q_B is truly a nonlinear function of T_B, the coefficients f_C and f_P will be obtained from T_B^* (our current estimate of T_B) and will have to be iteratively recalculated. We also require that f_P must be less than or equal to zero.

It will be our practice to linearize a nonlinear q_B~T_B relationship as

$$q_B = q_B^* + \left(\frac{dq_B}{dT_B}\right)^* (T_B - T_B^*) \qquad (2.86)$$

On comparing this with Eq. (2.50), we get

$$f_C = q_B^* - \left(\frac{dq_B}{dT_B}\right)^* T_B^* \qquad (2.87a)$$

$$f_P = \left(\frac{dq_B}{dT_B}\right)^* \qquad (2.87b)$$

The similarity of these relations with Eq. (2.85) should be obvious. Again. we shall not use Eq. (2.87) if it leads to a positive value of f_P. (Such physical situations are extremely rare. Almost invariably, you will get a naturally negative f_P.)

As an illustration, let us consider the following expression for the boundary heat flux q_B.

$$q_B = 5 (T_\infty - T_B) + 2 (T_\infty^4 - T_B^4) \qquad (2.88)$$

Here the two terms represent convective and radiative heat transfer respectively. The use of Eq. (2.87) leads to

$$f_C = 5 T_\infty + 2 T_\infty^4 + 6 T_B^{*4} \qquad (2.89a)$$

$$f_P = - (5 + 8 T_B^{*3}) \qquad (2.89b)$$

If T is regarded as the absolute temperature (which is consistent with the radiation term). T_B will always be positive; therefore. Eq. (2.89b) leads to a negative value of f_P. The coefficients f_C and f_P as given by Eq. (2.89) must be iteratively recalculated.

2.5-6   Underrelaxation

In an iterative solution of a nonlinear problem. as the temperature field changes every iteration, the discretization coefficients are recalculated. If the values of these coefficients change very rapidly from iteration to iteration, we may risk divergence. Therefore, in a highly nonlinear problem, it is desirable to slow down the changes in temperature from one iteration to the next. This process is called underrelaxation. It does not influence the temperature field in the final converged solution; it simply affects the *approach* to convergence.

Underrelaxation can be introduced in many different ways. We shall discuss only the particular practice employed in this book. Let us first consider Eq. (2.36) in absence of the source term. Then we can write

$$(a_E + a_W) T_P = a_E T_E + a_W T_W \qquad (2.90)$$

This can be generalized according to Eq. (2.38) as

$$(\Sigma a_{nb}) T_P = \Sigma a_{nb} T_{nb} \qquad (2.91)$$

This equation implies that T_P is a weighted average of the neighbor temperatures. If we use $\widetilde{T}_P$ to denote this average, we get

$$\tilde{T}_P = \frac{\Sigma a_{nb} T_{nb}}{\Sigma a_{nb}} \tag{2.92}$$

To slow down the change in T_P from iteration to iteration, we propose that the new value of T_P will be a combination of $\tilde{T}_P$ and the value T_P from the previous iteration. Thus,

$$T_P = \alpha \, \tilde{T}_P + (1 - \alpha) \, T_P^* \tag{2.93}$$

where α is the underrelaxation factor (between 0 and 1). If we now substitute Eq. (2.92) into (2.93) and rearrange, we get

$$(\Sigma a_{nb} + i) \, T_P = \Sigma a_{nb} T_{nb} + i \, T_P^* \tag{2.94}$$

where i is the so-called inertia given by

$$i = \frac{(1 - \alpha)}{\alpha} \, \Sigma a_{nb} \tag{2.95}$$

This underrelaxation scheme can now be generalized to the full discretization equation (2.38) with source terms and other complications. Our practice will be to introduce underrelaxation by replacing Eq. (2.38) by

$$(a_P + i) \, T_P = \Sigma a_{nb} T_{nb} + b + i \, T_P^* \tag{2.96}$$

where the inertia i is given by Eq. (2.95).

We must note that the introduction of the inertia does not alter the converged solution. When convergence is reached, T_P equals T_P^*; then the satisfaction of Eq. (2.96) also implies the satisfaction of the original equation (2.38).

The inertia equals zero when $\alpha = 1$. For smaller values of α, the role of the inertia is to hold the new T_P closer to T_P^*. The underrelaxation becomes heavier as α gets closer to zero. Of course, α should not be set equal to zero.

What is the best value of α in a given problem? Unfortunately, there are no general rules about this. An optimum value of α depends on the nature of the nonlinearity, number of grid points, boundary conditions, and other factors. You will have to find a suitable value of α by performing exploratory computations on a given problem. Still, it is nice to know that, for nonlinear problems, we can often combat possible divergence by the use of appropriate underrelaxation.

All this discussion pertains to the underrelaxation of the dependent variable T. In addition, we can usefully underrelax auxiliary quantities such as the conductivity k. If k depends on T, we can use during each iteration

$$k = \alpha \, k_{new} + (1 - \alpha) \, k_{old} \tag{2.97}$$

where k_{new} is the value obtained from the new temperature and k_{old} is the value used in the previous iteration. Again, α is the underrelaxation factor, which need not have the same value as in Eq. (2.95).

In the same manner, we can underrelax source terms, boundary values, and other quantities. Our aim is to arrange a controlled and gradual approach to the final converged solution, avoiding large jumps and oscillations that may otherwise occur.

2.5-7 Design of Control Volumes

While deriving the discretization equation in Section 2.4-1, we did not specify any precise location of the control-volume faces w and e in relation to the location of the grid points W, P, and E. In working out the sample problem in Section 2.4-5, we assumed that the control-volume faces were located exactly midway between the grid points. This is one possible practice of designing control volumes. We call it Practice A. There are other practices too. The practice that we shall employ in later chapters and in the computer program CONDUCT is called Practice B. These two practices will be described here. For your initial explorations with one-dimensional problems, you can use Practice A. However, all our later work with two-dimensional problems will be done with Practice B.

Details of Practice A. Figure 2.12 shows the locations of the control-volume faces obtained by the use of Practice A. In this method, we first place the grid points in the domain with any desired spacing, which may be nonuniform. A grid point is also placed on each boundary. Then the control-volume faces are located exactly *midway* between the grid points. This forms the regular control volumes around the internal grid points and the half control volumes near the two boundaries. For a nonuniform grid, whereas each control-volume face is *always* midway between the grid points, a grid point is not necessarily at the center of a control volume surrounding it.

Details of Practice B. General heat conduction problems may involve a discontinuity in the conductivity or heat generation rate at one or more locations

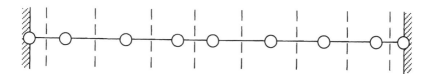

Fig. 2.12 Locations of grid points and control-volumes faces for Practice A

in the domain. In our calculation method, we treat these quantities as uniform over one control volume and allow discontinuities at the control-volume faces (as shown in Fig. 2.10 for the conductivity). Therefore, it is important that the control-volume faces are placed at the location of discontinuities. In Practice A, since we place the grid points first, it is often difficult to ensure that the resulting control-volume faces fall at the desired locations. Practice B is designed to overcome this shortcoming.

In Practice B, we first subdivide the entire calculation domain into a number of control volumes, which may have unequal widths. In doing this, we can ensure that the locations of discontinuity in conductivity or heat generation coincide with a control-volume face. Then we place a grid point in the geometrical center of each control volume. We also place a boundary grid point on each boundary. This arrangement is shown in Fig. 2.13.

For a nonuniform grid in Practice B, the control-volume faces may not be midway between the grid points; but each grid point is always at the center of the corresponding control volume. The control volumes used in Practice B are all regular control volumes; there are no half control volumes. This leads to some additional convenience.

You may, however, wonder how we can treat the boundary conditions without the half control volumes. We shall soon see that, even in Practice B, the half control volumes are present in a limiting form. Figure 2.14(a) shows a general half control volume similar to the one shown earlier in Fig. 2.8. In general, the control-volume face i can be located anywhere between the grid points B and I (thereby changing the thickness Δx of the half control volume). Now, as shown in Fig. 2.14(b), our Practice B can be interpreted as having the interface i located at B and making the thickness Δx zero. Thus, Practice B employs half control volumes of zero thickness. With this interpretation, all the equations in Section 2.4-3 are valid for Practice B as well. The only extra action required is to set $\Delta x = 0$ in these equations.

One aspect of Fig. 2.14 requires some comment. We calculate the heat flux q_i from the slope of the linear temperature profile between the grid points B and I. This approximation seems reasonable when i is located between B and I. When i coincides with B, the heat flux q_i (which now equals q_B) is still calculated from the linear profile between B and I. This one-sided formula for q_i is likely to make

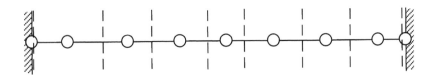

Fig. 2.13 Locations of grid points and control-volume faces for Practice B

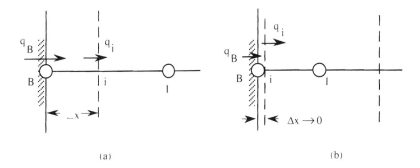

Fig. 2.14   (a) General half control volume
             (b) Zero-thickness half control volume for Practice B

the results of Practice B somewhat inaccurate. For this reason, we shall develop a
higher-order treatment for the boundary in Section 5.5-2. For the time being, you
may either overlook this shortcoming of Practice B or just use Practice A in your
one-dimensional explorations.

2.6   Unsteady Heat Conduction

After having completed the treatment of steady one-dimensional heat conduction,
let us extend the method to unsteady situations.

Preparation. The governing differential equation for unsteady one-
dimensional conduction is

$$\rho c \frac{\partial T}{\partial t} = \frac{\partial}{\partial x} \left(k \frac{\partial T}{\partial x} \right) + S \tag{2.98}$$

where the right-hand side is similar to the contents of Eq. (2.1). The quantities ρ
and c stand for the density and specific heat of the material respectively; the
product ρc can be thought of as the heat capacity per unit volume. There are now
two independent variables, the distance x and the time t. In an unsteady situation,
the initial distribution of temperature at time $t = 0$ is known. The task is to find the
temperature distributions at all subsequent instants of time.

For the numerical solution of an unsteady situation, time is discretized as
well. We consider a number of time steps and seek the temperature distribution at
various discrete values of t, which correspond to the end of each time step. Once
we learn how to perform the solution for one time step, we can repeat the same
process for any number of time steps. Therefore, the basic problem in solving

Eq. (2.98) can be stated as: given the temperature distribution at time t, find the temperature distribution at time t+Δt. Here Δt stands for the size of the time step.

Discretization method. There are a number of ways of deriving a discretization equation from Eq. (2.98). Three well-known methods are: explicit, Crank-Nicolson, and fully implicit. Here we shall not discuss these methods in detail. Their derivations can be found in many books on numerical analysis; alternatively, you may refer to Patankar (1980) for the details of these methods in the context of our control-volume formulation. In this book, we shall use the fully implicit method, because it is the only method (out of the three methods mentioned) that allows us to use any value of Δt without producing physically unrealistic results. Even the Crank-Nicolson method, which is often described as unconditionally stable, can exhibit unrealistic behavior when Δt exceeds a certain value. (Again, for further information on this topic, you may turn to Patankar, 1980 or Patankar and Baliga, 1978.) It is true that, for smaller Δt values, the Crank-Nicolson method can be more *accurate* than the fully implicit method. However, for our purposes, the guarantee of producing realistic results for any value of Δt is considered to be more important.

Discretization equation. Let T_P^0 denote the known value of T_P at time t. The quantity T_P (without the superscript) stands for the unknown temperature at time t+Δt. If the temperature T_P is considered to prevail over the control volume surrounding P, the term on the left side of Eq. (2.98) can be integrated over the control volume to give

$$a_P^0 (T_P - T_P^0) \tag{2.99}$$

where

$$a_P^0 = \frac{(\rho c)_P \, \Delta x}{\Delta t} \tag{2.100}$$

The other ingredients of the discretization equation can be taken from Sections 2.4-1 and 2.4-2; the final result is

$$a_P T_P = a_E T_E + a_W T_W + b \tag{2.101}$$

where

$$a_E = \frac{k_e}{(\delta x)_e} \tag{2.102a}$$

$$a_W = \frac{k_w}{(\delta x)_w} \tag{2.102b}$$

$$a_P^0 = \frac{(\rho c)_P \, \Delta x}{\Delta t} \tag{2.102c}$$

$$a_P = a_E + a_W + a_P^0 - S_P \, \Delta x \tag{2.102d}$$

$$b = S_C \, \Delta x + a_P^0 \, T_P^0 \tag{2.102e}$$

Here, it is interesting to note that, if $\Delta t \to \infty$, then $a_P^0 \to 0$; as a result, Eq. (2.102) reverts to the steady-state formulation given by Eq. (2.43).

In the derivation of Eq. (2.101), the fully implicit method expresses the heat conduction terms (involving T_E and T_W) and the temperature-dependent source term $(S_C + S_P \, T_P)$ by using the unknown temperatures at time $t+\Delta t$. In other words, the new (unknown) temperatures are considered to prevail throughout the time step.

Other details. All other matters considered in Sections 2.4 and 2.5 are directly applicable to the unsteady situation as well. Therefore, the treatment of boundary conditions, solution of algebraic equations, handling of the nonuniform conductivity, etc. are to be worked out in a very similar fashion. No further discussion seems to be necessary.

In fact, the only extra feature in the discretization equations for an unsteady situation is the presence of the coefficient a_P^0. Therefore, it is very convenient to write a computer program on the basis of the unsteady equations and use it for both steady and unsteady problems. When we want to use the program for a *steady* situation, all that needs to be done is to set the time step Δt to be a large number (thereby making a_P^0 almost zero).

Problems

2.1 Consider the steady one-dimensional conduction problem represented by three grid points that are uniformly spaced in the domain of length 2. The conductivity k and the heat generation S have constant values throughout the domain; k = 5, S = 150. The temperature T_1 equals 100, while at the grid point 3 heat is lost to a surrounding fluid at temperature T_∞ (= 20) with a convective heat transfer coefficient h (= 15). Using Practice A, write the discretization equations to determine the unknown temperatures T_2 and T_3. From these values, show that the overall heat balance is perfect, i.e., the heat generated in the entire domain equals the heat loss at the boundaries.

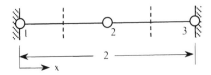

Problem 2.1

2.2 In a steady one-dimensional heat conduction situation, a rod of length 6 units has a uniform conductivity of 2.5. The source term is given by $S = 30 - 2T$. A numerical solution is to be obtained by using only three grid points as shown. Take $x_3 - x_2 = x_2 - x_1$ and use Practice A for locating the control-volume faces. The boundary conditions consist of: (a) a known heat flux $q = 15$ in maintained at $x = x_1$, while (b) at $x = x_3$, heat is exchanged with a surrounding fluid at $T_\infty = 30$ with a convective heat transfer coefficient $h = 5$.

Write the discretization equations to determine T_1, T_2, and T_3. Solve them by the TDMA to get the values of T_1, T_2, and T_3. From the value of T_3, obtain the heat loss at $x = x_3$. Also calculate the value of $S \Delta x$ for each control volume. Hence show that the overall heat balance is perfectly satisfied.

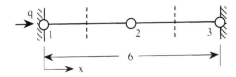

Problem 2.2

2.3 Steady one-dimensional heat conduction in a solid of nonuniform conductivity is governed by

$$\frac{d}{dx}\left(k\,\frac{dT}{dx}\right) + S = 0$$

where $S = 1$ everywhere. The boundary conditions are (i) at $x = 0$: $dT/dx = 0$, and (ii) at $x = 2$: $T = 0$. For the three-node grid shown for Problem 2.1, the grid-point values of conductivity k are given by $k = (2 + x)^3$. Construct the discretization equations for points 1 and 2. Solve these equations to obtain T_1 and T_2. Write the half control-volume equation for point 3. Use this equation to evaluate the heat flux at $x = 2$. Show that this heat flow equals the total amount of heat generated in the solid.

2.4 A boundary heat flux is given in terms of the boundary temperature by the expression $q_B = 10 - 6T_B^3$. Write the appropriate expressions for f_C and f_P in the linearized formula for the boundary flux.

2.5 A rod of conductivity k, diameter D and length L acts as a fin between the hot surfaces on the left and right at temperatures T_L and T_R. The surface of the rod loses heat to the surrounding fluid at T_f with a heat transfer coefficient h. Using a uniform, three-point grid of spacing L/2, derive the discretization equation based on Practice A for the control volume around the center grid point. Substitute the data given below and calculate the value of the center-point temperature. Use the half control volume equations to find the heat flow rates at the two ends of the rod. Finally, show that the heat flow rates at the ends and the heat loss to the surrounding fluid satisfy the overall heat balance. The given values are: $T_L = 150$, $T_R = 200$, $T_f = 20$, k = 120, h = 2.5, D = 0.1, and L = 1.

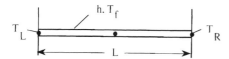

Problem 2.5

2.6 Steady one-dimensional heat conduction in a rod of nonuniform cross-sectional area is governed by

$$\frac{1}{A}\frac{d}{dx}\left(kA\frac{dT}{dx}\right) + S = 0$$

Derive the discretization equation by assuming that the values of the cross-sectional area A are available at the control-volume faces. (Multiply the equation by A and then integrate with respect to x.)

2.7 One-dimensional radial heat conduction in a cylindrical geometry is governed by

$$\frac{1}{r}\left(kr\frac{dT}{dr}\right) + S = 0$$

Derive the discretization equation resulting from this equation. (Multiply the equation by r and then integrate with respect to r.) Note that this is a special case of the general situation in Problem 2.6.

2.8 Use the derivation for Problem 2.7 to solve the following one-dimensional steady conduction problem in a hollow cylinder with uniform k and S. The inner and outer surfaces are maintained at known temperatures T_i and T_o. The radius ratio r_o/r_i equals 4. The source is given by $Sr_i^2/[k(T_o - T_i)] = 2.5$. Use only a few grid points (say 5) and compare the results with the exact solution. Calculate the boundary heat fluxes (made dimensionless if necessary) at the inner and outer surfaces and show that the heat balance is exactly satisfied even by the coarse-grid solution.

2.9 Steady one-dimensional heat conduction is a hollow sphere is governed by

$$\frac{1}{r^2}\frac{d}{dr}\left(r^2 k\frac{dT}{dr}\right) + S = 0$$

where k = 2 and S = 50. The boundary conditions are
(a) at the inner radius r = 1: T = 100
(b) at the outer radius r = 3: q = h (T – T_∞) with h = 4, T_∞ = 5.
Use a uniform three-point grid with grid points located at r = 1, r = 2, and r = 3. Write the control-volume equation for the grid point at r = 2. Calculate the numerical values of the coefficients in this equation. Also write the half-control-volume equations for the outer boundary. Solve these two equations to find the temperatures at r = 2 and r = 3.

2.10 In Problem 2.9, find the heat flow rates at r = 1 and r = 3. Also calculate the total amount of heat generation in the solid. From these numbers, show that the overall heat balance is perfectly satisfied.

2.11 In Problem 2.9, if the heat generation rate S is given by S = 50 – 0.3 $T^{1/2}$, write the control-volume equation for the grid point at r = 2. Using a current estimate of T' = 50 for the temperature at this point, find the numerical values of all the coefficients in the discretization equation.

2.12 A spherical object of radius R is embedded in an infinite solid material at a temperature T_0. The surface of the sphere is maintained at a temperature T_1. Calculate the steady-state temperature field outside the sphere by using 5 grid points. Compare the numerical solution with the exact solution. If Q is the total heat loss from the sphere, calculate the value of $Q/(kR(T_1 – T_0))$ from the numerical solution and compare it with the exact solution. (Place the outer boundary of the domain at a distance of 4R from the sphere.)

2.13 One-dimensional steady heat conduction in the triangular fin shown is to be calculated on the uniform three-point grid. The base width b equals 0.2. The fin cross section remains uniform in the direction normal to the diagram. For the depth D in that direction and a distance dx in the x direction, the fin surface area for heat transfer to the surrounding fluid can be taken as 2D dx. Use other data as: k = 50, T_1 = 120, T_∞ = 20, h =1. Write the discretization equations for T_2 and T_3 (use Practice A). Solve these equations to find the tip temperature T_3.

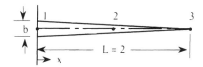

Problem 2.13

2.14 In Problem 2.13. find the heat flow rate at the base of the fin. Then verify that it exactly equals the total heat loss by convection from the entire fin.

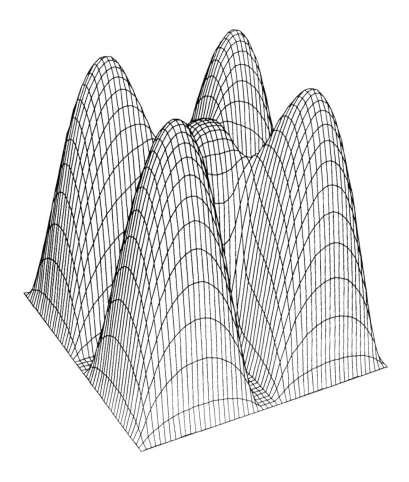

Axial velocity distribution in a square duct with four internal fins
(The fin height equals 0.35 times the side of the square.)

Chapter 3

General Mathematical Framework

Since this book is concerned with the simulation of conduction-like phenomena, we shall first examine the differential equation for heat conduction and then generalize it for use in other analogous processes. Let us assume that a detailed derivation of such equations is available to you from other sources. Here, our aim is to assemble and understand the equations required for the construction of the numerical method and the associated computer program.

3.1   The Heat Conduction Equation

The Fourier law of heat conduction is expressed as:

$$q_x = - k \frac{\partial T}{\partial x} \tag{3.1}$$

where q_x is the heat flux (i.e., the heat flow rate per unit area) in the x direction and k is the thermal conductivity of the material. The heat flux in a given direction is thus proportional to the temperature gradient (such as $\partial T / \partial x$) in that direction. The minus sign in Eq. (3.1) reflects our understanding of heat flow and temperature; that is, heat flows from a high temperature to a low temperature.

The application of the First Law of Thermodynamics to an infinitesimal control volume in a solid or a stationary medium leads to the well-known heat conduction equation. It contains terms, which, on a unit-volume basis, represent:
(a) the rate of change of stored internal energy,
(b) the net rate of outflow of heat by conduction, and
(c) the rate of energy generation (or "heat source") by processes such as nuclear or chemical reactions, radioactivity, and passage of electrical current through the material.

Denoting the three space dimensions by the Cartesian coordinates x, y, z and the time by t, we can write the heat conduction equation as:

$$\rho c \frac{\partial T}{\partial t} = \frac{\partial}{\partial x} \left(k \frac{\partial T}{\partial x} \right) + \frac{\partial}{\partial y} \left(k \frac{\partial T}{\partial y} \right) + \frac{\partial}{\partial z} \left(k \frac{\partial T}{\partial z} \right) + S_h \tag{3.2}$$

where ρ is the density, c is the specific heat, and S_h represents the heat source per unit volume.

In this equation, the quantities ρ, c, k, and S_h may depend on x, y, z and t, and may be functions of the temperature T. Our numerical method and computer program should be flexible enough to allow for such dependence.

The differential equations (2.1) and (2.98) that we used in Chapter 2 can now be seen as special cases of Eq. (3.2). The only extra feature of Eq. (3.2), compared to Eq. (2.98), is the presence of the heat conduction terms in all coordinate directions.

3.2   General Differential Equation

Analogous processes. The heat conduction equation is a direct result of the gradient-driven flux law expressed by Eq. (3.1) and of the conservation principle implied by the First Law of Thermodynamics. There are many other physical processes in which the relevant flux is also governed by a gradient law and for which a conservation principle exists. Then it follows that these processes will be governed by differential equations that have the same appearance as Eq. (3.2). Such processes include the diffusion of a chemical species, motion of charged particles in electromagnetic fields, flow in porous materials, potential flow, lubrication, heat and moisture transport in soil, and fully developed flow and heat transfer in ducts. Once a computational procedure is constructed for solving Eq. (3.2), it can also be used for any of the analogous processes simply by giving new meanings to the quantities T, k, S_h, etc. For example, in analyzing the diffusion of a chemical species, we may interpret T as the concentration, k as the diffusion coefficient, S_h as the chemical reaction rate, and so on. Alternatively, we can work with a general differential equation such that the equations for heat conduction and other analogous processes become its particular cases. The presentation in this book is based on such a general differential equation.

Differential equation for a general variable. Let ϕ denote the dependent variable of the general differential equation. The gradient of ϕ causes the corresponding diffusion flux. Thus, for the x direction, the diffusion flux J_x is given by:

$$J_x = - \Gamma \frac{\partial \phi}{\partial x} \tag{3.3}$$

where Γ is the generalized diffusion coefficient. The general differential equation representing the conservation of ϕ can now be written as:

$$\lambda \frac{\partial \phi}{\partial t} = \frac{\partial}{\partial x}\left(\Gamma \frac{\partial \phi}{\partial x}\right) + \frac{\partial}{\partial y}\left(\Gamma \frac{\partial \phi}{\partial y}\right) + \frac{\partial}{\partial z}\left(\Gamma \frac{\partial \phi}{\partial z}\right) + S \tag{3.4}$$

where λ is the storage capacity per unit volume (akin to the heat capacity ρc), and S is the generation rate per unit volume for the relevant physical quantity.

By comparing the heat conduction equation, Eq. (3.2), with the general differential equation, Eq. (3.4), we can conclude that, for the general equation to represent the heat conduction process, the following choices are appropriate.

$$\phi = T, \qquad \lambda = \rho c, \qquad \Gamma = k, \qquad S = S_h \tag{3.5}$$

In general, for each meaning of ϕ, there exist the corresponding meanings of λ, Γ, and S. (Indeed, it is useful to think of λ, Γ, and S as λ_ϕ, Γ_ϕ, and S_ϕ, although the subscript ϕ will normally be omitted for convenience.) In a simple problem, these quantities may be constant; but in general it is not uncommon for them to depend on the space coordinates and time, and on the variable ϕ itself.

Interlinkage between variables. A given physical problem can have more than one variable governed by the general differential equation. For example, in a mixture of many chemical species, the dependent variables are the concentrations of individual species. For the calculation of forced convection in a duct, we need to obtain both the axial velocity and the temperature by solving the appropriate forms of Eq. (3.4). In a plasma process, the dependent variable ϕ can stand for the electron temperature and also for the heavy-particle temperature. Often, these different manifestations of ϕ in a given situation are interlinked; that is, the quantities λ, Γ, and S for one ϕ may depend on the values of another ϕ. Therefore, the overall computational problem consists of the solution of a *system* of nonlinear and coupled differential equations of the form expressed by Eq. (3.4).

Tensor notation. The general differential equation, Eq. (3.4), is written in Cartesian coordinates. It can be expressed more compactly in the Cartesian-tensor notation as:

$$\lambda \frac{\partial \phi}{\partial t} = \frac{\partial}{\partial x_i} \left(\Gamma \frac{\partial \phi}{\partial x_i} \right) + S \tag{3.6}$$

where the summation convention is implied; that is, when a subscript is repeated in a term, a summation over all the space coordinates is to be taken. Thus, for a *three-dimensional* situation,

$$\frac{\partial}{\partial x_i} \left(\Gamma \frac{\partial \phi}{\partial x_i} \right) = \frac{\partial}{\partial x_1} \left(\Gamma \frac{\partial \phi}{\partial x_1} \right) + \frac{\partial}{\partial x_2} \left(\Gamma \frac{\partial \phi}{\partial x_2} \right) + \frac{\partial}{\partial x_3} \left(\Gamma \frac{\partial \phi}{\partial x_3} \right) \tag{3.7}$$

where x_1, x_2, and x_3 are the Cartesian coordinates (x, y, and z) in the three directions. We shall refer to the three terms in Eq. (3.6) as the unsteady term, the diffusion term, and the source term, respectively.

The diffusion flux expression given by Eq. (3.3) can now be written as:

$$J_i = - \Gamma \frac{\partial \phi}{\partial x_i} \tag{3.8}$$

where J_i represents the diffusion flux in the direction of the coordinate x_i. The diffusion term in Eq. (3.6) is related to J_i by:

$$\frac{\partial}{\partial x_i} \left(\Gamma \frac{\partial \phi}{\partial x_i} \right) = - \frac{\partial J_i}{\partial x_i} \tag{3.9}$$

Expressions in the three coordinate systems. In this book, we shall focus on *two-dimensional* problems formulated in one of the three coordinate systems introduced in Section 1.2. The expressions for $\partial J_1/\partial x_1$ in these systems are given by:

$$(x, y): \quad \frac{\partial J_1}{\partial x_1} = \frac{\partial J_x}{\partial x} + \frac{\partial J_y}{\partial y} \tag{3.10}$$

$$(x, r): \quad \frac{\partial J_1}{\partial x_i} = \frac{\partial J_x}{\partial x} + \frac{1}{r}\frac{\partial}{\partial y}\left(rJ_y\right) \tag{3.11}$$

$$(\theta, r): \quad \frac{\partial J_1}{\partial x_1} = \frac{1}{r}\frac{\partial J_\theta}{\partial \theta} + \frac{1}{r}\frac{\partial}{\partial y}\left(rJ_y\right) \tag{3.12}$$

In Eqs. (3.11) and (3.12), both y and r are coordinates in the radial direction. The only difference between y and r is that, whereas r must be measured from the axis of symmetry or the pole, the origin for y can be chosen arbitrarily. Thus,

$$y = r + \text{constant} \tag{3.13}$$

The fluxes J_x, J_y, J_θ appearing in Eqs. (3.10)–(3.12) are related to the gradients of ϕ by:

$$J_x = -\Gamma\frac{\partial \phi}{\partial x} \tag{3.14}$$

$$J_y = -\Gamma\frac{\partial \phi}{\partial y} \tag{3.15}$$

$$J_\theta = -\left(\frac{\Gamma}{r}\right)\frac{\partial \phi}{\partial \theta} \tag{3.16}$$

As a matter of convenience, the derivations in this book will often be given in only the xy coordinate system; their counterparts in the other two coordinate systems, however, should be easy to work out.

3.3 Boundary Conditions

For conduction-like problems, three types of boundary conditions are commonly encountered. At a boundary location, either the value of ϕ is specified, or the flux J_1 (normal to the boundary surface) is given, or a relation between the flux and the boundary value of ϕ is prescribed. In the context of one-dimensional heat conduction, we have encountered these boundary conditions in Chapter 2. In general, our calculation method and the computer program should be able to handle all these boundary conditions for each dependent variable.

3.4 Dimensionless Variables

The physical variables used in this chapter and in the computer program CONDUCT should, in general, be considered as *dimensional* quantities expressed in any *consistent* set of units. No conversion factors are embedded in the computer program. Therefore, although care is needed in using quantities in British units, no difficulty arises if standard S. I. units are used. The results of CONDUCT should be regarded as the values of the actual physical quantities, just like the data from a laboratory experiment.

In Chapter 2, we solved problems by giving simple numerical values to the physical parameters such as the conductivity k and the heat source S, without stating the units employed. For example, in Section 2.3 we used: $k = 1$, $S = 2$, $L = 5$, etc. We shall continue such a practice throughout the book. When the units are not mentioned, it is implied that the numbers represent the relevant physical quantities in any consistent set of units. In practical applications, the units are certainly important; but for the purpose of this book, there is no need to clutter our work with complicated units and numbers.

For many problems, however, the governing equations and their solutions are commonly expressed in terms of appropriate dimensionless quantities. This procedure is essential to determine the minimum number of dimensionless parameters that influence the situation and to present the solution in a generalized form. If you are used to analyzing problems in terms of dimensionless quantities, you may reasonably ask why CONDUCT does not have a dimensionless structure.

The reason is twofold. First, if you want to solve a practical problem directly without concerning yourself with the corresponding dimensionless generalization, you should be able to use CONDUCT simply by providing the actual geometry, temperatures, heat fluxes, etc. and receive the results in terms of various physical quantities. Second, since a single set of dimensionless variables is not applicable to all possible problems, a *general-purpose* computer program such as CONDUCT cannot be designed with *a priori* definitions of the required dimensionless variables.

Despite the foregoing comments, the program CONDUCT can indeed be used to *obtain* dimensionless solutions for a given class of problems. There are two ways of doing this. In the first method, the governing differential equation in *dimensionless* form is compared with Eq. (3.6) (which CONDUCT solves) and the quantities t, x, φ, Γ, S, etc. in the computer program are interpreted as the corresponding dimensionless variables. For example, the variable x may stand for the dimensionless distance x/L; φ may be taken to represent the dimensionless temperature $(T - T_1) / (T_2 - T_1)$; Γ may simply become unity; and so on. The second method is very similar to conducting a laboratory experiment. We perform the computations in terms of *dimensional* quantities, but print out the results in an appropriate dimensionless form. We can even verify the validity of the dimensionless formulation by checking whether the dimensionless results remain unchanged (as they should) when the dimensional values of the domain size, material properties, etc. are changed. It is very instructive to regard a computer run as akin to a laboratory experiment. One difference, however, is that, whereas an experiment must be conducted with practical sizes and available materials, any size of the domain and any material properties (even fictitious ones) can be used in a computation. Between the two methods of obtaining dimensionless solutions, I personally prefer the second method; but you can use whatever suits your taste.

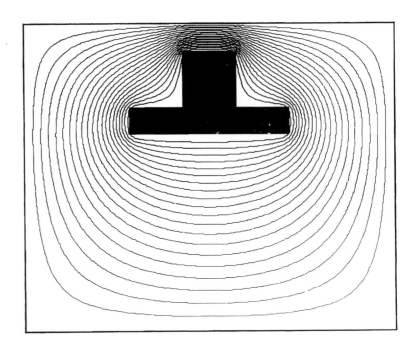

Temperature contours around a hot body embedded in a large solid

Chapter 4

Structure of the Computer Program

At this early point in the book, it is useful to develop some familiarity with the structure of the computer program CONDUCT. As already mentioned, the program is divided into two parts: the invariant part and the adaptation part. The invariant part embodies the general numerical procedure to be described in Chapter 5. The adaptation part provides the framework for the user-supplied information for the specific problem to be solved. A brief description of the program is given in this chapter; more complete details follow in the remainder of the book.

4.1 Flow Diagram

Figure 4.1 shows the flow diagram of the computer program. It indicates all the subprograms and their interrelations. Only the important "calls" are shown; double arrows imply that multiple calls are made to the particular subroutine. The subprograms enclosed within the dashed box constitute the invariant part. The subroutine ADAPT is meant for the problem-dependent details.

The subprograms such as DEFRD and ADAPT are actually assemblies of a number of member subroutines, whose names are shown in the appropriate boxes in Fig. 4.1. These assemblies are created by using the ENTRY statement in a novel manner. In such an assembly, all the member subroutines start with an ENTRY statement and can be used, for all practical purposes, as independent subroutines. The assemblies are themselves never "called" by their names; only the member routines are called. Thus, the statement CALL ADAPT is never used in the program; but CALL BEGIN provides access to the appropriate member of ADAPT. Except VALUES, no subroutine has any arguments; all the required information is transferred through COMMON statements. The expedient of using the ENTRY statement to stack a number of member subroutines in one nominal subprogram reduces the repetition of COMMON statements, enables the member routines to share variables without passing them through COMMON, and keeps the related physical information or mathematical operations together.

4.2 Subprograms in the Invariant Part

The MAIN subprogram controls the sequence of important operations by calling a number of subroutines as shown in Fig. 4.1. Some routines are called only once; these calls constitute the getting-ready phase of the computation. The remaining operations are arranged in a loop; each pass through the loop represents one iteration for a steady problem or one time step for an unsteady problem.

"invariant" part

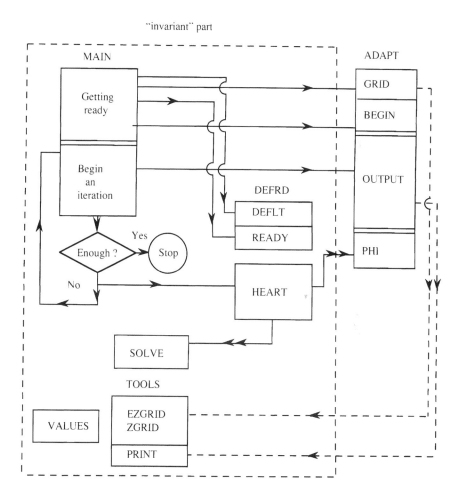

Fig. 4.1 Flow diagram of the computer program

In the subroutine DEFLT, we assign the "default values" to a number of important variables. It will be helpful for you to know these default values well. Then, while specifying the details of your problem, you need not assign values to certain parameters if their default values are acceptable. For example, if your problem does not involve internal heat generation, you need not specify the value of the source term; it will remain zero by default. Incidentally, all the default values are listed in Appendix C.

The subroutine READY is designed for the calculation of numerous geometrical quantities that are frequently needed in the rest of the program. HEART is the most important subroutine: it prepares the numerical counterparts of the general differential equation given in Chapter 3. To obtain the numerical solution, we need to solve a set of simultaneous algebraic equations; this is accomplished in SOLVE.

The subprogram TOOLS is actually not an essential part of the computer program. It is, however, provided as an aid to you in designing the adaptation part of the program. The member routines, namely EZGRID, ZGRID, and PRINT, incorporate some commonly required operations. These utilities are expected to be called from the adaptation part if and when the need for them arises. The subprogram VALUES performs a similar utility function. It enables us to assign values to a number of variables in a concise manner.

Since you would not normally make any changes in the invariant part of the program, it would be a good idea to compile this part once and simply use the compiled version (i.e., the object code) in all the applications of CONDUCT.

4.3   Subprograms in the Adaptation Part

The subprogram ADAPT contains four member subroutines; they are: GRID, BEGIN, OUTPUT, and PHI. Of these, GRID and BEGIN are called only once and provide grid-related geometrical information and initial values respectively. OUTPUT is called once per iteration. The required printout is specified in OUTPUT.

PHI is the most frequently called part of ADAPT; it is also the most important member subroutine. In each iteration, it is called several times—once for each dependent variable. Its primary function is to specify the appropriate information about λ, Γ, and S in Eq. (3.6) for each ϕ; also, some boundary condition details for each ϕ are given in PHI.

This brief overview of CONDUCT should provide the necessary background for the remaining chapters of the book, where the actual details of numerical analysis, program nomenclature, and its application are given.

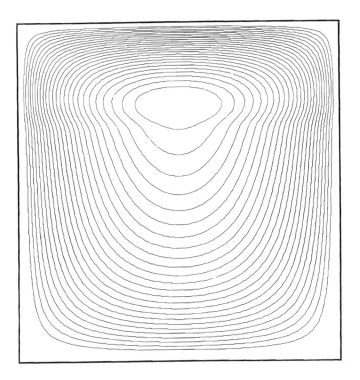

Axial velocity contours for two immiscible fluids of different viscosities
(The viscosity of the fluid in the top one-fourth layer is one-fourth of the
viscosity of the other fluid.)

Chapter 5

Numerical Scheme and Its Implementation

The purpose of this chapter is to describe in detail the numerical scheme that is embodied in CONDUCT. This scheme is a logical extension of the one-dimensional procedure we developed in Chapter 2. Interwoven with the description here are other matters such as the nomenclature used in the program, the numbering scheme, sign conventions, etc. Also, occasional references are made to the subroutines in the computer program.

5.1   Grid and Control Volumes

As we have seen in Chapter 2, the differential equations such as Eq. (3.6) are solved by converting them into algebraic equations known as discretization equations. Discretization equations contain as unknowns the values of ϕ at chosen discrete locations. These locations are arranged on a grid and called the grid points. A small control volume is constructed around each grid point, and the discretization equations are formed by integrating Eq. (3.6) over such control volumes.

The grid design. The general principles of designing the control volumes were discussed in Section 2.5-7 in the one-dimensional context. The scheme that we called Practice B is employed in CONDUCT. For a two-dimensional situation, the basic construction of the control volumes and the grid is shown in Fig. 5.1. At first, the calculation domain is divided into control volumes; the dashed lines denote the boundaries of the control volumes. Then, grid points are placed at the geometrical centers of the control volumes. In Fig. 5.1, the solid lines are the grid lines, and the dots denote the grid points. A typical control volume is shown shaded. It can be seen that a given grid point communicates with the four neighboring grid points through the four faces of the control volume. For a near-boundary control volume, one of its faces coincides with the boundary of the calculation domain, and a boundary grid point is placed at the center of the control-volume face. You may find it convenient to imagine a control volume of infinitesimal thickness associated with the boundary point.

For most purposes, the value of a variable at a grid point will be assumed to prevail over the control volume surrounding it. Similarly, the value at a boundary grid point will be considered to prevail over the associated control-volume face. It, therefore, follows that the control-volume boundaries should be located such that they coincide with the discontinuities in material properties, source terms, boundary conditions, etc.

With reference to Fig. 5.1, the grid locations in the x and y directions are denoted by I and J respectively. Of course, I increases with x, and J with y. The value $I = 1$ refers to the grid line at the left boundary, while $I = L1$ indicates the grid line at the right boundary. Similarly, $J = 1$ and $J = M1$ stand for the bottom

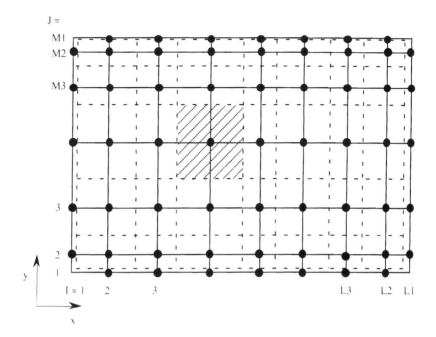

Fig. 5.1 Control volumes and grid points

and top boundary grid lines respectively. Indeed. in many places in the program. the left and bottom boundaries of the domain are called the I1 and J1 boundaries respectively. and the right and top boundaries as the L1 and M1 boundaries: the logic for this nomenclature should now be obvious. For convenience. some additional variable names are introduced: they are defined by:

$$L2 = L1 - 1, \qquad L3 = L1 - 2 \qquad\qquad (5.1)$$

$$M2 = M1 - 1. \qquad M3 = M1 - 2 \qquad\qquad (5.2)$$

Thus $I = 2$ and $I = L2$ represent the first internal locations adjacent to the domain boundaries. and $I = 3$ and $I = L3$ are the internal neighbors. Similar meanings apply to M2 and M3. (By the way. our introduction of the variable names such as L1. L2. and L3 does not mean that we are also numbering all grid points again from right to left. We simply have L1. L2. and L3: there are no L4. L5. etc. Often. there is a need to make a reference to the last three grid locations in each direction. The names like L1. L2. and L3 are useful for this purpose.) Thus. the range $I=2$. L2 and $J=2$. M2 specifies all the *internal* grid points. A dependent variable such as the temperature at the grid point (I, J) will be denoted by

$T(I,J)$. The boundary values of T are contained in $T(1,J)$, $T(L1,J)$, $T(I,1)$, and $T(I,M1)$.

Corner values. You should note that, in general, the values of any dependent variable stored at the four corners of the calculation domain do not have any meaningful role to play. Thus, $T(1,1)$, $T(L1,1)$, $T(1,M1)$, and $T(L1,M1)$ are meaningless. These values are printed out by the utility routine PRINT (because it is too much trouble to suppress their printout), but you are urged to ignore them as meaningless numbers having no influence whatsoever on the solution of the problem. (This statement may not be repeated elsewhere in the book. When you see the results of our examples, please do not be puzzled by the values at the corner grid points, which may often appear surprising.)

You may think that this exclusion of the corner values from the calculation scheme is inconvenient and even undesirable. However, you will experience that the practice adopted here is actually both convenient and desirable. While calculating any quantities related to a boundary, such as a mean temperature for the boundary, you need not give any weightage to the corner point. For example, the mean temperature along the top boundary will be an average of $T(I,M1)$ with $I=2,L2$. (If the other boundary points are supposed to have a control volume of infinitesimal thickness associated with them, a corner point carries a control volume of infinitesimal dimensions in both x and y directions.) Often, the corner locations happen to be the places of discontinuity in the specified boundary conditions. For example, if the bottom boundary has a temperature T_1 and the left boundary is at T_2, the corner point temperature $T(1,1)$ is not properly defined. It is then very convenient that our procedure does not require or produce a value for $T(1,1)$. If for any reason (such as plotting the results) you want meaningful values for the variables at the corner grid points, you are free to assign them by a suitable extrapolation. This action will be simply cosmetic and will not have any effect on the solution produced by CONDUCT.

The three coordinate systems. As mentioned earlier, CONDUCT has provision for three coordinate systems. The value of the integer variable MODE indicates a particular coordinate system.

MODE = 1:   Cartesian coordinates (x, y)

MODE = 2:   Axisymmetric system (x, r)

MODE = 3:   Polar coordinates (θ, r)

The appearance of the grid for MODE = 1 or 2 is as shown in Fig. 5.1. For MODE = 2, x is the axial coordinate, while y is measured in the radial direction.

The relationship between y and r was explained in Eq. (3.13). The grid for MODE = 3 is shown in Fig. 5.2. Here, the angle θ (measured in radians) takes the place of the coordinate x. In the computer program, the variable X(I) is to be interpreted as the distance x for MODE = 1 and 2, and as the angle θ in radians for MODE = 3.

A unified treatment for the three coordinate systems is made possible by the use of an array R(J) in addition to the y coordinate Y(J). R(J) is set equal to unity for MODE = 1 and is given the local value of radius r for MODE = 2 or 3. Also, since X(I) has a different meaning for MODE = 3, a general scale factor SX(J) is introduced for the x direction. The x-direction *length* between grid points (I,J) and (I+1,J) is, in general, given by SX(J)*(X(I+1)-X(I)). It is then obvious that SX(J) should be unity for MODE = 1 or 2 and SX(J) should be equal to r for MODE = 3.

In a two-dimensional computer program, the dependent variables do not change in the third coordinate direction. Still, it is convenient to assume a certain "depth" of the domain in the third coordinate. This depth is taken as unity for MODE = 1 or 3. For MODE = 2, the third coordinate is θ; the domain is considered to have one radian extent in the θ direction.

These generalized geometrical quantities are summarized in the following table.

MODE	X(I)	R(J)	SX(J)	"depth"
1	x	1	1	1
2	x	r	1	1 radian
3	θ	r	r	1

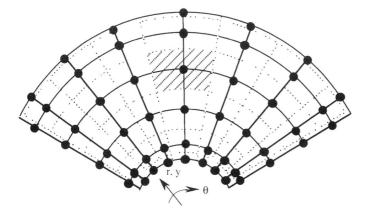

Fig. 5.2 Control volumes and grid locations for the θr system

Other related quantities that are stored for each control volume are: $XCV(I)$ and $YCV(J)$, which are the appropriate widths of the control volumes: $YCVR(J)$, which stands for $r\Delta y$ of the control volume; and $ARX(J)$, which is the area of the control-volume face normal to the x direction.

You may wonder why the area normal to the y direction is not stored as an array. Also, what about storing quantities like the volume of a control volume (to be given in Eq. (5.9) later)? Our policy here is to provide storage for only those geometrical quantities that can be stored *one-dimensionally*: the quantities that would have to be stored as two-dimensional arrays are not stored, but are computed as required. This is a reasonable compromise between the use of computer memory and execution time.

5.2   Interface-Related Quantities

There are a number of geometrical quantities that refer to the control-volume faces rather than to the grid points. The x and y coordinates of the interfaces are given by $XU(I)$ and $YV(J)$. The numbering convention used here assumes that the interface I lies between the grid locations $I-1$ and I. In other words, an interface has the same subscript as the grid point on the positive side of it, i.e., in the direction of increasing I or J. This is illustrated in Fig. 5.3 for the numbering in the x direction. An identical pattern is used in the y direction. A particular consequence of the control-volume design and the numbering scheme is that the interface $I = 2$ and the grid location $I = 1$ both coincide with the left boundary of the domain. At the right boundary, the interface and the grid location are both given by $I = L1$. Similar relations hold for the y direction. As a result,

$$XU(2) = X(1), \qquad XU(L1) = X(L1). \qquad (5.3)$$

$$YV(2) = Y(1), \qquad YV(M1) = Y(M1). \qquad (5.4)$$

and $XU(1)$ and $YV(1)$ are meaningless.

In the specification of the grid, the values of $XU(I)$ and $YV(J)$ denoting the locations of the control-volume faces are specified by the user. The coordinates $X(I)$ and $Y(J)$ and all other geometrical quantities are then calculated in the subroutine READY.

For the interfaces normal to the y direction, in addition to $YV(J)$, the radius $RV(J)$ is stored. The values of the radius at the bottom boundary is given both by $R(1)$ and $RV(2)$. Similarly, $R(M1)$ and $RV(M1)$ both denote the radius at the top boundary.

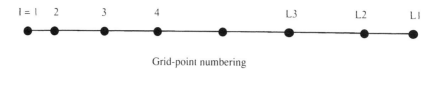

Grid-point numbering

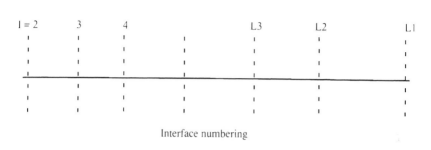

Interface numbering

Fig. 5.3 Numbering scheme for grid points and interfaces

5.3 General Discretization Equation

From the foregoing description, it can be seen that the general dependent variable ϕ will be stored at the grid points. The array $F(I,J,NF)$ will be used to store the values of the different ϕ's. where I and J denote the grid-point location. while NF identifies a particular kind of ϕ such as the temperature. velocity. turbulence kinetic energy. etc. A discretization equation relates the value of $F(I,J,NF)$ at one grid point to the values at the four neighboring grid points. The equation is derived by integrating Eq. (3.6) over the control volume around (I,J).

Conservation over the control volume. A typical control volume is shown in Fig. 5.4. The integration of Eq. (3.6) over the control volume gives:

$$\lambda_P \frac{\Delta V}{\Delta t} (\phi_P - \phi_P^0) = J_w A_w - J_e A_e + J_s A_s - J_n A_n + \overline{S} \Delta V \tag{5.5}$$

Here the superscript 0 denotes the known value of ϕ at the beginning of the time step Δt. the J's are the diffusion fluxes across the control-volume faces. $\overline{S}$ is the average source term over the control volume. ΔV is the volume of the control volume. and the A's are the areas of the control-volume faces.

For all the three coordinate systems. the areas and the volume can be calculated in terms of the geometrical quantities stored in the program. Thus.

$$A_e = A_w = ARX(J) \tag{5.6}$$

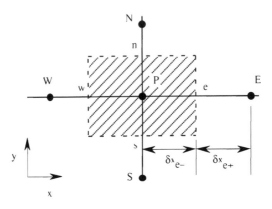

Fig. 5.4 A typical control volume

$$A_n = RV(J+1)*XCV(I) \tag{5.7}$$

$$A_s = RV(J)*XCV(I) \tag{5.8}$$

$$\Delta V = YCVR(J)*XCV(I) \tag{5.9}$$

where it is assumed that point P in Fig. 5.4 corresponds to (I,J).

Diffusion flux. The diffusion fluxes at the control-volume faces e and w can be calculated as

$$J_e A_e = D_e(\phi_P - \phi_E) \tag{5.10}$$

$$J_w A_w = D_w(\phi_W - \phi_P) \tag{5.11}$$

The quantity D_e is the diffusion conductance of the link PE and is calculated from the Γ values specified at P and E. With reference to the distances $(\delta x)_{e-}$ and $(\delta x)_{e+}$ shown in Fig. 5.4, D_e is given by:

$$D_e = A_e \left[\frac{(\delta x)_{e-}}{\Gamma_P} + \frac{(\delta x)_{e+}}{\Gamma_E} \right]^{-1} \tag{5.12}$$

The definitions of D at other interfaces are to be constructed in a similar manner. Formulas similar to Eqs. (5.10) and (5.11) apply to the fluxes J_n and J_s.

This formulation of the diffusion fluxes can now be recognized as a straightforward generalization of Eqs. (2.33), (2.34), and (2.80). The conductance such as D_e now contains the area A_e of the control-volume face; in Eq. (2.78), we used a unit cross-sectional area for the one-dimensional case.

Source-term linearization. Since the source term S may often depend on ϕ itself, it is desirable to include a (nominally) linear dependence of S on ϕ. For this purpose, $\overline{S}$ is written as

$$\overline{S} = S_C + S_P \, \phi_P \tag{5.13}$$

where S_P is the coefficient of ϕ_P, and S_C is the part of $\overline{S}$ that does not explicitly depend on ϕ_P.

Once again, Eq. (5.13) follows from the source-term treatment expressed in Eq. (2.41). Also, all the considerations given in Section 2.5-4 about the source-term linearization apply here as well. When the linearization is not desired, you should simply set S_P equal to zero and equate S_C to $\overline{S}$. In any case, S_P should never be a positive quantity. When the S~ϕ relationship is not truly linear, the nominally linear behavior given in Eq. (5.13) can be formulated along the lines of Eq. (2.85). In such a case, S_C and S_P depend on ϕ_P; they are to be calculated from the current estimate of ϕ_P and iteratively updated as the solution progresses towards convergence.

Final discretization equation. The substitution of the foregoing expressions for the J's and for $\overline{S}$ into Eq. (5.5) leads to the final form of the discretization equation. It is written as:

$$a_P \phi_P = a_E \phi_E + a_W \phi_W + a_N \phi_N + a_S \phi_S + b \tag{5.14}$$

where

$$a_E \;=\; D_e \tag{5.15}$$

$$a_W \;=\; \qquad\qquad\qquad\qquad\qquad\qquad D_w \tag{5.16}$$

$$a_N \;=\; D_n \tag{5.17}$$

$$a_S \;=\; D_s \tag{5.18}$$

$$b \;=\; S_C \Delta V + a_P^0 \, \phi_P^0 \tag{5.19}$$

$$a_P^0 = \frac{\lambda_P \, \Delta V}{\Delta t} \tag{5.20}$$

$$a_P = a_E + a_W + a_N + a_S + a_P^0 - S_P \Delta V \tag{5.21}$$

Here the diffusion conductances D_e, D_w, D_n, and D_s are defined along the lines of Eq. (5.12). The coefficient a_P^0 results from the unsteady term. When Δt equals a very large number, a_P^0 becomes negligible and the formulation reduces to the steady-state situation. In CONDUCT, the default value of Δt is set equal to 1.E20, which is appropriate for steady-state problems. For unsteady problems, you should give a proper value to Δt. If you use a finite Δt for steady problems, it has the effect of introducing underrelaxation into the discretization equation.

For compactness, it is often useful to write Eq. (5.14) as:

$$a_P \phi_P = \sum a_{nb} \phi_{nb} + b \tag{5.22}$$

where nb denotes a neighbor grid point of P, and the summation is taken over the four neighbors.

5.4 Related Fortran Names

The coefficients in Eq. (5.14) for the grid point (I, J) are given the following Fortran names.

$$AP(I,J) = a_P \qquad CON(I,J) = b$$

$$AIP(I,J) = a_E \qquad AIM(I,J) = a_W$$

$$AJP(I,J) = a_N \qquad AJM(I,J) = a_S$$

The neighbor-point coefficient names are based on the fact that E is $(I+1, J)$, S is $(I, J-1)$, and so on.

The quantities λ, Γ, S_C, and S_P are specified in the arrays ALAM(I, J), GAM(I, J), SC(I, J), and SP(I, J) for *internal* grid points. Please note that the arrays SC and SP are made equivalent to, and therefore occupy the same storage locations as, CON and AP respectively. The names SC and SP are used only in the subprogram ADAPT for the convenience of the user. In the rest of the program, it is understood that S_C and S_P have been supplied in the arrays CON and AP.

Although the dependent variables ϕ are stored as F(I, J, NF), the subscript NF is not used for ALAM, GAM, SC, and SP. These arrays and also the coefficient arrays are used, over and over again, for all the successive ϕ's. Thus, at any given stage in the computation, ALAM, GAM, SC, and SP contain the values of λ, Γ, S_C

and S_p for the *current* ϕ under consideration. The default values of SC and SP are zero.

Incidentally. it is our practice to use the implicit type declarations of the Fortran variables. Thus, the variable names that begin with I, J, K, L, M, or N are treated as integer variables, while the others are considered to be real. This is why we use for λ the variable name ALAM(I,J) and not LAM(I,J). You will see later that, for the conductivity k and viscosity μ, the names AK and AMU are used, instead of K and MU.

5.5 Treatment of Boundary Conditions

For every near-boundary grid point, the corresponding boundary grid point acts as one of the neighbors. Therefore, either a value or an equation must be available for the ϕ at the boundary node. Since exactly the same treatment is used for all the four boundaries of the calculation domain, it is sufficient here to describe the treatment for the left boundary. The translation to other boundaries should be straightforward.

The treatment of boundary conditions was described in Section 2.4-3 for the one-dimensional situation. The same practices are applicable here. Since we use Practice B for the control-volume design, the half-control-volume thickness, as pointed out in Section 2.5-7, approaches zero. This leads to some inaccuracy in the determination of the boundary flux. Therefore, two kinds of boundary treatment are given here. The lower-order treatment is a logical consequence of the formulation described in Section 5.3 (or in Section 2.4-3). In addition, a more accurate higher-order treatment has also been worked out. It is this higher-order treatment that is recommended for the general ϕ equation in CONDUCT; but the lower-order treatment is also available for use.

5.5-1 Lower-Order Treatment

Figure 5.5 shows the situation near the boundary grid point (1,J). The subscript J is dropped in the following derivations for convenience. The control-volume face for I = 2. which coincides with the left boundary of the domain can be considered to lie *between* the locations of ϕ_1 and ϕ_2 if a control volume of an infinitesimal thickness is imagined around the location of ϕ_1. Then the total flux J_2 at the boundary is still given by Eq. (5.10). It is convenient to write this as:

$$J_2 = AIP(1,J) * (\phi_1 - \phi_2) \tag{5.23}$$

where

$$AIP(1,J) = \frac{\Gamma_2}{\delta} \tag{5.24}$$

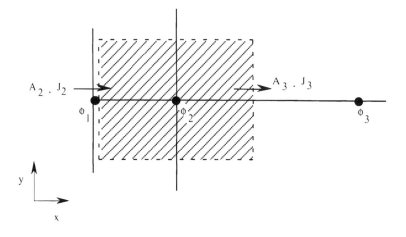

Fig. 5.5 A control volume near the left boundary

It should be noted that, whereas the coefficients such as $AIP(I,J)$ for the control volumes contain the area of the control-volume face, the boundary coefficient $AIP(1,J)$ is defined on a unit area basis. This is done for two reasons. First, the input or output for the boundary flux is normally available or desired per unit area. Second, for $MODE = 2$ or 3, if the lower boundary of the domain coincides with the axis ($r = 0$), the interface area there is zero. Then the area-integrated quantities are not as useful as the per-unit-area quantities. Incidentally, $AIP(1,J)$ and similar coefficients will henceforth be abbreviated as $AIP(1)$.

The practice of calculating the boundary flux J_2 from ϕ_1 and ϕ_2 can be seen from Fig. 5.5 as a kind of one-sided scheme since the interface is not located midway between ϕ_1 and ϕ_2. This practice would give somewhat inaccurate results. Since the boundary treatment has a strong effect on the whole solution and since the boundary fluxes are often a significant outcome of the computation, a more accurate formula for the boundary fluxes is desirable. This is described next.

5.5-2 Higher-Order Treatment

The total flux formulas in Eqs. (5.10) and (5.23) are derived from the piecewise-linear profile which implies that the flux J remains constant between two neighboring grid points. A higher-order formula can be obtained if the diffusion flux is regarded as being linear in distance between the two opposite faces of a control volume. This concept when applied to the boundary control volume leads to the flux profile shown in Fig. 5.6. There the J distribution is linear with x

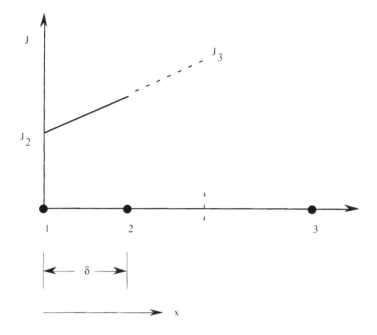

Fig. 5.6 Assumed profile for the total flux J in the higher-order treatment

between J_2 at the left face and J_3 at the right face of the control volume. The assumed J distribution between grid points 1 and 2 is thus given by:

$$J = -\Gamma \frac{d\phi}{dx} = J_2 + J'(x - x_1)$$ (5.25)

where

$$J' = \frac{J_3 - J_2}{2\,\delta}$$ (5.26)

Equation (5.25) can be integrated with respect to x to give a profile expression for ϕ. This integration pertains only to the region between the grid points 1 and 2, where Γ has a uniform value. The constant of integration is obtained from the condition:

at $x - x_1 = 0$: $\phi = \phi_1$ (5.27)

A further condition to be satisfied by the profile is:

at $x - x_1 = \delta$: $\phi = \phi_2$ (5.28)

The use of this condition leads. after some algebraic manipulation, to:

$$J_2 = \left(\frac{4}{3}\right)\left[\left(\frac{\Gamma_2}{\delta}\right)(\phi_1 - \phi_2)\right] - \left(\frac{1}{3}\right)J_3 \tag{5.29}$$

A generalized form of this equation is:

$$J_2 = \beta\left[\left(\frac{\Gamma_2}{\delta}\right)(\phi_1 - \phi_2)\right] - (\beta - 1)J_3 \tag{5.30}$$

If the factor β is set equal to 4/3. we get Eq. (5.29). For $\beta = 1$. the equation reverts to the lower-order treatment given by Eqs. (5.23)–(5.24).

An expression for J_3 can be written along the lines of Eq. (5.10) as:

$$J_3 A_3 = D_3 (\phi_2 - \phi_3) \tag{5.31}$$

where D_3 is the appropriate diffusion conductance for the link connecting ϕ_2 and ϕ_3.

The discretization equation for the near-boundary control volume $(2,J)$ should now be based on Eq. (5.30) for the flux J_2 through its left face. rather than on Eq. (5.11). As a result. the coefficients $AIM(2)$ and $AIP(2)$ would be given by

$$AIM(2) = \beta\left(\frac{\Gamma_2}{\delta}\right)A_2 \tag{5.32}$$

$$AIP(2) = D_3 + (\beta - 1)A_2\left(\frac{D_3}{A_3}\right) \tag{5.33}$$

The formula for $AP(2)$ is still given by Eq. (5.21) provided that the neighbor coefficients appearing in the expression for a_p are modified as indicated here.

Along the lines of Eq. (5.23). an expression for the diffusion flux J_2 can now be written as:

$$J_2 = AIP(1) * (\phi_1 - \phi_2) + AIM(1) * (\phi_3 - \phi_2) \tag{5.34}$$

where

$$AIP(1) = \beta\left(\frac{\Gamma_2}{\delta}\right) \tag{5.35}$$

$$AIM(1) = (\beta - 1)\left(\frac{D_3}{A_3}\right) \tag{5.36}$$

Here the use of the coefficient name $AIM(1)$ is somewhat unusual. For the node at $I = 1$. there is an $I+1$ neighbor. but there is no $I-1$ neighbor. Thus, the coefficient $AIM(1)$ has no proper role to play. In the higher-order formula for J_2, however, there is a need for an extra coefficient: $AIM(1)$ is used for that purpose.

To summarize, Eqs. (5.32)–(5.36) can be used for both types of boundary condition treatment with appropriate values of β given by :

higher-order treatment: $\beta = \dfrac{4}{3}$ (5.37)

lower-order treatment: $\beta = 1$ (5.38)

In CONDUCT, the appropriate level of boundary condition practice can be chosen by setting the parameter KORD. It can take the values 1 or 2, which correspond to the lower and higher-order treatments respectively. The default value of KORD is 2, which you would rarely need to change.

5.5-3   Boundary-Condition Indicators

The type of the boundary condition at the left, right, bottom, and top boundaries of the domain is specified by the indicators KBCI1(J), KBCL1(J), KBCJ1(I), and KBCM1(I) respectively. Again, all these indicators refer to the current F(I,J,NF) under consideration and must be correctly reset for every NF.

The boundary indicators KBC can take the values 1 or 2. When the value is 1, it implies that the boundary value of ϕ is known. Thus, when KBCI1(J) = 1, the value of F(1,J,NF) is considered as given. When KBC equals 2, the diffusion flux at the boundary is assumed to be given in the form:

$$J_B = f_C + f_P \phi_B$$ (5.39)

where the subscript B refers to the boundary. You will recall that we used a similar expression in Eq. (2.50) for the boundary flux. The sign convention used for Eq. (5.39) is that, for any boundary, J_B denotes the diffusion flux entering the calculation domain. Thus, in the case of a heat flux, J_B will be positive when heat is supplied to the domain and negative when there is a heat loss from the domain. When the diffusion flux is wholly known, f_C represents the known value of the flux and f_P is set equal to zero. When the diffusion flux is known as a linear function of ϕ_P, both f_C and f_P will, in general, be nonzero. When the flux is a nonlinear function for ϕ_B, Eq. (5.39) would imply that f_C and f_P themselves depend on ϕ_P and should be iteratively updated. Such linearization of the boundary condition was discussed in Section 2.5-5.

The similarity between Eqs. (5.39) and (5.13) is worth noting. The boundary flux J_B acts as a source term for the infinitesimal control volume around the boundary point. Whereas S_C and S_P are calculated per unit volume, f_C and f_P are to be specified on a unit-area basis. For a properly specified problem, f_P must be zero or negative. If the boundary heat flux J_B is given by a convective heat transfer coefficient h such that:

$$J_B = h \, (\phi_\infty - \phi_B) \tag{5.40}$$

where ϕ_∞ stands for the temperature of the surrounding fluid, the expressions for f_C and f_P will be:

$$f_C = h \, \phi_\infty \tag{5.41a}$$

$$f_P = - h \tag{5.41b}$$

You need to supply the values of f_C in the arrays $FLXCI1(J)$, $FLXCL1(J)$, $FLXCJ1(I)$, and $FLXCM1(I)$ for the four boundaries of the domain. The values of f_P are given in $FLXPI1(J)$, $FLXPL1(J)$, etc. Thus, for $KBCI1(J) = 2$, the values of $FLXCI1(J)$ and $FLXPI1(J)$ should be given.

The default values of the KBC indicators are all 1, while those of FLXC and FLXP are zero. Because of these default specifications, your work of specifying the boundary conditions increases with the complexity of the boundary conditions. Consider the following four cases:

(a) When the boundary value of ϕ_B is known, *no action* is needed. (Of course, you should have put the known value of ϕ_B in the appropriate boundary locations of $F(I,J,NF)$. But after having done this, you need to do nothing extra to state that the boundary value ϕ_B is considered to be known.)

(b) For adiabatic boundaries and symmetry surfaces, the diffusion flux at the boundary is zero. For such a zero-flux boundary condition, *all* that you do is to set the appropriate KBC equal to 2. (The *value* of the flux will remain zero by default.)

(c) When the boundary flux J_B is given, you perform two actions: set the KBC equal to 2 and the corresponding FLXC equal to the known value of the boundary flux. Here, FLXC should be positive if the flux enters the domain and negative if the flux leaves the domain.

(d) When the unknown flux J_B depends on the unknown value ϕ_B, three actions are needed: set the KBC equal to 2 and put the values of f_C and f_P into FLXC and FLXP respectively. The most common situation for this type of boundary condition is the case of a given convective heat transfer coefficient. Then, the expressions given in Eq. (5.41) for f_C and f_P should be used.

This describes how a user should introduce the boundary-condition information in the problem-dependent part of the program. The further processing of this information in the invariant part of the program is outlined in the next two subsections.

5.5-4   Treatment for KBC=1

When the boundary value ϕ_B of the dependent variable is known, all that needs to be done is to include the influence of the boundary point in the constant term b of the discretization equation. Thus, if $KBCI1(J) = 1$,

$$CON(2) = CON(2) + AIM(2) * F(1,J,NF) \qquad (5.42a)$$

$$AIM(2) = 0. \qquad (5.42b)$$

would be the required modification. Again, the subscript J for CON and AIM is omitted for convenience.

5.5-5   Treatment for KBC=2

If the diffusion flux J_2 is given by

$$J_2 = f_C + f_P \phi_1 \qquad (5.43)$$

Eq. (5.34) provides the required algebraic equation for the unknown ϕ_1. It can be written as:

$$AP(1) \phi_1 = AIP(1) \phi_2 + AIM(1) * (\phi_2 - \phi_3) + CON(1) \qquad (5.44)$$

where $AIP(1)$ and $AIM(1)$ have already been defined by Eqs. (5.35) and (5.36) and

$$AP(1) = AIP(1) - f_P \qquad (5.45)$$

$$CON(1) = f_C \qquad (5.46)$$

With the help of Eq. (5.44), the unknown ϕ_1 is eliminated from the discretization equation for the control volume $(2,J)$. In the final form of this equation, the coefficient $AIM(2)$ of the unknown value ϕ_1 will be zero. After the set of the discretization equations is solved for the ϕ values at the internal grid points, the unknown value ϕ_1 can be calculated from Eq. (5.44).

5.5-6   Calculation of the Boundary Flux

Once the solution of the discretization equation is obtained, the diffusion flux at all the boundary points is calculated from expressions like Eq. (5.34). These flux values may form an important outcome of the calculation. For this reason, they are stored in the arrays `FLUXI1(J,NF)`, `FLUXL1(J,NF)`, `FLUXJ1(I,NF)`, and `FLUXM1(I,NF)` for the left, right, bottom, and top boundaries respectively. We shall often use these flux values to obtain the total heat flow from a boundary of the domain, the heat transfer coefficient for a duct flow, etc. Once again, the sign convention is: if the value of the boundary flux is positive, it implies a heat flow *into* the domain; a negative value denotes a heat loss.

5.6   Solution of the Discretization Equations

For each control volume around an internal grid point, a discretization equation such as (5.14) has been formulated. When the foregoing modifications are made in the equations for the near-boundary control volumes, the ϕ values at the boundary grid points do not explicitly appear in the set of equations. The algebraic equations are, at least nominally, linear and there are exactly as many equations as unknowns. Therefore, the equation set can be solved by any convenient solution algorithm.

In Section 2.4-4, we used the TriDiagonal-Matrix Algorithm (TDMA) for the solution of the one-dimensional equations. For the two-dimensional equations such as Eq. (5.14), the TDMA cannot be easily extended. The standard *direct* methods for the two-dimensional equations require very large amounts of computer memory and computer time. Therefore, we employ an iterative method for the solution of these linear algebraic equations. You will see that, in this iterative method, the TDMA is used as an important unit. Our solution procedure is a combination of the line-by-line method and a block-correction scheme.

Line-by-line method. If, in the discretization equation (5.14), the y-direction neighbors ϕ_N and ϕ_S are considered to be tentatively known, then the equation would have only three unknowns, ϕ_P, ϕ_E, and ϕ_W. If such three-unknown equations are formed along an x-direction line, their form will be

$$a_i \phi_i = b_i \phi_{i+1} + c_i \phi_{i-1} + d_i \qquad \text{for i = 2, L2} \tag{5.47}$$

where ϕ_i, ϕ_{i+1}, and ϕ_{i-1} stand for ϕ_P, ϕ_E, and ϕ_W respectively, and the coefficients are related to those in Eq. (5.14) by

$$a_i = a_P, \quad b_i = a_E, \quad c_i = a_W, \quad d_i = a_N \phi_N' + a_S \phi_S' + b \tag{5.48}$$

Here ϕ' denotes an estimated value. Because of the boundary modifications, the coefficients c_2 and b_{L2} are zero. Equation (5.47) has the same form as Eq. (2.56) and therefore, can be solved by the TDMA. The details of the algorithm are rewritten here for completeness.

First, new coefficients P_i and Q_i are calculated from the recurrence relations

$$P_i = \frac{b_i}{(a_i - c_i P_{i-1})}$$ (5.49)

$$Q_i = \frac{d_i + c_i Q_{i-1}}{a_i - c_i P_{i-1}}$$ (5.50)

for $i = 2$, L2. Since c_2 is zero, the recurrence process can be started by assuming any values for P_1 and Q_1. Finally, the coefficient P_{L2} will be zero, since b_{L2} is zero. The values of ϕ_i are obtained by using

$$\phi_i = P_i \phi_{i+1} + Q_i$$ (5.51)

in the reverse order, i.e. for $i = $ L2, L2 − 1, L2 − 2....., 4, 3, 2. For $i = $ L2, the value of ϕ_{L2+1} is not needed since P_{L2} is zero.

The line-by-line method consists of employing the TDMA along all the lines in the x direction. This is then repeated along the lines in the y direction. As the new values of ϕ along a line are calculated, they are used as the estimates ϕ^* in the line solution for the next line. The sequence in which the lines are chosen can be arbitrary. The practice employed in CONDUCT is as follows. First, the line traverse is made along the x-direction line just above the bottom boundary. Then, the successive parallel lines up to the top boundary are visited. The same lines are then traversed in a top-to-bottom sweep direction. This is followed by the TDMA traverses along the y-direction lines using a left-to-right sweep followed by a right-to-left sweep.

This practice is intended to bring the influence of all the boundary values quickly into the interior of the calculation domain. The speed of convergence of the line-by-line method is further enhanced by a block-correction scheme, which is described next.

Block-correction scheme. The scheme employed here is based on the additive-correction strategy of Settari and Aziz (1973). The actual details of the scheme are taken from Patankar (1981). Let the discretization equation (5.14) be expressed as

$$a\phi_{i,j} = b\phi_{i+1,j} + c\phi_{i-1,j} + d\phi_{i,j+1} + e\phi_{i,j-1} + f$$ (5.52)

for $i = 2$, L2 and $j = 2$, M2. The subscripts (i,j) for the coefficients a, b, c, d, e, and f have been omitted for convenience. For the near-boundary control volumes,

$$b_{L2,j} = 0, \quad c_{2,j} = 0, \quad d_{i,M2} = 0, \quad e_{j,2} = 0$$ (5.53)

The central idea of the block-correction scheme is that an unconverged field $\phi^*_{i,j}$ obtained from prior iterations is corrected by adding *uniform* corrections $\bar{\phi}_i$ along lines of constant i. Thus

$$\phi_{i,j} = \phi^*_{i,j} + \bar{\phi}_i \tag{5.54}$$

The corrections $\bar{\phi}_i$ are chosen such that the *integral* conservation over the control-volume blocks defined by each constant-i line is exactly satisfied. The equation governing $\bar{\phi}_i$ is obtained by substituting Eq. (5.54) into Eq. (5.52) and adding such equations for all values of j. The result is

$$BL(I)\,\bar{\phi}_i = BLP(I)\,\bar{\phi}_{i+1} + BLM(I)\,\bar{\phi}_{i-1} + BLC(I) \tag{5.55}$$

where

$$BL(I) = \sum (a - d - e) \tag{5.56}$$

$$BLP(I) = \sum b \tag{5.57}$$

$$BLM(I) = \sum c \tag{5.58}$$

$$BLC(I) = \sum (b\phi^*_{i+1,j} + c\phi^*_{i-1,j} + d\,\phi^*_{i,j+1} + e\,\phi^*_{i,j-1} + f - a\phi^*_{i,j}) \tag{5.59}$$

The summations in these expressions are taken over $j = 2, M2$. The equation set (5.55) written for $i = 2, L2$ can be conveniently solved by the TDMA. Since Eq. (5.53) implies that $BLM(2)$ and $BLP(L2)$ will be zero, the boundary values $\bar{\phi}_1$ and $\bar{\phi}_{L2+1}$ are not needed.

It should be noted that $BLC(I)$ represents the integral residual for the block around a constant-i line. The corrections $\bar{\phi}_i$ reduce all the integral residuals to zero. In other words, when the block corrections are made, the corrected field of ϕ implies perfect integral conservation (of heat, mass, momentum, etc.) over each block. The forgoing description applies to the block correction along lines of constant i. A similar procedure can be worked out for lines of constant j.

Although the block-correction scheme is, in general, found to enhance the speed of convergence quite significantly, it can occasionally lead to unrealistic solutions and even divergence. The reason is that the corrections implied by Eq. (5.54) are rather indiscriminate. If $\phi^*_{i,j}$ is highly nonuniform, the uniform corrections $\bar{\phi}_i$, although appropriate for adjusting the *mean* value to the correct level, can lead to *local* values of $\phi_{i,j}$ that lie outside the reasonable range for the variable ϕ. Because of this potential of the block-correction scheme, a provision is made in CONDUCT for omitting the block correction. An index KBLOC(NF) is used for this purpose. When $KBLOC(NF)$ is zero, the block correction is omitted. The default value of $KBLOC(NF)$ is 1.

Repetitions of the solution algorithm. Since the solution procedure described so far is iterative, it should be repeated a number of times to obtain the converged solution. How we control the number of these repetitions is described here.

The solution algorithm is incorporated in subroutine SOLVE. It consists of the block corrections for the i and j directions followed by the four sweeps of the line-by-line method. This complete set of operations is repeated a number of times. The number of repetitions is controlled by two quantities, $NTIMES(NF)$ and $CRIT(NF)$, which are supplied by the user for every value of NF. $NTIMES(NF)$ denotes the *maximum* number of repetitions that are allowed. At a minimum, one pass through the solution algorithm is always made.

$CRIT(NF)$ controls the *actual* number of repetitions. When, in a given pass through the solution algorithm, the local relative error ε is found to be everywhere less than or equal to $CRIT(NF)$, subsequent repetitions of the algorithm are halted. The local relative error ε is defined as

$$\varepsilon = |(RES)/(TERM)| \tag{5.60}$$

where RES is the residual of the discretization equation given by

$$RES = b\phi_{i+1,j} + c\phi_{i-1,j} + d\phi_{i,j+1} + e\phi_{i,j-1} + f - a\phi_{i,j} \tag{5.61}$$

and $TERM$ is the largest of all the terms in the discretization equation.

The default values of $NTIMES(NF)$ and $CRIT(NF)$ are 10 and 1.E–5 respectively. What the optimum values of $NTIMES$ and $CRIT$ are depends on the experience with a given type of problem, but some general advice will be given here.

To get an absolutely perfect convergence, $NTIMES$ should be very large and $CRIT$ very small. However, for practical purposes, it is seldom necessary to set $CRIT$ below 1.E–6. Also, the word length of your computer may not carry enough significant figures for a very tight convergence criterion. On machines with 32-bit words, since the real numbers are stored to only about 7 significant figures, $CRIT$ values below 1.E–6 do not serve any useful purpose. When you use a large number of grid points, it generally takes more repetitions of the solution algorithm to bring the error ε below the value of $CRIT(NF)$. You should then use a larger value of $NTIMES(NF)$. We could have always used a very large value of $NTIMES$, for example, 1000; but then, if you happen to specify the problem incorrectly so that the set of algebraic equations have no converged solution, the computer would be unnecessarily making 1000 repetitions of the algorithm A reasonable value of $NTIMES(NF)$ is a safety feature that is useful when something goes wrong.

The *actual* number of repetitions performed in SOLVE are stored as $NTC(NF)$; this number may be printed out to see whether the algorithm converged (to the specified $CRIT$ value) before reaching the maximum limit set by $NTIMES$.

5.7 Nonlinearity and Underrelaxation

Treatment of nonlinearity. We have already seen that the problems to be solved by CONDUCT can be nonlinear. Then, the coefficients in a given discretization equation themselves depend on ϕ. Further, since ϕ can stand for a number of physical quantities, the coefficients for one meaning of ϕ may be influenced by some of the other ϕ's.

Because of these interlinkages and nonlinearities, the final solution is obtained by iteration. At any given stage, the discretization coefficients are calculated from the current estimates of all the ϕ values. The solution of the discretization equations gives an improved estimate, which is used to recalculate the coefficients. When, after many repetitions of this process, all the ϕ values cease to change, the final converged solution is reached. This matter was discussed in Section 2.5-3 for the one-dimensional case. The same ideas apply to the two-dimensional situation as well.

Use of underrelaxation. It does not always follow that successive iterations would lead to a converged solution. At times, the values of ϕ oscillate or drift continuously. Such divergence of the iterative process must be avoided. Although, for linear equations, the line-by-line method used in CONDUCT is guaranteed to converge, no such certainty exists for nonlinear and coupled problems. It is true that various ingredients of the calculation scheme have been designed to minimize the chances of divergence; but additional techniques are often needed to promote convergence. One such strategy is to slow down the changes in the coefficients from iteration to iteration by controlling the changes in the ϕ values. This practice is known as underrelaxation.

Once again, this concept was explained in Section 2.5-6. It is generalized here to the discretization equation for ϕ. The practice used in CONDUCT is to write the general discretization equation

$$a_p \phi_p = \sum a_{nb} \phi_{nb} + b \tag{5.62}$$

as

$$(a_p + i)\phi_p = \sum a_{nb}\phi_{nb} + b + i\,\hat{\phi}_p \tag{5.63}$$

where i is the so-called inertia. Further, the inertia is chosen as

$$i = \frac{(1-\alpha)}{\alpha} \Sigma a_{nb} \tag{5.64}$$

where α is the relaxation factor, stored as $RELAX(NF)$ for all the dependent variables. When $\alpha = 1$, no underrelaxation is introduced. As α gets closer to zero, the changes in ϕ are greatly slowed down. If α is chosen to be greater than unity, the effect is overrelaxation, i.e., the changes in ϕ are actually exaggerated. In general, $\alpha > 1$ is not recommended. Incidentally, the default values of $RELAX(NF)$ in CONDUCT are all set equal to 1.

As already mentioned, the unsteady-term coefficient a_p^0 in Eq. (5.21) plays the same role as the inertia i. Thus, the use of a finite time step Δt for steady-state problems also introduces underrelaxation. Which underrelaxation practice should be used depends on the nature of the problem, experience, and personal taste.

In the fifteen applications of CONDUCT that are included in this book, it has not been necessary to use the underrelaxation through $RELAX(NF)$. This is because the nonlinearities in our example problems are probably moderate enough to be handled without underrelaxation. However, as you apply CONDUCT to increasingly complex problems, you may need to use underrelaxation of the dependent variables. In that case, all that you have to do is to set the appropriate $RELAX(NF)$ equal to a number less than 1.

Underrelaxation of auxiliary variables. In addition to the dependent variables, other quantities can be underrelaxed with advantage. Although no provision for this has been made in the invariant part of CONDUCT, you can introduce the desired underrelaxation in the adaptation part. The formula for underrelaxing an auxiliary variable such as the diffusion coefficient Γ is

$$\Gamma = \alpha \Gamma_{new} + (1 - \alpha) \Gamma_{old} \tag{5.65}$$

In the same manner, the source terms and boundary conditions can be underrelaxed. You may recognize that this topic too was introduced in Section 2.5-6. Whereas the underrelaxation through $RELAX(NF)$ is not illustrated in our fifteen examples, we do use the underrelaxation expressed in Eq. (5.65) in Examples 11 and 13 in Chapter 11.

Iterations for unsteady problems. If an unsteady problem is nonlinear, CONDUCT does not make a provision by which, in a given time step, the discretization coefficients are continually updated and multiple iterations are performed. It is assumed that the coefficient values obtained form the known values of ϕ at time t are sufficiently accurate for the entire time step. It then follows that, for nonlinear problems, the time step Δt should be made sufficiently small. Further, the relaxation factors RELAX(NF) must always be unity for an unsteady problem. Otherwise, the unsteady solution will represent the outcome of the artificially modified equation, Eq. (5.63), rather than that of the original equation, Eq. (5.62).

Inner and outer iterations. At this stage, the concepts of iterations, time steps, and the repetitions in SOLVE should be clearly understood. Let us first discuss only steady-state problems. The problem may be linear or nonlinear. For a linear problem, the coefficients a_P, a_E, a_W, etc. in Eq. (5.14) are constant. Then, all we need to do is to solve this set of equations in SOLVE and obtain the final solution. The procedure we use in SOLVE is iterative and therefore multiple repetitions are needed to get a satisfactory solution. These are controlled by NTIMES(NF) and CRIT(NF). The repetitions in SOLVE can be thought of an *inner* iterations. The *outer* iteration loop shown in the flow diagram (Fig. 4.1) in Chapter 4 is meant for handling the nonlinearities by recalculating the discretization coefficients every iteration. These outer iterations are counted by the variable ITER; at any stage, ITER gives the number of iterations that have been completed. In each outer iteration, we call HEART, which in turn calls SOLVE for each dependent variable considered. Every time we go through HEART, the discretization coefficients are recalculated. When we use the term "iteration", we normally imply the *outer* iteration, which is provided for handling nonlinearity by updating the coefficients. The inner iterations in SOLVE are generally described as repetitions of the solution algorithm. To summarize:

> The repetitions in SOLVE give the solution of linear algebraic equations; no coefficient recalculation is done in SOLVE. The (outer) iterations provide for the recalculation of the coefficients to account for nonlinearity.

From this discussion, it follows that, for a linear problem, we need to perform only one iteration since no coefficient recalculation is necessary. Of course, we must use NTIMES(NF) sufficiently large and CRIT(NF) sufficiently small so that SOLVE produces a good solution of the algebraic equations. Even though one iteration is sufficient, our practice in this book will be to perform three iterations for linear problems. Then, by printing out some representative quantities after each iteration, we get a chance to *confirm* that the solution has in fact converged. Also, if occasionally the number of repetitions in SOLVE performed during the first iteration turn out to be insufficient, the additional work in the second and third iterations would help to produce a well-converged solution.

The number of iterations required for a nonlinear problem cannot be predetermined. One may decide this number from some exploratory computations or introduce an appropriate convergence criterion in the problem-dependent part of the program. The only termination condition built into the invariant part of CONDUCT is that the computations stop once the number of iterations completed equals the value of the variable LAST, to which you can give any desired value. How a problem-dependent convergence criterion can be introduced in the adaptation part is illustrated in Example 2 in Chapter 8.

Now let us turn our attention to the unsteady problems. In the computation of unsteady situations, performing *each* time step is, strictly speaking, equivalent to solving a full steady-state problem; for the next time step, a new problem begins. If the problem is nonlinear, there should be a number of iterations within a time step. However, no provision is made in CONDUCT to perform multiple iterations within a time step. One outer iteration for a steady-state problem is treated as one time step for an unsteady problem. For most applications, this aspect of CONDUCT should not cause any difficulty. However, if you wish to modify this feature of CONDUCT, some suggestions are made in Section 12.4.

5.8   Relative Dependent Variable

For certain dependent variables, the specified boundary conditions may lead to a situation that ϕ and $\phi + C$ (where C is an arbitrary constant) are both acceptable solutions. This happens for a steady heat conduction in the presence of *given* heat fluxes at all the boundaries. We shall also encounter this in duct flows with heat-flux boundary conditions. In such cases, the absolute values of the variable ϕ are not relevant; only the differences between the ϕ values at any two grid points are meaningful, and these are not altered by an arbitrary constant added to the ϕ field. Such variables are called relative dependent variables.

If the absolute values of such variables are not unique, would the computations converge to a definite solution? Fortunately, the iterative method used in SOLVE does converge to a solution, the absolute values being indirectly decided by the initial guesses. A direct method would have failed by encountering a singular matrix of the coefficients.

There is, however, one ingredient of SOLVE that is essentially a direct method. In the block-correction scheme, the pseudo one-dimensional problem for $\bar{\phi}_I$ is solved by a direct TDMA. When ϕ happens to be a relative variable, a singular matrix is indeed encountered in the block-correction scheme. In SOLVE, a provision is made to check for this singularity; when it is detected, the block correction in one block is set equal to zero and the other corrections are calculated relative to it.

When the printout for a relative variable is obtained, it is desirable to eliminate the arbitrariness in the absolute values of ϕ by some convenient device. For example, if the temperature T over a duct cross section behaves as a relative variable, it would be more meaningful to print out $T - T_b$ (where T_b is the bulk temperature), or $T - T_c$ (where T_c is the centerline temperature), and so on.

Otherwise, if the absolute values of T happen to be very large numbers, the significant information about their differences would probably be all lost.

In this chapter, the complete numerical technique has been described. Along with this description, numerous references have been made to the subroutine and variable names. With this background, the stage is now set for the complete description of the computer program; the next two chapters are devoted to this activity.

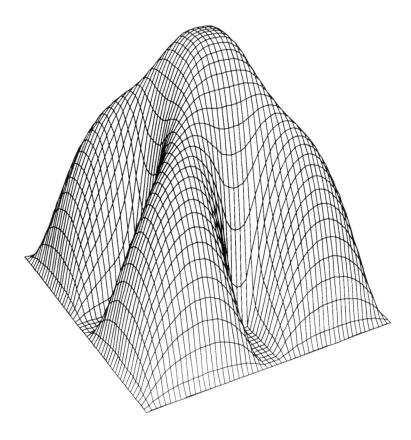

Axial velocity distribution in a square duct with four internal fins
(The fin height equals 0.25 times the side of the square.)

Chapter 6

Invariant Part Of the Computer Program

As already mentioned. the computer program CONDUCT is constructed in two parts: the invariant part contains the general calculation procedure for conduction-type problems and the adaptation part is a user-designed subroutine for the specification of a particular problem. The details of the invariant part are given in this chapter: the adaptation part will be described in Chapter 7.

6.1   Important Fortran Names

All the important Fortran variables in the invariant part of the program are included in COMMON statements. which are used in various subprograms and also in the adaptation part of CONDUCT. To avoid the repetition of these statements in every subprogram. all the COMMON statements (and some related declarations) are written in a file named "COMMON". which is inserted via an INCLUDE statement into the relevant subprograms. The syntax of the INCLUDE statement used in the program listing in this book is the one required for the Microsoft FORTRAN Compiler for the IBM PC. If you use a different compiler. you should modify the syntax as needed. If your compiler does not allow the INCLUDE statement. you should replace each occurrence of the INCLUDE statement by the contents of the file COMMON.

This file begins with a PARAMETER statement: it defines the values of the constants NI. NJ. NFMAX. and NZMAX which are used for dimensioning various arrays. These values decide the size of arrays. for which computer memory is allocated: they do not specify the actual number of grid points used in the calculation.

The dependent variables ϕ are all stored in the array $F(I,J,NF)$. where I and J denote the locations in the x and y directions. and the value of NF identifies a particular dependent variable such as the temperature. the velocity. etc. The F array is dimensioned as $F(NI,NJ,NFMAX)$. Thus. the largest values that I and J can take are given by NI and NJ respectively. It then follows. that the actual number of grid locations in each direction. namely $L1$ and $M1$. cannot exceed NI and NJ respectively. Incidentally. $L1$ and $M1$ cannot be less than 4 either. The constant $NFMAX$ represents the maximum number of different dependent variables that we can store. The constant $NZMAX$ in the PARAMETER statement is the maximum number of zones allowed in connection with the utility ZGRID to be described later. The values used in the PARAMETER statement are considered more than sufficient for our purposes in the book: but if you wish to use finer grids. more dependent variables. or more zones. you can easily alter the PARAMETER statement as needed.

The meanings of $F(I,J,NF)$ for each value of NF are not predetermined. You can associate any desired meaning with a particular value of NF and provide the appropriate λ. Γ. S_C. S_P. and boundary conditions. Depending on your chosen meaning for a particular value of NF, you can also introduce a more meaningful array name for that part of the F array. Thus, if you wish to use NF = 2 for temperature, the array name T(I , J) can be introduced in subroutine ADAPT and made equivalent to F(I , J , 2). We shall use this practice in all our example problems.

The NFMAX dependent variables may not be needed in every problem. Often, it is convenient to use the storage space in the F array for auxiliary variables, which are obtained from simple algebraic relationships rather than by solving the general differential equation for ϕ. For example, we could store in the F array, in addition to the temperature, a corresponding *dimensionless* temperature. Further, in a given iteration cycle, it may not be desired to include the solution of all the relevant dependent variables. To make provision for all these alternatives, an array KSOLVE(NF) is defined. If KSOLVE(NF) is zero, the differential equation is not solved for F(I , J , NF). Thus, we can distinguish between the true dependent variables in the F array and the auxiliary variables. The value of KSOLVE(NF) may be changed at every iteration. Thus, it is possible to arrange that, as the iterations proceed, the equations for certain variables are switched on or off. Many imaginative uses of the KSOLVE facility are possible.

There are other arrays that are directly related to the members of the F array. Of these, KBLOC(NF), NTIMES(NF), CRIT(NF), and NTC(NF) are used in conjunction with the solution of the discretization equations and have been defined in Section 5.6. The array RELAX(NF) is used for the underrelaxation factor α as indicated in Section 5.7. The printout of the two-dimensional field of F(I , J , NF) is arranged in subroutine PRINT if the corresponding indicator KPRINT(NF) is nonzero; otherwise the printout is suppressed. In this printout, an 18-character title for the variable F(I , J , NF) is used: this should be supplied in the array TITLE(NF). To prepare a file for plotting purposes, a variable KPLOT(NF) is used; its function is similar to KPRINT(NF). When KPLOT(NF) is nonzero, the corresponding F(I , J , NF) is included in the plotfile. Two important auxiliary variables, λ and Γ, are stored as ALAM(I , J) and GAM(I , J).

The Fortran names of all the useful geometrical quantities have been explained in Sections 5.1-5.2. The discretization coefficient names were introduced in Section 5.4. The boundary condition indicators KBCI1(J), KBCL1(J), etc. were defined in Section 5.5-3. Section 5.5-6 makes reference to the boundary-flux arrays such as FLUXI1(J , NF), to be used for printout purposes.

The variable ITER indicates the number of iterations that have been completed at any given stage. For an unsteady situation, ITER counts the number of time steps that have been completed. DT is used for the time step Δt, while the cumulative time t is stored as TIME. The indicator KSTOP is used to decide whether the computation should be terminated. KSTOP is initially set equal to zero; when it becomes nonzero, further iterations are halted. The only termination criterion used in the invariant part of the program is that ITER equals LAST, where LAST is the total number of iterations (or time steps) specified by the user. However, you are free to introduce any termination condition in ADAPT by making KSTOP nonzero when the solution has progressed to the desired state.

The variables SMALL and BIG represent a very small and a very big number respectively. We set SMALL equal to 1.E–20 and BIG equal to 1.E20. If your computer has a different range and does not accept these numbers, you may appropriately modify them. We often add SMALL to the denominator of an expression to avoid division by zero. BIG is used as an approximation to infinity.

With this general background about the main variables, the different subroutines in the invariant part of the program can now be described. While following this description, you would find it useful to refer to the flow diagram given in Fig. 4.1 and to glance at the listing of the invariant part in Appendix A. The Fortran variables in the program and their meanings are listed in Appendix B.

6.2 Main Program

The Main program is so short that it is virtually self-explanatory. Initially, it calls DEFLT to set the default values for many important variables. This will be described shortly. After DEFLT, the routine GRID is called to receive the user-specified geometrical information about the grid. Then a call is made to READY to process the geometrical quantities. The initial phase ends with a call to BEGIN, where you supply the initial values of the dependent variables. The iteration loop begins with CALL OUTPUT. Depending on the value of KSTOP, either the computation is terminated or the subroutine HEART is called, where one (outer) iteration cycle on all the relevant variables is performed.

6.3 Subroutine DEFRD

If you take a quick look at the listing in Appendix A, you will notice that DEFRD is actually an assembly of two member subroutines, DEFLT and READY. This is also indicated in Fig. 4.1. What we do is use ENTRY statements to create member routines. The structure of this arrangement can be expressed as follows.

```
SUBROUTINE DEFRD
ENTRY DEFLT
...
...
RETURN
ENTRY READY
...
```

```
...
RETURN
END
```

Now it is easy to see that DEFLT and READY are, for all practical purposes, two separate subroutines. In this particular instance, this arrangement does not have any overwhelming advantages; however, we shall use the same arrangement in the adaptation part of the program, where its benefits are very significant.

Routine DEFLT. When a general-purpose program such as CONDUCT contains so many important variables and indicators, their correct specification for every problem appears to be a formidable task. However, in CONDUCT, this task is made much easier by the judicious use of default values for many of these variables. In general, the default values chosen are the logical or commonly encountered values for each variable. It is a good idea to study the default values inserted in the program, so that the specification in ADAPT could be limited to only those quantities that must depart from their default values. A complete list of default values is given in Appendix C. Incidentally, all the default values are initially set in the routine DEFLT.

Routine READY. The function of READY is to calculate and store the values of numerous geometrical quantities that were introduced in Sections 5.1–5.2. Only the one-dimensional geometrical quantities are stored. Quantities such as $YCVR(J)*XCV(I)$, although useful in later work, are not stored as they would require two-dimensional storage. The unified treatment of the three coordinate systems for $MODE = 1, 2$, and 3 becomes possible because of the appropriate definitions of the generalized geometrical quantities employed in READY. Almost nowhere else in the program is there any need to use IF statements to implement different formulas for each coordinate system.

READY also performs another function. It prints a general title to indicate the coordinate system used. You will find this output very valuable; because if you inadvertently used the wrong coordinate system, you would immediately be alerted. Further, READY prints out the character variable $HEADER$, which contains the 64-character heading for the problem, supplied by the user.

6.4 Subroutine HEART

The most important calculation machinery of CONDUCT is contained in the subroutine HEART. It is here that the discretization coefficients are calculated and the boundary condition modifications are made. A glance at the entire subroutine HEART will reveal an outer DO loop for $N=1, NFMAX$ for handling all the relevant dependent variables. Inside the loop, a call is made to the subroutine PHI, where the user supplies the appropriate values of $ALAM(I,J)$, $GAM(I,J)$, $SC(I,J)$, $SP(I,J)$, and the boundary indicators KBC. The subroutine SOLVE is finally called to obtain an iterative solution of the algebraic equations.

At the end of HEART, the values of $TIME$ and $ITER$ are incremented, and the value of $KSTOP$ is set equal to 1 if the value of $ITER$ has reached its limit given by $LAST$.

As you become more familiar with CONDUCT, you should study the subroutine HEART in detail. Only then would you fully understand the actual calculation of the discretization coefficients.

6.5   Subroutine SOLVE

The task of the subroutine SOLVE is to perform the iterative solution of the linear algebraic equations like (5.14). The solution algorithm embodied in SOLVE has been described in Section 5.6. After the desired repetitions of the algorithm are performed, an additional function is carried out at the end of SOLVE. It involves the calculation of the unknown boundary values of ϕ when the boundary indicator KBC is set equal to 2. It should be noted that the solution algorithm in SOLVE operates on the ϕ values at only the internal grid points. Once the solution is completed, the unknown boundary values such as ϕ_1 are obtained from equations like (5.44). Also, the diffusion fluxes at the boundaries are calculated and stored in the arrays $FLUXI1(J,NF)$, $FLUXL1(J,NF)$, $FLUXJ1(I,NF)$, and $FLUXM1(I,NF)$. These are available to you for printout and other processing. The sign convention used is that a flux is positive if it *enters* the calculation domain.

As you know, we assign default values to a number of variables in the routine DEFLT. However, certain variables need to be continually restored to their default state after each use. Such variables include the source terms $SC(I,J)$ and $SP(I,J)$, and all boundary condition parameters KBC, $FLXC$, and $FLXP$. These variables are reset to their default values at the end of SOLVE. (Note that SC and SP are made equivalent to CON and AP respectively. In the invariant part of the program, the names SC and SP are not used.)

6.6   Subroutine TOOLS

The subprograms described so far constitute the essential routines in the invariant part of CONDUCT. The subroutine TOOLS is a collection of a few optional routines that are provided to facilitate the work of the user. These routines are not called from any point in the invariant part. You may call them from appropriate locations in ADAPT. The utility subroutine TOOLS is an assembly of four member routines EZGRID, ZGRID, PRINT, and PLOT.

EZGRID. The user has the task of specifying the grid-related information in terms of the locations of the control-volume faces $XU(I)$ and $YV(J)$. In general, the grid can be nonuniform and the actual distribution of the values of $XU(I)$ and $YV(J)$ would depend on the details of the problem. However, often it is desired to specify a uniform grid such that all the control volumes would have a uniform width in the x direction and also in the y direction. This task can be accomplished by calling EZGRID from the GRID part of ADAPT.

The input required by EZGRID consists of the values of the overall domain lengths XL and YL in the x and y directions respectively, and the desired number of control-volume widths in each direction. $NCVLX$ denotes the number of control-volume widths into which XL is subdivided. $NCVLY$ is a similar number for the y direction. (Incidentally, for $MODE = 3$, XL represents the θ-direction extent of the domain in radians.) The output of EZGRID is the values of $L1$, $M1$, and the locations $XU(I)$ and $YV(J)$ for all control-volume faces.

EZGRID can also be used to introduce certain simple nonuniformities in the grid spacing. If we use the formula

$$\frac{XU(I)}{XL} = \left(\frac{I-2}{L1-2}\right)^n \qquad (6.1)$$

where n is a positive constant, we can get a corresponding nonuniform grid for different values of n. The variation given by Eq. (6.1) is plotted in Fig. 6.1. When n > 1, the grid is fine near the left boundary and becomes progressively coarser toward the right; near the right boundary, the grid tends to be coarse and uniform.

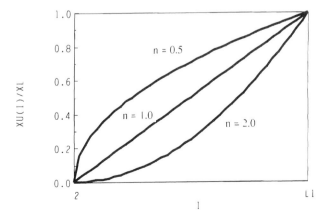

Fig. 6.1 Locations of control-volume faces given by Eq. (6.1)

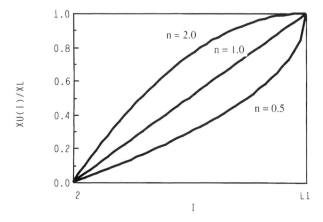

Fig. 6.2 Locations of control-volume faces given by Eq. (6.2)

When n < 1, the grid is coarse at the left end and becomes uniformly fine near the right end.

These behaviors at the left and right ends can be reversed if we use

$$\frac{XU(I)}{XL} = 1 - \left(1 - \frac{I-2}{L1-2}\right)^n \tag{6.2}$$

This variation is shown in Fig. 6.2.

In EZGRID, the locations $XU(I)$ are calculated from Eqs. (6.1) and (6.2). For n = 1, which is our default value, we get a uniform grid. Other values of n give the corresponding nonuniformity. The variable POWERX contains the value of n for the x direction. Although the constant n in Eqs. (6.1) and (6.2) should always be positive, we use the sign of POWERX to decide whether Eq. (6.1) or (6.2) is used. If POWERX > 0, EZGRID uses Eq. (6.1); if POWERX < 0, Eq. (6.2) is used (where n is set equal to $-$POWERX). The nonuniformity in the y direction is treated in the same manner: the variable POWERY is used to specify the exponent for the y-direction grid.

POWERX and POWERY are set equal to 1 by default. So, if you do not change these values, you get a uniform grid. Normally, it is not desirable to set the magnitudes of POWERX and POWERY greater than 2 or less than 0.5; because then we get excessive nonuniformity of the grid.

ZGRID. An extended version of EZGRID is provided in ZGRID, in which the x- and y-direction lengths are divided into different zones, and for each zone a grid design similar to EZGRID is provided.

In our calculation procedure. it is important to ensure that the control-volume faces coincide with the locations of discontinuity in conductivity, source terms, boundary conditions, etc. With arbitrary locations of these discontinuities, it is not always possible to capture them either with a uniform grid or a grid based on Eqs. (6.1)–(6.2). In these cases, we can subdivide the x-direction length (and similarly the y-direction length) into different zones such that the zone boundaries coincide with the discontinuities. Then the number of control-volume widths and the index n can be supplied separately for each zone. ZGRID provides this type of grid design.

The quantities required as input to ZGRID are: the number of zones NZX and NZY for the x and y directions respectively; and for each zone: the physical length, number of control-volume widths, and the index n. These quantities are given as

    (XZONE(NZ),NCVX(NZ),POWRX(NZ),NZ=1,NZX)
and
    (YZONE(NZ),NCVY(NZ),POWRY(NZ),NZ=1,NZY)

With this information. ZGRID calculates the values of XU(I). YV(J). L1, and M1.

The variables POWRX(NZ) and POWRY(NZ) are treated like POWERX and POWERY in EZGRID. their default values being unity. The number of zones NZX and NZY must not exceed NZMAX. which is specified in the PARAMETER statement in the "include" file COMMON.

PRINT. When the solution to a physical problem is calculated. it is desirable to print out the grid locations and the two dimensional fields of the relevant dependent variables. Such a printout is obtained by calling PRINT. Although normally PRINT is called when the converged solution is reached. it can in principle be called at any time. The logical place for a call to PRINT is the OUTPUT part of ADAPT. where, in addition to the other problem-dependent output. the printout provided by PRINT is usually desirable.

The printout from PRINT includes the values of X(I) and Y(J). followed by the two-dimensional fields of those F(I,J,NF) for which KPRINT(NF) is nonzero. The eighteen-character title TITLE(NF) supplied by the user is printed out to identify the field of F for each NF.

If you call PRINT more than once in the ADAPT routine (for example. to get the temperature fields at different instants of time in an unsteady problem). you may not want to print out the same X(I) and Y(J) values every time. The variable KPGR is provided for this purpose. Normally. KPGR will be equal to 1 and then you will get the printout of the grid locations. When KPGR = 0. the grid printout is omitted. If you call PRINT several times. you may wish to set KPGR = 0 after the first call to PRINT.

PLOT. If the results of the calculation are to be written to a file for plotting or other purposes, the utility PLOT is used. It writes on a file the grid-related data and the fields of those F(I,J,NF) values for which KPLOT(NF) is nonzero.The name of the plotfile is specified by the user as a value of the character variable PLOTF.

The contents of the PLOT routine in CONDUCT are designed to produce a plotfile for a color-graphics program microGRAPHICS, which is available from the publisher of this book, Innovative Research, Inc. The color pictures on the cover of this book are examples of the color displays created by microGRAPHICS. The graphics package can produce displays of the grid geometry, color contour lines and color-shaded maps for any scalar variable, vectors, and profiles of any variable in any coordinate direction. All these displays can be generated for the whole calculation domain or for a smaller region on which you choose to zoom in. If the calculation domain contains any blockages, such as the solid fins in the cross section of a duct, they are drawn by microGRAPHICS if you specify the value of IBLOCK(I,J) equal to 1 in the corresponding control volume(s). By default, the value of IBLOCK(I,J) is zero, which indicates no blockage. microGRAPHICS is designed to operate on the IBM PC and compatible computers and requires the EGA or VGA card and the corresponding color monitor.

If you use a graphics package other than microGRAPHICS, you may be able to modify the PLOT routine accordingly. If you do not intend to use any plotting facility, you may simply ignore all references to the PLOT routine in this book. For this reason, in our description of the fifteen examples, no mention of the PLOT utility is made, although in the ADAPT routine we do include a call to PLOT and set the IBLOCK(I,J) array wherever necessary.

6.7 Subroutine VALUES

Another utility routine provided in CONDUCT is the subroutine VALUES. It is called from DEFLT and, therefore, cannot be termed as nonessential. Yet, its function could have been performed by simple arithmetic statements, READ statements, or possibly DATA statements. The purpose of VALUES is to assign numerical values to several variables in a concise manner. Its justification lies in the convenience it provides to the user; but the use of VALUES cannot be considered as computationally efficient.

A quick glance at the listing of VALUES will reveal an interesting use of ENTRY statements, which enable us to enter this routine with a variable number of arguments. Since an argument must be of a designated type, two sets of entry points are provided. The entries labeled DATA9, DATA8, etc. are to be used for real variables, and INTA9, INTA8, etc. are for integer variables.

A call such as

```
CALL DATA4(A,2.5,G1,7.8,HG,1.2,B,5.0)
```

accomplishes the task of the following four statements:

```
A=2.5
G1=7.8
HG=1.2
B=5.0
```

Alternatively, this task can sometimes be performed by a DATA statement such as:

```
DATA A,G1,HG,B/2.5,7.8,1.2,5.0/
```

However, there are a number of restrictions on the use of the DATA statements. It is not an executable statement and initializes data only at the compilation stage. You cannot use DATA to insert values for variables in the unlabeled COMMON. For variables in a labeled COMMON, the corresponding DATA statements should normally be placed in a BLOCK DATA subprogram. Finally, different compilers put additional restrictions on the use of BLOCK DATA subprograms. The subroutine VALUES is our attempt to get much of the convenience of the DATA statement without having to accept its limitations.

The numerical part in the ENTRY names such as DATA6, or INTA4 indicates the number of variables that are to be assigned values. Although the entries DATA1 and INTA1 are provided for completeness, there is obviously no benefit in using them. The task of

```
CALL DATA1(Q,3.7)
```

can be done more easily by

```
Q=3.7
```

The even-numbered arguments of the entries in VALUES can be constants, variables, or expressions. For example, we can use

```
CALL DATA3(RHO,1.2,U,22.,CP,127.)
CALL DATA2(RHOCP,RHO*CP,USQ,U*U)
```

Warnings. In the entries like DATA9, DATA8 etc., all the arguments must be real; in INTA9, INTA8, etc., they must be integer. Statements such as

```
CALL DATA2(A,12,B,3.5)
CALL INTA3(I,1,J,10.,K,5)
```

would lead to incorrect values for A and J. Further, the numerical part of each entry name must exactly match the number of variables to which values are assigned. For example, you cannot call DATA4 and assign values to more than four variables. The statements like

```
CALL DATA3(A,1.2,B,1.5)
CALL DATA3(A,1.2,B,1.5,C,27.,D,108.)
```

are incorrect; the compiler may not complain about them but their effects are unpredictable.

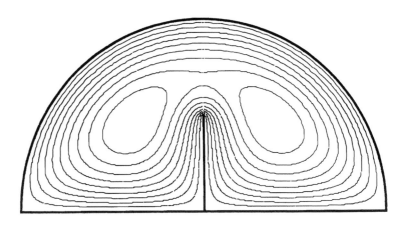

Contours of axial velocity in a semicircular duct with an internal fin

Adaptation Part of the Computer Program

Unlike the invariant part of CONDUCT, which is complete in all respects, the adaptation part is provided only as a skeleton. Its contents are to be filled by the user according to the nature of the problem and the information sought from its solution. Even the examples of various adaptations of CONDUCT that are included in this book are only illustrative and are not meant to be general, efficient, or optimized. Their purpose is simply to assist you in learning to use the computer program. Before we begin to examine these examples, the required background information about the subroutine ADAPT is given in this chapter.

7.1 General Framework of ADAPT

The subroutine ADAPT is an assembly of four member subroutines: GRID, BEGIN, OUTPUT, and PHI. ADAPT communicates with the invariant part through the COMMON statements in the "include" file COMMON. Once again, we use the ENTRY-RETURN combination, explained in Section 6.3, to stack several member routines inside one general subroutine. In this manner, the COMMON statements need not be repeated in each member routine. More importantly, any new variable names introduced by you need not be put in COMMON, since they are automatically available to all the parts of ADAPT. This feature is the overriding reason for making ADAPT an assembly of four routines instead of writing four separate subroutines. Also, this arrangement keeps all the problem-dependent information physically in one place.

The physical arrangement of the four member subroutines in ADAPT follows the structure in the flow diagram shown in Fig 4.1. Thus, in designing any member routine, it is convenient to know that, when the execution of that routine begins, all the routines appearing before it have been executed at least once.

It is important to study Fig 4.1 to identify the frequency of calls to the various members of ADAPT. The routines GRID and BEGIN are called only once. OUTPUT is called once during each iteration. PHI is called in each iteration once for every NF for which KSOLVE(NF) is nonzero.

This knowledge of the frequency of calling is useful in making imaginative use of the various routines beyond their nominal stated function. The member subroutines represent strategic points at which the workings of the invariant part can be intercepted and influenced by the user. For example, the quantities that would remain unchanged throughout the computation can be conveniently introduced in GRID or BEGIN. On the other hand, it would be inappropriate to include in GRID or BEGIN those operations that depend on the changing values in the iteration cycle.

7.2   Initial Declarations

As mentioned earlier, convenient and meaningful names can be given to the members of the F array through appropriate DIMENSION and EQUIVALENCE statements. For example, if F(I,J,1) and F(I,J,2) are supposed to stand for the axial velocity w and the temperature T in a duct flow, the names W(I,J) and T(I,J) can be introduced through:

DIMENSION W(NI,NJ),T(NI,NJ)

EQUIVALENCE (F(1,1,1),W(1,1)), (F(1,1,2),T(1,1))

Making a two-dimensional array equivalent to a certain part of a three-dimensional array may be a novel concept to many of you. This is possible because of the certain manner in which Fortran handles multidimensional arrays. In storing F(I,J,NF), all the elements of F(I,J,1) are stored first, those of F(I,J,2) come next, and so on. Thus, the three-dimensional array is stored as a collection of many two-dimensional arrays, to which we can give equivalent names such as T(I,J) or W(I,J). That NF is the third subscript in F(I,J,NF) makes this practice possible. If we had used F(NF,I,J) or F(I,NF,J) as the subscript sequence, we could not have made T(I,J) equivalent to the appropriate part of the F array.

If you plan to use additional subscripted variables in ADAPT, they should be dimensioned in the beginning. In order to conserve computer storage, it is best to use the available arrays before declaring new ones. In a given problem, certain members of the F array may be unused; the storage they occupy can be used for other purposes, possibly by making them equivalent to some other array names.

7.3   GRID

The function of the routine GRID is to supply the grid-related information. In particular, the values of MODE, L1, M1, (XU(I),I=2,L1), and (YV(J),J=2,M1) are required. If the value of MODE is 2 or 3, the value of R(1), i.e., the radius at the bottom boundary of the domain must be additionally specified. The coordinates XU(I) and YV(J) of the control-volume faces need not be uniformly spaced; however, the values of XU(I) must increase with I, and those of YV(J) with J. For MODE = 2 or 3, the values of YV(J) must increase in the radially *outward* direction. XU(2) or YV(2) need not be zero. The value of R(1) can be zero, but never negative.

If a uniform grid is desired, you should supply, instead of XU(I), YV(J), L1, and M1, the values of the domain lengths XL and YL, and the number of control-volume widths NCVLX and NCVLY, and then call EZGRID. You can even introduce some simple nonuniformity by giving values to POWERX or POWERY, as explained in Section 6.6. For more complex domains, the utility ZGRID may be used for generating the grid-related information. The input required for ZGRID has been described in Section 6.6.

The use of a nonuniform grid is a powerful way of making efficient deployment of a given number of grid points. Also, discontinuities in boundary conditions, material properties, and generation rates can be conveniently handled by locating the control-volume faces to coincide with the lines of discontinuity. Often, exploratory solutions with coarse grids are useful in providing guidance for proper design of a nonuniform grid, for which you may use EZGRID or ZGRID, or design the GRID routine independently.

In designing GRID, you should remember that no portion of the invariant part of the program (except the DEFLT routine) has been executed when the statements in GRID are encountered. Thus, none of the geometrical quantities nor the integers like L2 or M2 have yet been calculated.

That GRID is the very first routine that is called suggests many other uses for it. You can print out a descriptive title for your problem, open any files, perform some preliminary operations, and READ in the input data. Our practice in the illustrative examples in this book is to give in GRID the desired values to the character variables HEADER, PRINTF, and PLOTF, which stand for the problem heading, the name of the printfile, and the name of the plotfile.

The use of the printfile requires some explanation. These days, the default output device on most computers is the monitor screen. Therefore, ordinary output will scroll on the screen and be lost. In order to save the output on a file, a special provision is made in CONDUCT. First, PRINTF gives the name of the printfile used for this purpose. This file is opened at the beginning of READY to UNIT = 7. The default output device (screen) corresponds to UNIT = 6. Now you have three options: you can send the output to the screen only, to the printfile only, or to both. These options are specified by setting the value of the variable KOUT equal to 1, 2, or 3 respectively. The default value of KOUT is 3, which gives output on the printfile and on the screen. If you need to change the value of KOUT, this must be done in GRID, since KOUT is used immediately afterwards in READY.

All the output statements in CONDUCT are placed inside a DO loop such as

```
      DO 100 IUNIT=IU1,IU2
      WRITE(IUNIT,110)---
  110 FORMAT(---)
  100 CONTINUE
```

The values of IU1 and IU2 are assigned at the beginning of READY on the basis of the specified value of KOUT.

7.4 BEGIN

The primary function of BEGIN is to supply the initial values of F(I,J,NF) for all the relevant NF values. For a steady-state problem, the initial values represent only a starting guess. For an unsteady problem, the initial values refer to the known values at time t = 0. If any boundary values of F(I,J,NF) are known, it is best to fill here the correct values in the boundary locations of the F(I,J,NF) array. These values will be left unaltered by the program provided the relevant KBC indicator is kept equal to 1 in subroutine PHI.

BEGIN is a convenient place also for introducing any other initialization and once-for-all information. A number of constants required for material properties, heat flow rates, boundary conditions, governing parameters, etc. can be specified in BEGIN. For an unsteady problem, DT should be given the value of the time step Δt. However, if it is desired to change DT at every time step, the final specification of DT must be placed in OUTPUT.

Our practice in the illustrative examples is to start BEGIN by giving the values of TITLE(NF), KSOLVE(NF), KPRINT(NF), etc. for the relevant dependent variables, then specify various constants needed for material properties and boundary conditions, and then proceed to fill the F(I,J,NF) array.

7.5 OUTPUT

Whereas the other parts of ADAPT must supply certain minimum information to the invariant part of the program, no such demand is placed on OUTPUT. The program would often function smoothly even if OUTPUT were left blank. However, from the user's point of view, the most important function is performed by OUTPUT. The primary function of OUTPUT is to arrange any desired printout. The user has complete freedom in designing the details of the output.

It is usually desirable to provide a short (one or two lines) output after every iteration and to arrange the two-dimensional field printout of all the relevant ϕ's after the final iteration. The short output per iteration is useful in observing the progress of the steady-state solution towards convergence or the time evolution of the unsteady solution. The field printout can be obtained by calling the utility routine PRINT provided for this purpose.

OUTPUT is a logical place for much of the post-processing of the results of the computation. It is here that one can calculate a Nusselt Number, a friction factor, total heat flow rate, mean temperature, net force on a wall, and so on. The dependent variables F(I,J,NF) can be turned, if so desired, into appropriate dimensionless variables before they are printed out by PRINT.

Any convergence criterion for terminating the computations can also be introduced in OUTPUT. One possibility is to monitor the changes in some significant quantity of interest (such as friction factor, heat flow, maximum temperature in the domain, etc.) and set KSTOP to be nonzero when the changes between two successive iterations are acceptably small. Another practice is to watch the changes in the values of $F(I,J,NF)$ at chosen grid points.

Often it is desirable to perform an overall heat balance for the whole domain. For this, we calculate the total heat loss from all the boundaries and the total internal heat generation. It is convenient to perform these calculations in OUTPUT and print out the results of the heat balance every iteration. As the solution attains convergence, the heat balance would become perfect (within the roundoff error of the computer).

Since OUTPUT is called once every iteration, it is a proper place for updating any quantities that change with the iteration counter or the time step. If you wish to change the KSOLVE(NF) values to switch on or off the solution of different φ's, this should be done in OUTPUT. For an unsteady problem, any time-dependent boundary temperatures and a variable time step Δt should be specified in OUTPUT.

7.6 PHI

Perhaps the most significant information about the problem is contained in the routine PHI. For problems incorporating complex mathematical models, PHI can be very extensive. On the other hand, many commonly encountered problems require rather simple versions of PHI.

The invariant part of the program has the provision for solving the general differential equation for a number of different φ's. The meanings of $F(I,J,NF)$ are not predetermined. How does the program know the meaning and physical behavior of each relevant φ? This happens, not through TITLE(NF) or other informative printout labels, but via the information provided by PHI for each value of NF.

In subroutine HEART, the value of KSOLVE(NF) is examined for each value of NF. If KSOLVE(NF) is nonzero, the calculation of the discretization coefficients for the relevant $F(I,J,NF)$ is initiated. At the start of this calculation, a call is made to subroutine PHI.

The task of PHI is to supply λ, Γ, S_C, S_P, and the boundary-condition information. The λ and Γ values are given in ALAM(I,J) and GAM(I,J) arrays for the internal grid points $I=2,L2$ and $J=2,M2$. The values of λ are needed only for unsteady problems.

PHI also supplies the values of the source-related quantities S_C, S_P, which were defined in Eq (5.13). These are given in the arrays SC(I,J) and SP(I,J) for $I=2,L2$ and $J=2,M2$. The units of S_C and of $S_P \phi_P$ should correspond to a generation rate *per unit volume*. Further, S_p must be zero or negative. The default values of S_C and S_p are zero: thus, if the dependent variable does not have a source term, no action is needed.

The final task of PHI is to specify the boundary conditions through the indicators KBCI1(J), KBCL1(J), KBCJ1(I), and KBCM1(I), and the associated values for FLXC and FLXP. This specification has been fully discussed in Section 5.5-3, and no further elaboration seems necessary. The default values of all the KBC's are 1, and those of FLXC and FLXP are zero.

There is one easy method to decide whether a problem is linear or nonlinear. If, in the specifications of λ, Γ, S_C, S_p, and boundary conditions, the unknown dependent variable (F(I,J,NF) or equivalent) appears, then the problem is nonlinear: if the quantities in PHI can be calculated without knowing F(I,J,NF), then the problem is linear.

7.7 Treatment of Irregular Geometries

The computer program can be used for two-dimensional domains of any arbitrary geometry. However, when the domain shape is irregular, additional work is needed in the adaptation part of the program. The treatment of different types of boundary conditions on irregular geometries gets somewhat involved. For this reason, you may wish to skip this section in your initial reading of this chapter. However, by the time you study Example 3 in Chapter 8, you should have gone through this section.

7.7-1 Main Concept

CONDUCT and irregular geometry. The calculation method in CONDUCT is based on a grid drawn in one of the three coordinate systems shown in Fig. 1.1. As a result, the program is directly applicable to only to those geometries that neatly fit in these coordinates. Thus, if the shape of the domain is a rectangle for MODE = 1 and 2 and an annular or circular sector for MODE = 3, the program can be readily used. Any other shape must be regarded as an irregular geometry with reference to the framework of the program.

Yet, it is possible to apply CONDUCT to irregular geometries by using the techniques to be described here. You should understand that these techniques are computational tricks devised to apply a regular-geometry program to irregularly shaped domains. Therefore, some inelegance and wastefulness are inevitable. Because of our use of regular coordinates, the calculation method and the computer program offer considerable simplicity, convenience, and efficiency. The other side of the coin is that the program is somewhat cumbersome and inelegant for its use in irregular domains. If arbitrary geometries were our primary focus, we could have built the program using a curvilinear nonorthogonal grid or finite-element discretization. But then the program would have been much more complicated to construct, understand, and use. For an introductory book, aimed at physical understanding, our choice of the regular grid seems quite appropriate.

Having considered all this, let us note that CONDUCT can in fact adequately handle a number of irregular geometries. The required computational tricks are based on physical principles and should be used with full understanding. Finally, if the domain is very highly irregular, you may decide that CONDUCT is unsuitable for that application.

Active and inactive control volumes. The program can nominally handle domains that have the shape of a rectangle, a circular sector, or an annular sector. For convenience, we shall discuss this topic in terms of a rectangular shape (MODE = 1 or 2), but the same ideas can be easily used for circular or annular sectors (MODE = 3). Given an irregular domain shape, we begin by drawing a nominal rectangular domain around it. This procedure is shown in Fig. 7.1. Now, some boundaries of the real domain fall inside the nominal domain. We should discretize the nominal domain into control volumes in such a way that the boundaries of the real domain coincide with the control-volume faces. The control volumes that lie inside the real domain are called active control volumes; the solution there is of interest. The control volumes that lie outside the real domain (in the shaded area in Fig. 7.1) are considered inactive; we carry them along to form a nominally regular domain, but we do not seek a solution there.

For the geometry given in Fig. 7.1(a), it is easy to design control volumes such that they fall wholly inside or wholly outside the real domain. For curved or inclined boundaries as shown in Fig. 7.2, we need to approximate them by staircase-like zigzag lines. Although we shall compute the solution on the approximated geometry, the resulting error in the solution—for a moderately fine grid—is usually surprisingly small. The staircase approximation to a smooth line

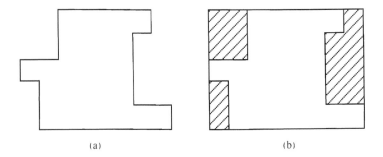

(a)   (b)

Fig. 7.1 Treatment of an irregular boundary:

(a) real domain, (b) nominal domain enclosing the real domain

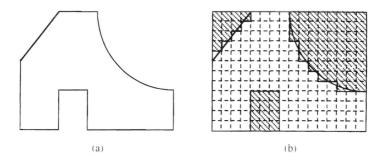

Fig. 7.2 Treatment of inclined and curved boundaries:
(a) real domain, (b) nominal domain with active and inactive control volumes

looks worse than it really is in terms of the accuracy of the solution. Also, this kind of approximation in duct flow problems does not have the effect of replacing a smooth wall by a rough wall. Finally, you can always use a finer grid to improve the geometrical approximation and to reduce the solution error. When you begin to use CONDUCT, you will have many occasions to evaluate the accuracy of this procedure, for example, by computing a circular domain on a rectangular grid.

Since our objective is to obtain the solution in the active control volumes, we must handle the inactive control volumes in such a manner that the given boundary conditions at the boundaries of the real calculation domain are felt by the active control volumes. Exactly how we do this depends on the nature of the boundary condition. Some procedures for this will be discussed here; if you recognize that these are basically computational tricks, you may be able to invent additional procedures of this kind if needed. Most of the description given here is in terms of heat conduction. However, the general variable ϕ is used in the equations, and the treatment is , of course, applicable to any dependent variable.

7.7-2   Formulation for Different Boundary Conditions

Known value of ϕ. Consider the situation shown in Fig. 7.3, where the value of the temperature T is given on a boundary of the real domain. Since this boundary becomes an *internal* location in the nominal domain, we cannot directly specify the temperature there. However, as shown in Fig. 7.3(b), we can prescribe the known temperature as 150 at the *outer* boundary of the shaded region (which coincides with the boundary of the nominal domain) and set the conductivity, i.e., GAM(I , J) in the shaded region equal to a very large number such that 1.E20. The result will be that the temperature throughout the high-conductivity shaded region will be equal to the value 150 prescribed on the boundary. Thus, the correct boundary condition will be felt by the real domain.

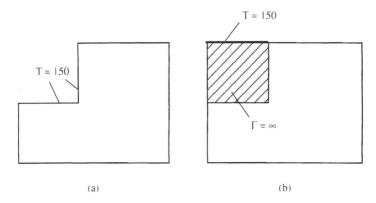

Fig. 7.3 Given temperature boundary condition
(a) real boundary condition, (b) simulated boundary condition on the nominal boundary

In duct flow problems involving fins or other solid projections from the duct wall, the control volumes in the solid material can be treated similarly. For the velocity equation, the velocity at the boundary of the nominal domain is set equal to zero, and the GAM(I , J) in the solid material is made very large. As a result, the velocity over the entire region will be uniform and equal to the zero velocity at the boundary. In this manner, the zero values of the velocity in the solid provide the appropriate boundary conditions for the fluid flow in the other control volumes. For the velocity equation, GAM(I , J) equals viscosity. It is interesting to note that our practice amounts to treating the solid material as a fluid of very large viscosity.

The practice of setting GAM(I , J) equal to a large number is useful when the inactive region is adjacent to the boundary of the nominal domain. Only then the given value of ϕ at the nominal boundary gets established throughout the high-Γ region. Sometimes, the inactive region forms, as shown in Fig. 7.4, an island totally isolated from the nominal boundaries. Then it is not sufficient to set GAM(I , J) equal to a large number in the inactive region. In addition, we must directly establish the known value of ϕ at one or more grid points on the island. This can be done by specifying the source terms S_C and S_P for the selected grid point(s) according to:

$$S_C = M \, \phi_{desired} \tag{7.1a}$$

$$S_P = - M \tag{7.1b}$$

where M is a very large number such as 1.E20 and the known value of ϕ is denoted by $\phi_{desired}$. The source terms given by Eq. (7.1) are generally very large

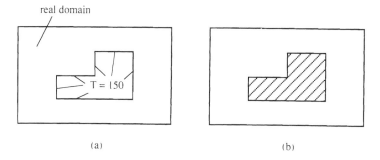

real domain

T = 150

(a)   (b)

Fig. 7.4 An internal boundary with known temperature:
(a) real domain. (b) inactive region forming an island

when compared with the other terms in the discretization equation (5.14). When they are substituted into the coefficient expressions in Eqs. (5.15)–(5.21), the discretization equation reduces to

$$\phi_P = \phi_{desired} \tag{7.2}$$

In this manner, the known value of ϕ can be established at a grid point on the island. Since the control volumes on the island have a large value of GAM(I , J), the values of ϕ on the island will be uniform and equal to $\phi_{desired}$. (The grid point at which the source terms given by Eq. (7.1) are used should not have the value of GAM(I , J) also equally large. Otherwise, the source terms alone would not dominate the equation and the outcome would not conform to Eq. (7.2). Although GAM(I , J) should be large, it should not be so large as to overpower M. If M equals 1.E20, GAM(I , J) can be set equal to 1.E15.)

Given flux. Suppose that, for the domain shown in Fig. 7.3(a), the boundary condition is, instead of a specified temperature, a specified zero flux. This is very easy to handle; all that we have to do is set GAM(I , J) in the inactive region equal to *zero*. Since no heat flux goes through a zero-conductivity material, we have established the given zero flux boundary conditions at the boundaries of the real domain.

Now consider the situation in Fig. 7.5, where a nonzero heat flux q_B is specified. We can handle this in two steps. First, we set GAM(I , J) equal to zero in the inactive shaded region. This will establish a *zero* flux at the surface in question. Then we provide the given flux q_B as an extra source term for the near-boundary control volumes shown in Fig. 7.5(b). These are the control volumes into which the boundary flux q_B would normally enter. The extra source term for these control volumes is given by

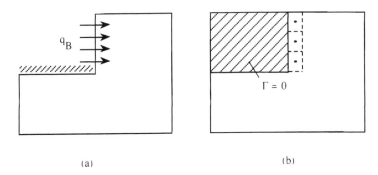

(a)   (b)

Fig. 7.5 Heat flux boundary condition

(a) real domain. (b) nominal domain showing the affected control volumes in the inactive region

$$S_{C.extra} = \frac{q_B A}{\Delta V} \tag{7.3}$$

where A is the area of the control-volume face through which the heat flux q_B enters, and ΔV is the volume of the control volume. The division by ΔV is needed because the source term S is defined as the generation rate per unit volume. Of course, the value of S_C gets multiplied by ΔV in Eq. (5.19); so the heat flow rate $q_B A$ finally enters the discretization equation.

Convective boundary condition. Now let us consider the most general boundary condition shown in Fig. 7.6. where the convective heat transfer coefficient h and the surrounding fluid temperature ϕ_∞ are prescribed. Once again our strategy would be to set GAM(I,J) equal to zero in the inactive region (thus stopping the heat flow through the boundary face considered) and then to bring the convective heat flux into the near-boundary control volumes via the source term. An enlarged view of a typical near-boundary control volume is shown in Fig. 7.7. For the convective boundary. we can express the heat flux q_B as

$$q_B = h\,(\phi_\infty - \phi_B) \tag{7.4}$$

The heat conduction from B to P in Fig. 7.7 gives

$$q_B = \frac{\Gamma}{\delta}\,(\phi_B - \phi_P) \tag{7.5}$$

where δ is the distance shown in Fig. 7.7. and Γ is the conductivity in the near-boundary control volume. If we eliminate ϕ_B from Eqs. (7.4) and (7.5), we get

$$q_B = \left[\frac{1}{h} + \frac{\delta}{\Gamma}\right]^{-1}(\phi_\infty - \phi_P) \tag{7.6}$$

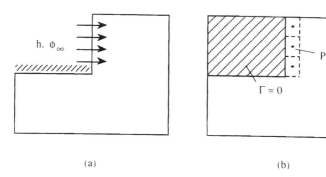

(a)   (b)

Fig. 7.6 Convective boundary condition
(a) real domain. (b) nominal domain showing the affected control volumes

The expression in the square brackets can be recognized as the overall resistance to heat transfer between the surrounding fluid and the grid point P. If A is the area through which the heat flux q_B enters the control volume. and ΔV is the volume of the control volume, the extra heat source due to the boundary condition, when expressed on a unit-volume basis, becomes

$$S_{extra} = \frac{A}{\Delta V}\left[\frac{1}{h} + \frac{\delta}{\Gamma}\right]^{-1}(\Phi_\infty - \Phi_P) \tag{7.7}$$

The form of this expression suggests that the extra source term for the control volume would contain both S_C and S_P parts as given by

$$S_{C.extra} = \frac{A}{\Delta V}\left[\frac{1}{h} + \frac{\delta}{\Gamma}\right]^{-1}\Phi_\infty \tag{7.8a}$$

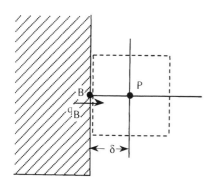

Fig. 7.7 Enlarged view of the near-boundary control volume

$$S_{P,extra} = -\frac{A}{\Delta V}\left[\frac{1}{h} + \frac{\delta}{\Gamma}\right]^{-1} \tag{7.8b}$$

Known value of φ: another approach. If the heat transfer coefficient h in Fig. 7.6 becomes very large, then the temperature of the boundary surface will approach ϕ_∞. This provides another procedure for handling a known temperature at a boundary of the real domain. Of course, for the situation shown in Fig. 7.3, the high-Γ procedure is quite satisfactory; but it will not be suitable for more complex boundary conditions such as those shown in Fig. 7.8. In these cases, we need to provide extra source terms for the near-boundary control volumes. For the typical control volume shown in Fig. 7.7, if the temperature at B is to be specified as $\phi_{desired}$, the required extra source terms are

$$S_{C,extra} = \left(\frac{A}{\Delta V}\right)\left(\frac{\Gamma}{\delta}\right)\phi_{desired} \tag{7.9a}$$

$$S_{P,extra} = -\left(\frac{A}{\Delta V}\right)\left(\frac{\Gamma}{\delta}\right) \tag{7.9b}$$

These expressions result from Eqs. (7.8) by setting the heat transfer coefficient h equal to a very large number.

7.7-3   Final Comments and Reminders

The techniques described here for the treatment of irregular geometry should be studied with great care. Some of the subtle points and precautions are summarized here.

(a) Our particular practice of obtaining the overall conductance between the grid

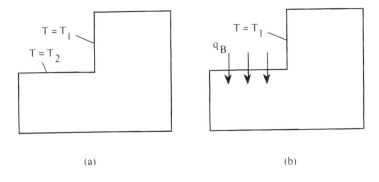

(a)   (b)

Fig. 7.8 Two examples of boundary conditions that cannot be handled by the large conductivity approach

points, as expressed by Eqs. (2.78) and (5.12). enables us to accommodate large discontinuities in the distribution of GAM(I,J). This is why we can easily set GAM(I,J) equal to zero or to a very large number and get the appropriate consequences.

(b) The inclusion of inactive regions in the calculation domain through large Γ's or large source terms can have a serious effect on the block-correction scheme described in Section 5.6. If a block contains some inactive control volume, the block equations would normally predict a zero value of $\overline{\phi}_I$ there. Thus, the block corrections get virtually eliminated if most blocks include inactive control volumes. A remedy for this is a selective block-correction scheme, which omits the inactive control volumes from the block-correction calculation. This is incorporated in subroutine SOLVE by making sure that discretization-coefficient values in excess of 1.E10 do not participate in the block-correction scheme.

(c) The treatment of irregular geometries described here is also suited for conjugate heat transfer problems, in which the conduction in a solid is treated along with the convection in the adjacent fluid. This procedure has been incorporated in some of the adaptations of the program to duct flow problems.

(d) Our output of the two-dimensional field of ϕ will contain the values at all grid points, including the ones in the inactive control volumes. Usually, there is no need to pay any attention to these values especially if GAM(I,J) has been set equal to zero in the inactive region. This action *decouples* the values of ϕ in the active region from those in the inactive region. Even if the computed values of ϕ in the inactive region seem absurd, they do not influence the solution in the active region.

(e) Our printout contains the values of ϕ at the grid points, not at the control-volume faces. The internal boundary locations such as B in Fig. 7.7 coincide with the control-volume faces, and the ϕ values there are not automatically printed out. If you need the value of ϕ_B, you will have to calculate it from Eqs. (7.4) and (7.5).

(f) When we approximate an inclined or curved boundary by staircase-like steps shown in Fig. 7.2, the surface area of the original boundary can be significantly smaller than the area represented by the staircase steps. Therefore, care is needed in the use of the approximate area in connection with heat fluxes, heat transfer coefficients, etc. For example, if the total heat flow from an inclined boundary is given, then in the calculation of the *average* heat flux, it will be better to use the actual area rather than the artificially increased area of the staircase approximation. If the heat flux normal to the inclined boundary is given, we should use its components q_x and q_y in the x and y directions as the prescribed fluxes at the control-volume faces normal to the x and y directions respectively. Remember that the computational tricks for handling the irregular geometry perform well when used with full understanding and care.

(g) When by the use of Eq. (7.1) you establish a desired temperature at one or more grid points on an island, such as the one shown in Fig. 7.4, you still need to set GAM(I,J) at all points on the island equal to a large number. This would ensure that the desired temperature prevails over the island all the way up to the control-volume faces that coincide with the outer surfaces of the island. Then the correct boundary condition is felt by the control volumes in the real calculation domain.

(h) Note that Eqs. (7.3) and (7.9) give the *extra* source terms for the near-boundary control volume, needed for the treatment of the boundary conditions. These should be *added* to the regular source terms that the control volume may have. A common mistake is to use Eqs. (7.3) or (7.9) directly to fill the SC(I,J) and SP(I,J) arrays, thereby wiping out the real heat generation that may occur in the control volume. This point is illustrated in Example 4 in Section 8.4.

(i) By now, we have seen many interesting ways of using the source terms S_C and S_p. If our calculation procedure did not use the linear source-term expression given by Eqs. (2.41) and (5.13), many of these convenient practices would not be available to us.

(j) Finally, the use of inactive regions to represent irregular geometry is admittedly a wasteful process; because storage is provided for, and arithmetic operations are performed on, the meaningless values of the variables in the inactive region. Still, that a program that normally enjoys the benefits of the regular-geometry structure can also be used for irregular geometries represents a significant practical advantage.

7.8 Before You Use the Program

Now that the description of CONDUCT has been completed, you are in a position to embark upon an unlimited number of possible adaptations of the computer program. In this undertaking, it is very important to acquire sufficient experience of using the program. Thus, the need for initial trial explorations with the program cannot be too strongly emphasized. Such explorations are helpful in learning about the behavior of the physical problem as well as about the nature of the program. These can be usefully carried out on coarse grids, which save computer time and facilitate examination of the results.

For any problem, please DO NOT write the subroutine ADAPT from scratch. You should always start from one of the available examples of ADAPT and modify it step-by-step until the desired problem is programmed. The adaptation part should be tested after every few modifications. Then, if the results show undesirable behavior, it becomes easy to detect which changes must be responsible for the trouble. The temptation to go directly to the final complex application should be avoided. Subroutine ADAPT and your experience with it should gradually grow and culminate in the successful solution of the physical problem. Even here, "Walk before you run" continues to be a wise policy.

In our illustrative examples, the grids used are rather coarse. This has been done to avoid voluminous output. In your initial work with CONDUCT, you should also use coarse grids. However, the reasonable-looking results produced on coarse grids should not be regarded as accurate solutions of the differential equations. If accurate numerical values are required, you must use a sufficiently fine grid. In fact, for a given type of problem, you should determine how fine the grid needs to be so that any further grid refinement does not appreciably change the solution. For unsteady problems, the time step Δt should also be made sufficiently small for an accurate solution.

As a final note, you should not lose sight of the fact that a computer program such as CONDUCT is simply a device for solving a set of differential equations by a numerical method. If the equations are a satisfactory description of the reality, the computed results are very useful. However, when the mathematical model is questionable, the results of the computations can be only as good as the underlying equations. In such cases, it is appropriate to undertake some experimental validation, which often leads to further refinement and adjustment of the mathematical model. On the numerical side, there is no unconditional guarantee that the iterative technique used in CONDUCT would converge for all types of nonlinearities and interlinkages. You can nevertheless derive hope from the empirical evidence that the program has already been used with success for a very large number of complex problems. Similar success is now awaiting you.

7.9 Introduction to Adaptation Examples

Chapters 8, 10, and 11 include a number of adaptations of CONDUCT to steady and unsteady heat conduction, to fully developed flow and heat transfer in ducts, and to other physical situations. Here we shall discuss some general characteristics of all these adaptations.

The adaptations of CONDUCT given in this book illustrate how the general-purpose computer program can be used for various specific problems. The adaptations are designed to assist your initial learning process; thus, they serve the same purpose as do the solved problems in most textbooks. Each problem is chosen to emphasize a few special features and practices. Other complications are intentionally kept to a minimum. Since simplicity is essential for the ease of learning, the programming in the subroutine ADAPT is often retained in a crude, non-general, and somewhat inefficient form. Optimum, efficient, and overly general passages are avoided as they frequently tend to become difficult to unravel. For the same reason, we do not use READ statements but insert the numerical data directly into the ADAPT routine. This means that the subroutine must be recompiled every time we need to change the data. This is true; but in the illustrative adaptations, we are after clarity and simplicity, not efficiency.

The physical problems chosen for the illustrative adaptations, and the values, units, fluid properties, etc. used therein, may not have direct practical relevance. Often, the problems are artificially contrived to illustrate a computational feature. The assumptions made about the physical processes may not always be valid. The aim, in most cases, is to show how to incorporate a given problem description into the program rather than to present problem descriptions that are valid in engineering practice. In the heat conduction problems in Chapter 8, the problem description and results are in terms of *dimensional* quantities. For the duct flow problems in Chapter 10, the objective is to obtain the *dimensionless* solution. When values are given to physical quantities such as the conductivity or the source term, we use simple numbers (for example, $k = 2$, $S = 5$) without specifying the units. In a practical application of CONDUCT, you should use the actual properties of materials in some convenient set of units. However, for the purpose of this book, no particular advantage is gained by specifying that the material is aluminum with a thermal conductivity of 204 W/m K.

On the numerical side, the solutions in the adaptation problems are obtained with one particular grid. No grid refinement is made to ensure that the numerical solution is sufficiently close to the exact solution. In most adaptations, no convergence criterion is used to terminate the sequence of the iterations; instead, a predetermined number of iterations are performed and the approach to convergence is observed. It is not implied here that grid refinement and use of appropriate convergence criteria are unimportant matters; but these matters are left to you for exploration on your own.

All the adaptations in the book should be studied carefully. The description in Chapters 8, 10, and 11 includes all main points, but you should try to understand every line in ADAPT routines. The first few adaptations are explained in great detail; subsequently, only the extra features are discussed. After studying the relevant adaptations, you should experiment with the program by making small changes in the physical or numerical parameters. Such initial exploration of the behavior of the problem and the program is very helpful. You should resist the temptation to proceed directly to the complex application of practical interest.

Once the learning phase is over, there is no need to continue with the easy-to-learn and perhaps inefficient forms of the adaptation routines. For a given class of problems, more general and efficient routines can be written. If you are going to perform many runs of the program for a particular physical problem, it will be desirable to construct a special automatic version of ADAPT for that problem, which READs the required data interactively or from a data file.

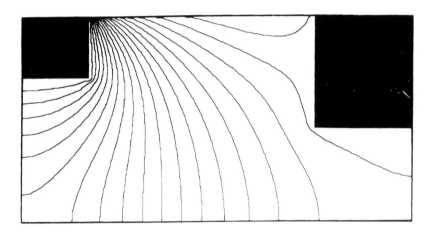

Temperature distribution around underground structures

Chapter 8

Adaptation Examples: Heat Conduction

Now that we have seen the details of the invariant part of CONDUCT and the general framework of the adaptation routine. we can begin to design specific versions of ADAPT for solving particular problems. The examples of ADAPT included here serve to illustrate certain aspects of problem formulation and associated programming. For each example. we shall look at the description of the problem. the corresponding design of the ADAPT subroutine. a list of new Fortran names introduced. an actual listing of ADAPT. the computer output for the problem. and some comments on the results. While reading the description of ADAPT. you should frequently refer to the listing of the subroutine. After you complete a careful study of these adaptation examples. you will be able to build your own adaptations to other problems.

8.1 Steady Conduction with Heat Generation (Example 1)

8.1-1 Problem Description

We begin with a very simple problem of steady heat conduction in a plane solid of square shape. As shown in Fig 8.1. the boundaries are at a prescribed temperature T_w. The solid has a uniform conductivity k and a constant heat generation rate S. We shall use the values

$$T_w = 0. \qquad k = 1. \qquad S = 5 \tag{8.1}$$

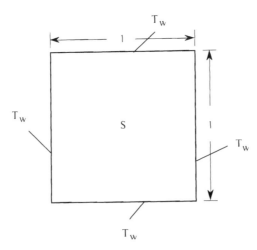

Fig. 8.1 Steady conduction with heat generation

Our aim is to calculate the steady temperature distribution.

Here we make ample use of the default specifications; therefore, the resulting ADAPT routine is quite short.

8.1-2 Design of ADAPT

GRID. We construct a uniform grid by prescribing the domain lengths XL and YL, and choosing 5 control-volume widths in each coordinate direction. Here, MODE equals 1 by default. Also, the default values of POWERX and POWERY are used. As a result, a uniform grid is created by calling EZGRID.

BEGIN. After making the necessary declarations for our chosen variable NF = 1, we provide the numerical values specified in Eq. (8.1). Then we fill the T(I,J) array by TW, which serves as the given boundary value at the boundary points and a convenient initial guess at the internal points.

OUTPUT. For each iteration, we print out the values of the temperature at three representative locations. This output would help us in verifying the convergence of the solution. At the final iteration, we get the field printout of T(I,J) by calling PRINT.

PHI. Here we specify the values of Γ and S_C. By default, S_P remains zero. Also, since all boundaries have a given temperature, no KBC values need to be set.

8.1-3 Additional Fortran Names

COND	conductivity k. Eq. (8.1)
SOURCE	heat generation rate S. Eq. (8.1)
T(I,J)	temperature T
TW	boundary temperature T_w. Eq. (8.1)

8.1-4 Listing of ADAPT for Example 1

```
CCCCCCCCCCCCCCCCCCCCCCCCCCCCCCCCCCCCCCCCCCCCCCCCCCCCCCCCCCCCCCCCCCCCCCC
      SUBROUTINE ADAPT
C
C     EXAMPLE 1 -- STEADY CONDUCTION WITH HEAT GENERATION
C
$INCLUDE:'COMMON'
C*********************************************************************
      DIMENSION T(NI,NJ)
```

```
      EQUIVALENCE (F(1,1,1),T(1,1))
C*-*-*-*-*-*-*-*-*-*-*-*-*-*-*-*-*-*-*-*-*-*-*-*-*-*-*-*-*-*-*
      ENTRY GRID
      HEADER='STEADY CONDUCTION WITH HEAT GENERATION'
      PRINTF='PRINT1'
      PLOTF='PLOT1'
      CALL DATA2(XL,1.,YL,1.)
      CALL INTA2(NCVLX,5,NCVLY,5)
      CALL EZGRID
      RETURN
C*-*-*-*-*-*-*-*-*-*-*-*-*-*-*-*-*-*-*-*-*-*-*-*-*-*-*-*-*-*-*
      ENTRY BEGIN
      TITLE(1)='  TEMPERATURE '
      CALL INTA4(KSOLVE(1),1,KPRINT(1),1,KPLOT(1),1,LAST,3)
      CALL DATA3(TW,0.,COND,1.,SOURCE,5.)
      DO 100 J=1,M1
      DO 100 I=1,L1
        T(I,J)=TW
  100 CONTINUE
      RETURN
C*-*-*-*-*-*-*-*-*-*-*-*-*-*-*-*-*-*-*-*-*-*-*-*-*-*-*-*-*-*-*
      ENTRY OUTPUT
      DO 200 IUNIT=IU1,IU2
        IF(ITER.EQ.0) WRITE(IUNIT,210)
  210   FORMAT(2X,'ITER',3X,'T(2,2)',4X,'T(4,2)',4X,
     1   'T(6,3)')
        WRITE(IUNIT,220) ITER,T(2,2),T(4,2),T(6,3)
  220   FORMAT(2X,I2,2X,1P3E10.2)
  200 CONTINUE
      IF(ITER.EQ.LAST) THEN
        CALL PRINT
        CALL PLOT
      ENDIF
      RETURN
C*-*-*-*-*-*-*-*-*-*-*-*-*-*-*-*-*-*-*-*-*-*-*-*-*-*-*-*-*-*-*
      ENTRY PHI
      DO 300 J=2,M2
      DO 300 I=2,L2
        GAM(I,J)=COND
        SC(I,J)=SOURCE
  300 CONTINUE
      RETURN
```

```
        END
CCCCCCCCCCCCCCCCCCCCCCCCCCCCCCCCCCCCCCCCCCCCCCCCCCCCCCCCCCCCCCCCCCCCCCCCCC
```

8.1-5   Results for Example 1

```
RESULTS OF CONDUCT FOR CARTESIAN COORDINATE SYSTEM
****************************************************
```

```
.  .  .  .  .  .  .  .  .  .  .  .  .  .  .  .  .  .  .  .  .  .  .  .  .  .  .  .  .  .  .  .  .  .  .  .  .  .  .  .  .  .  .  .  .  .  .
STEADY CONDUCTION WITH HEAT GENERATION
.  .  .  .  .  .  .  .  .  .  .  .  .  .  .  .  .  .  .  .  .  .  .  .  .  .  .  .  .  .  .  .  .  .  .  .  .  .  .  .  .  .  .  .  .  .  .
```

```
ITER    T(2,2)    T(4,2)    T(6,3)
 0     0.00E+00  0.00E+00  0.00E+00
 1     6.81E 02  1.46E 01  1.29E 01
 2     6.81E 02  1.46E 01  1.29E 01
 3     6.81E 02  1.46E 01  1.29E 01
```

```
I =     1        2        3        4        5        6        7
X = 0.00E+00 1.00E 01 3.00E 01 5.00E 01 7.00E 01 9.00E 01 1.00E+00
```

```
J =     1        2        3        4        5        6        7
Y = 0.00E+00 1.00E 01 3.00E 01 5.00E 01 7.00E 01 9.00E 01 1.00E+00
```

```
******        TEMPERATURE        ******
              .  .  .  .  .  .  .  .  .  .  .  .  .  .  .  .
```

```
I =    1        2        3        4        5        6        7
J
7   0.00E+00 0.00E+00 0.00E+00 0.00E+00 0.00E+00 0.00E+00 0.00E+00
6   0.00E+00 6.81E 02 1.29E 01 1.46E 01 1.29E 01 6.81E 02 0.00E+00
5   0.00E+00 1.29E 01 2.71E 01 3.13E 01 2.71E 01 1.29E 01 0.00E+00
4   0.00E+00 1.46E 01 3.13E 01 3.63E 01 3.13E 01 1.46E 01 0.00E+00
3   0.00E+00 1.29E 01 2.71E 01 3.13E 01 2.71E 01 1.29E 01 0.00E+00
2   0.00E+00 6.81E 02 1.29E 01 1.46E 01 1.29E 01 6.81E 02 0.00E+00
1   0.00E+00 0.00E+00 0.00E+00 0.00E+00 0.00E+00 0.00E+00 0.00E+00
```

8.1-6   Discussion of Results

As expected, the solution can be seen to have converged well in one iteration. The final printout shows that the effect of heat generation is to create a high temperature in the center of the solid. Also note that the temperature distribution is symmetrical about the vertical and horizontal centerlines of the solid.

The exact solution for this problem with the data in Eq. (8.1) gives the *maximum* temperature as 0.368. This is to be compared with 0.363 given by our numerical solution. For the very small number of grid points we have used, this agreement is very good indeed.

8.1-7   Final Remarks

Here you have seen the first application of CONDUCT. You can now try different variations of this problem. Change the number of grid points, the values of k and S, or the dimensions of the domain. For a square domain of side L, the dimensionless temperature $k(T-T_w)/(SL^2)$ is simply a function of the dimensionless coordinates x/L and y/L You can verify this assertion by solving the problem with different numerical values of k, S, L, etc. and observing that the dimensionless temperature remains unaffected. The exact solution gives the maximum value of this dimensionless temperature as 0.0737.

Such simple explorations of CONDUCT are useful in becoming familiar with the program. By studying the results, you will also develop a better understanding of the relationship between the temperature, the conductivity, the heat generation rate, etc.

8.2   Steady Conduction with Mixed Boundary Conditions (Example 2)

8.2-1   Problem Description

Now we consider a fairly complex situation. The problem is designed to illustrate different boundary conditions, discontinuous and variable conductivity, and a nonuniform source term. The physical situation is shown in Fig. 8.2. The numerical values and property relations are given by:

$$T_{w1} = 100, \qquad T_{w2} = 20, \qquad\qquad T_\infty = 5 \qquad\qquad (8.2a)$$

$$k_1 = 5, \qquad\qquad k_2 = 1 + 0.01\,T, \qquad h_e = 20, \qquad q_w = 800 \qquad (8.2b)$$

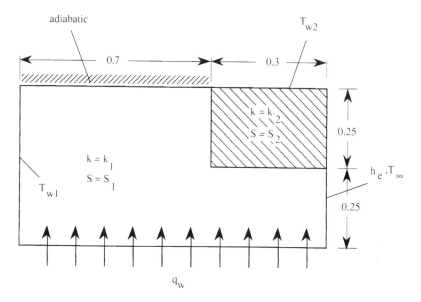

Fig. 8.2 Steady conduction with mixed boundary conditions

S = 0 in the shaded area, and
S = a – b T³ with a = 1000 and b = 4.0E–5 in the unshaded area.   (8.2c)

8.2-2   Design of ADAPT

GRID. Although there are discontinuities in the conductivity distribution and the boundary conditions, the given dimensions are simple numbers. Therefore, we can use a suitable uniform grid and still capture all the discontinuities at the control-volume faces. For this purpose, we choose 10 control-volume widths in the x direction and 6 in the y direction.

BEGIN. Since the problem is nonlinear, we need multiple iterations. Here we shall develop a convergence criterion and terminate the calculation when the criterion is met. Still, it will be desirable to perform at least a minimum number of iterations (ITRMIN) and allow a certain maximum number of iterations (LAST). These quantities are specified in BEGIN, along with the numerical values of T_{w1}, T_{w2}, etc. given in Eq. (8.2). Finally, the T(I,J) array is filled so as to have T_{w1} as the initial guess at all internal grid points and to set T_{w1} or T_{w2} as the known boundary values at the appropriate boundary locations.

OUTPUT. In addition to arranging the desired printout, we incorporate a suitable convergence criterion here. We can judge the convergence by changes in the $T(I,J)$ values from iteration to iteration or examine the change in some related quantity. Here, we shall use the total heat flow through the left boundary of the domain as the basis of our convergence criterion. This quantity, denoted by $HTFLX$, is first calculated by a summation over all control-volume faces on that boundary. In this calculation, we use the flux quantity $FLUXI1(J,1)$ produced by the invariant part of CONDUCT. Note that the first subscript J denotes the location along the left boundary, while the second subscript is the value of NF for the variable $T(I,J)$ under consideration.

At every iteration, we print out a few typical values of $T(I,J)$ and the value of $HTFLX$. The previous-iteration value of $HTFLX$ is stored as $HTFLXO$. The convergence criterion is designed to require that the relative change in $HTFLX$ is less than a very small number. (Note the use of the absolute-value function here. We want the *magnitude* of the change to be small. If we do not use the absolute value, the criterion will be satisfied even when the relative change in $HTFLX$ is large but negative.)

This particular example was chosen to illustrate the construction and use of a convergence criterion. Such criteria should be employed in all nonlinear problems. However, in order to keep things simple, we shall not use a convergence criterion in the remaining examples in this book. Our practice would be simply to set $LAST$ equal to the desired number of iterations. (This number can be decided from some trial runs.)

Another useful practice that is illustrated only in this example is the verification of the overall heat balance. Performing such balances is very useful and you are encouraged to include them in the adaptations you design. For a properly converged solution, we expect the overall heat balance to be perfect (within the computer roundoff error).

We perform the overall balance calculation after the convergence criterion is met. For this calculation, we obtain the total heat flows through the four boundaries of the domain. Also, the given source term is integrated over the unshaded region. Finally, the difference between the total heat outflow and the total heat generation is calculated. We expect this difference to be zero (or very small).

The final printout consists of the local heat fluxes on the four boundaries of the domain, the heat balance quantities, and the field of $T(I,J)$, which is obtained by calling PRINT. To terminate the computation, we set $KSTOP = 1$.

PHI. The shaded area is identified through an IF statement. Then, the conductivity values are calculated according to the specification in Eq. (8.2b). The source term in the unshaded area is linearized in accordance with the procedure developed in Section 2.5-4. Here, the appropriate expressions for S_C and S_P are

$$S_C = a + 2\,b\,T_P^{\cdot\,3} \tag{8.3a}$$

$$S_P = -3 \, b \, T_P^{'2}$$

(8.3b)

The source term in the shaded area remains zero by default. Finally, the boundary conditions are specified through the appropriate values of KBC, FLXC, and FLXP.

8.2-3   Additional Fortran Names

AK1	thermal conductivity k_1, Eq. (8.2b)
DIFF	relative difference for convergence criterion
GEN	total heat generated
HE	heat transfer coefficient h_e for the right boundary, Eq. (8.2b)
HTBAL	heat balance quantity
HTFLX	total heat flow through the left boundary
HTFLXO	previous-iteration value of HTFLX
HTB	total heat flow at the bottom boundary
HTL	total heat flow at the left boundary
HTOUT	total heat outflow from the domain
HTR	totatl heat flow at the right boundary
HTT	totatl heat flow at the top boundary
ITRMIN	minimum number of iterations to be performed
QW	constant wall heat flux q_w for the bottom boundary
T(I,J)	temperature T
TINF	surrounding temperature T_∞, Eq. (8.2a)
TW1,TW2	boundary temperatures T_{w1} and T_{w2}, Eq. (8.2a)

8.2-4   Listing of ADAPT for Example 2

```
CCCCCCCCCCCCCCCCCCCCCCCCCCCCCCCCCCCCCCCCCCCCCCCCCCCCCCCCCCCCCCCCCCCCCCCCCC
      SUBROUTINE ADAPT
C . . . . . . . . . . . . . . . . . . . . . . . . . . . . . . . . . . . . .
C     EXAMPLE 2 -- STEADY CONDUCTION WITH MIXED BOUNDARY CONDITIONS
C . . . . . . . . . . . . . . . . . . . . . . . . . . . . . . . . . . . . .
$INCLUDE:'COMMON'
C**********************************************************************
      DIMENSION T(NI,NJ)
      EQUIVALENCE (F(1,1,1),T(1,1))
C*-*-*-*-*-*-*-*-*-*-*-*-*-*-*-*-*-*-*-*-*-*-*-*-*-*-*-*-*-*-*-*-*-*-*
      ENTRY GRID
      HEADER='STEADY CONDUCTION WITH MIXED BOUNDARY CONDITIONS'
      PRINTF='PRINT2'
      PLOTF='PLOT2'
```

```
      CALL DATA2(XL,1.,YL,0.5)
      CALL INTA2(NCVLX,10,NCVLY,6)
      CALL EZGRID
      RETURN
C*.*.*.*.*.*.*.*.*.*.*-*.*.*.*.*.*.*.*-*.*.*.*.*.*.*.*.*.*.*.*.*-*.*.*.*
      ENTRY BEGIN
      TITLE(1)='  TEMPERATURE '
      CALL INTA5(KSOLVE(1),1,KPRINT(1),1,KPLOT(1),1,ITRMIN,3,LAST,15)
      CALL DATA6(AK1,5.,TW1,100.,TW2,20.,TINF,5.,HE,20.,QW,800.)
      DO 100 J=1,M1
      DO 100 I=1,L1
         T(I,J)=TW1
  100 CONTINUE
      DO 110 I=2,L2
         IF(X(I).GT.0.7) T(I,M1)=TW2
  110 CONTINUE
      HTFLXO=0.
      RETURN
C*.*-*-*.*.*.*.*.*.*.*-*.*.*.*.*-*.*.*.*.*.*.*.*.*.*.*.*.*.*-*.*.*.*.*.*
      ENTRY OUTPUT
      HTFLX=0.
      DO 200 J=2,M2
         HTFLX=HTFLX+ARX(J)*FLUXI1(J,1)
  200 CONTINUE
      DO 210 IUNIT=IU1,IU2
         IF(ITER.EQ.0) WRITE(IUNIT,220)
  220    FORMAT(2X,'ITER',3X,'T(3,3)',4X,'T(5,4)',4X,'T(10,7)'
     1    ,4X,'HEAT FLOW (LEFT FACE)')
         WRITE(IUNIT,230) ITER,T(3,3),T(5,4),T(10,7),HTFLX
  230    FORMAT(2X,I2,2X,1P3E10.2,8X,1PE11.3)
  210 CONTINUE
CREATE A CONVERGENCE CRITERION
      IF(ITER.LT.ITRMIN) RETURN
      DIFF=ABS((HTFLX-HTFLXO)/(HTFLX+SMALL))
      HTFLXO=HTFLX
      IF(DIFF.LE.1.E-5.OR.ITER.EQ.LAST) THEN
CALCULATE QUANTITIES FOR OVERALL HEAT BALANCE
      HTR=0.
      DO 231 J=2,M2
         HTR=HTR+ARX(J)*FLUXL1(J,1)
  231 CONTINUE
      HTL=HTFLX
```

```
      HTB=QW*XL
      HTT=0.
      DO 232 I=2,L2
        IF(X(I).GT.0.7) HTT=HTT+XCV(I)*FLUXMl(I,1)
  232 CONTINUE
      HTOUT= (HTL+HTR+HTB+HTT)
      GEN=0.
      DO 233 J=2,M2
      DO 233 I=2,L2
        IF(X(I).GT.0.7.AND.Y(J).GT.0.25) GO TO 233
        GEN=GEN+(1000. 4.E 5*T(I,J)**3)*XCV(I)*YCV(J)
  233 CONTINUE
      HTBAL=HTOUT-GEN
CONSTRUCT FINAL PRINTOUT
          DO 240 IUNIT=IU1,IU2
            WRITE(IUNIT,250)
  250       FORMAT(1X,/,'  J',8X,'Y(J)',5X,'FLUX(LEFT)',4X,
      1     'FLUX(RIGHT)')
            DO 260 J=M2,2, 1
              WRITE(IUNIT,270) J,Y(J),FLUXI1(J,1),FLUXL1(J,1)
  270         FORMAT(1X,I2,5X,1PE9.2,3X,1PE9.2,5X,1PE9.2)
  260       CONTINUE
            WRITE(IUNIT,280)
  280       FORMAT(1X,/,'  I',8X,'X(I)',5X,'FLUX(BOTTOM)',2X,
      1     'FLUX(TOP)')
            DO 290 I=2,L2
              WRITE(IUNIT,270) I,X(I),FLUXJ1(I,1),FLUXM1(I,1)
  290       CONTINUE
            WRITE(IUNIT,291) HTOUT,GEN,HTBAL
  291       FORMAT(/1X,'OVERALL HEAT BALANCE'/1X,20('*')/1X,
      1     'HEAT OUTFLOW',5X,'GENERATION',6X,'DIFFERENCE'/2X,
      2     1PE10.3,6X,1PE10.3,5X,1PE10.3)
  240     CONTINUE
          CALL PRINT
          CALL PLOT
          KSTOP=1
      ENDIF
      RETURN
C*-*-*-*-*-*-*-*-*-*-*-*-*-*-*-*-*-*-*-*-*-*-*-*-*-*-*-*-*-*-*-*-*
      ENTRY PHI
      DO 300 J=2,M2
      DO 300 I=2,L2
```

```
         IF(X(I).GT.0.7.AND.Y(J).GT.0.25) THEN
           GAM(I,J)=1.+0.01*T(I,J)
         ELSE
           GAM(I,J)=AK1
           TP2=4.E-5*T(I,J)**2
           SC(I,J)=1000.+2.*TP2*T(I,J)
           SP(I,J)=-3.*TP2
         ENDIF
  300 CONTINUE
COME HERE TO SPECIFY BOUNDARY CONDITIONS
       DO 310 I=2,L2
         KBCJ1(I)=2
         FLXCJ1(I)=QW
         IF(X(I).LT.0.7) KBCM1(I)=2
  310 CONTINUE
       DO 320 J=2,M2
         KBCL1(J)=2
         FLXCL1(J)=HE*TINF
         FLXPL1(J)=-HE
  320 CONTINUE
       RETURN
       END
CCCCCCCCCCCCCCCCCCCCCCCCCCCCCCCCCCCCCCCCCCCCCCCCCCCCCCCCCCCCCCCCCCCCCCCCC
```

8.2-5   Results for Example 2

```
RESULTS OF CONDUCT FOR CARTESIAN COORDINATE SYSTEM
******************************************************


. . . . . . . . . . . . . . . . . . . . . . . . . . . . . . . . . . . . . . . . . .
STEADY CONDUCTION WITH MIXED BOUNDARY CONDITIONS
. . . . . . . . . . . . . . . . . . . . . . . . . . . . . . . . . . . . . . . . . .
```

ITER	T(3,3)	T(5,4)	T(10,7)	HEAT FLOW (LEFT FACE)
0	1.00E+02	1.00E+02	1.00E+02	0.000E+00
1	1.29E+02	1.38E+02	3.38E+01	-4.948E+02
2	1.30E+02	1.41E+02	3.70E+01	-5.158E+02
3	1.30E+02	1.41E+02	3.70E+01	-5.126E+02
4	1.30E+02	1.41E+02	3.70E+01	-5.127E+02
5	1.30E+02	1.41E+02	3.70E+01	-5.127E+02

```
  6      1.30E+02  1.41E+02  3.70E+01           5.127E+02
  7      1.30E+02  1.41E+02  3.70E+01           5.127E+02


  J       Y(J)      FLUX(LEFT)     FLUX(RIGHT)
  7      4.58E 01    6.77E+02      2.45E+02
  6      3.75E 01    7.15E+02      4.33E+02
  5      2.92E 01    7.97E+02      6.83E+02
  4      2.08E 01    9.42E+02      1.20E+03
  3      1.25E 01    1.21E+03      1.34E+03
  2      4.17E 02    1.81E+03      1.51E+03


  I       X(I)      FLUX(BOTTOM)   FLUX(TOP)
  2      5.00E 02    8.00E+02      1.55E 04
  3      1.50E 01    8.00E+02      4.52E 04
  4      2.50E 01    8.00E+02      4.49E 04
  5      3.50E 01    8.00E+02      5.99E 04
  6      4.50E 01    8.00E+02      5.94E 04
  7      5.50E 01    8.00E+02      2.21E 05
  8      6.50E 01    8.00E+02      2.62E 04
  9      7.50E 01    8.00E+02      1.50E+03
 10      8.50E 01    8.00E+02      5.83E+02
 11      9.50E 01    8.00E+02      1.60E+02


OVERALL HEAT BALANCE
********************

HEAT OUTFLOW      GENERATION      DIFFERENCE
  3.885E+02        3.885E+02       1.434E 03



 I =     1        2        3        4        5        6        7
 X = 0.00E+00 5.00E 02 1.50E 01 2.50E 01 3.50E 01 4.50E 01 5.50E 01

 I =     8        9       10       11       12
 X = 6.50E 01 7.50E 01 8.50E 01 9.50E 01 1.00E+00



 J =     1        2        3        4        5        6        7
 Y = 0.00E+00 4.17E 02 1.25E-01 2.08E-01 2.92E 01 3.75E 01 4.58E-01

 J =     8
 Y = 5.00E 01
```

```
******        TEMPERATURE        ******
              . . . . . . . . . . . . . . . . . . . .
```

I =	1	2	3	4	5	6	7
J							
8	1.00E+02	1.06E+02	1.17E+02	1.25E+02	1.29E+02	1.28E+02	1.21E+02
7	1.00E+02	1.06E+02	1.18E+02	1.25E+02	1.29E+02	1.28E+02	1.21E+02
6	1.00E+02	1.07E+02	1.19E+02	1.27E+02	1.31E+02	1.30E+02	1.25E+02
5	1.00E+02	1.08E+02	1.21E+02	1.30E+02	1.35E+02	1.35E+02	1.31E+02
4	1.00E+02	1.09E+02	1.24E+02	1.35E+02	1.41E+02	1.42E+02	1.38E+02
3	1.00E+02	1.11E+02	1.30E+02	1.42E+02	1.49E+02	1.51E+02	1.47E+02
2	1.00E+02	1.16E+02	1.39E+02	1.52E+02	1.60E+02	1.61E+02	1.58E+02
1	1.00E+02	1.22E+02	1.45E+02	1.59E+02	1.66E+02	1.68E+02	1.64E+02

I =	8	9	10	11	12
J					
8	1.07E+02	2.00E+01	2.00E+01	2.00E+01	1.00E+02
7	1.08E+02	5.48E+01	3.70E+01	2.60E+01	1.73E+01
6	1.14E+02	8.80E+01	6.42E+01	4.12E+01	2.67E+01
5	1.22E+02	1.06E+02	8.48E+01	5.88E+01	3.92E+01
4	1.30E+02	1.17E+02	9.87E+01	7.66E+01	6.48E+01
3	1.39E+02	1.26E+02	1.08E+02	8.50E+01	7.21E+01
2	1.50E+02	1.37E+02	1.19E+02	9.50E+01	8.07E+01
1	1.56E+02	1.43E+02	1.25E+02	1.01E+02	1.00E+02

8.2-6   Discussion of Results

In this problem, the nonlinearity of k is rather mild, and the source term has been linearized in an optimum manner. Therefore, the convergence of the iterations can be seen to be very rapid.

The printout of the local fluxes gives us the values of the known and unknown fluxes along the boundaries of the domain. For the adiabatic portion of the top boundary, the printed flux values are very small, but not quite zero. The small numbers are a result of the computer roundoff error.

In the temperature field, local temperatures in excess of the given boundary temperatures are created by internal heat generation and the heat-flux boundary condition. At the bottom boundary, the temperature is quite high so as to transfer the prescribed heat flux into the calculation domain.

8.2-7   Final Remarks

In this problem, we have seen the treatment of variable conductivity, nonlinear source term, and a variety of boundary conditions. With this background, you should be able to apply CONDUCT to a large number of steady heat conduction problems. The treatment of irregular geometry and of nonlinear boundary conditions will be illustrated in the next two examples.

8.3   Steady Conduction in Irregular Geometry (Example 3)

8.3-1   Problem Description

Figure 8.3 shows a square solid with some rectangular cut-outs. The values of the temperatures on the inner surfaces of the cut-outs are prescribed. The bottom boundary exchanges heat with the surroundings by convection and radiation. The heat flux at the bottom boundary is given by

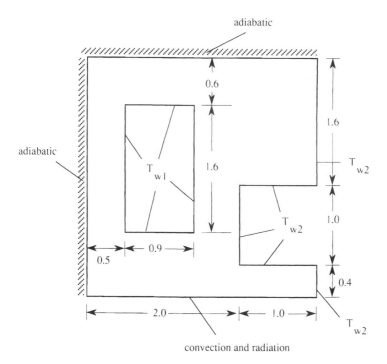

Fig. 8.3 Steady conduction in irregular geometry

$$q_B = a\,(T_\infty - T_B) + b\,(T_\infty^4 - T_B^4) \tag{8.4}$$

where T_∞ and T_B stand for the temperature of the surroundings and of the boundary respectively.

The geometry of the problem is shown in Fig. 8.3. The numerical values for other parameters are

$$T_{w1} = 400, \qquad T_{w2} = 500, \qquad T_\infty = 300 \tag{8.5a}$$

$$k = 12, \qquad a = 50, \qquad b = 2.0E{-}8 \tag{8.5b}$$

Because of the radiation term in Eq. (8.4), all temperatures will be taken to be absolute temperatures.

8.3-2   Design of ADAPT

GRID. This problem is a good candidate for the use of ZGRID. We can define zones such that their boundaries coincide with different irregularities in the geometry of the domain. From Fig. 8.3, we can see that 4 zones in the x direction and 5 zones in the y direction will be appropriate. Then, each zone can be subdivided into any desired number of control-volume widths. All this information is prescribed in GRID before calling ZGRID. Since we have left POWRX(NZ) and POWRY(NZ) at their default values of unity, we shall get a uniform grid within each zone.

BEGIN. Here we assign numerical values to T_{w1}, T_{w2}, k, etc. according to the information in Eq. (8.5). The only temperature known on the outer boundary of the domain is T_{w2}; therefore, we fill the entire T(I,J) array with this value.

OUTPUT. Because of the nonlinear boundary condition, this problem requires multiple iterations. To observe the approach to the converged solution, we print out, after every iteration, the values of some typical temperatures and the total heat flow HTFLY at the bottom boundary. This quantity is calculated by a summation over the bottom boundary; here, XCV(I) is used as the area (per unit depth) of each control-volume face along the boundary.

After the final iteration, a field printout of T(I,J) is arranged by calling PRINT.

PHI. The main handling of the irregular geometry is done in PHI through appropriate values of GAM(I,J). The two cut-outs are represented as regions of high conductivity, so that a uniform temperature prevails throughout each region.

This will be our way of specifying the temperatures T_{w1} and T_{w2} at the surfaces of the cut-out regions.

In the first part of PHI, we set GAM(I,J) equal to the uniform conductivity AK, but we overwrite it by a large number when the grid points fall in the cut-out regions. The IF statements are designed to identify these special regions.

Just the GAM(I,J) specification is sufficient for the treatment of the cut-out on the right boundary. This is because this region is adjoining the boundary and hence the specified temperature T_{w2} at the nominal boundary of the domain will get established throughout this high-conductivity region. For the cut-out on the left, there is no adjoining boundary and hence the known temperature T_{w1} for the region cannot be specified through any outer boundary of the domain. Therefore, as explained in Section 7.6, we need to set the temperature at one or more grid points in the region equal to T_{w1}. Here, we employ Eq. (7.1) to set T(4,6) equal to T_{w1} through the use of large source terms.

One other subtle point deserves mention here. For the cut-out on the left, we use large conductivity and large source terms. However, the technique of employing large source terms will be successful only if they overpower all other terms in the discretization equation. If the conductivity values are equally large, then the large source terms will not establish the desired temperature at the chosen grid point. For this reason, whereas we set the conductivity equal to 1.0E12, the source terms are based on BIG, which has a value of 1.0E20.

The nonlinear boundary condition at the bottom boundary is linearized according to the practice described in Section 2.5-5. The expressions for f_C and f_P are

$$f_C = a\,T_\infty + b\,T_\infty^4 + 3\,b\,T_B^{'4} \tag{8.6a}$$

$$f_P = -(a + 4\,b\,T_B^{'3}) \tag{8.6b}$$

The appropriate boundary conditions are expressed through the values of KBC, FLXC, and FLXP.

8.3-3 Additional Fortran Names

AK	thermal conductivity k
AQ,BQ	constants a and b, Eq. (8.5b)
HTFLY	total heat flow at the bottom boundary
T(I,J)	temperature T
TINF	surrounding temperature T_∞, Eq. (8.5a)
TW1,TW2	boundary temperatures T_{w1} and T_{w2}, Eq. (8.5a)

8.3-4 Listing of ADAPT for Example 3

```
CCCCCCCCCCCCCCCCCCCCCCCCCCCCCCCCCCCCCCCCCCCCCCCCCCCCCCCCCCCCCCCCCCCCCCC
      SUBROUTINE ADAPT
C- - - - - - - - - - - - - - - - - - - - - - - - - - - - - - - - - - - - - - - - -
C- - - - -EXAMPLE 3 -- STEADY CONDUCTION IN IRREGULAR GEOMETRY
C- - - - - - - - - - - - - - - - - - - - - - - - - - - - - - - - - - - - - - - - -
$INCLUDE:'COMMON'
C*********************************************************************
      DIMENSION T(NI,NJ)
      EQUIVALENCE (F(1,1,1),T(1,1))
C*-*-*-*-*-*-*-*-*-*-*-*-*-*-*-*-*-*-*-*-*-*-*-*-*-*-*-*-*-*-*-*-*-*-*
      ENTRY GRID
      HEADER='STEADY CONDUCTION IN IRREGULAR GEOMETRY'
      PRINTF='PRINT3'
      PLOTF='PLOT3'
      CALL INTA5(NZX,4,NCVX(1),2,NCVX(2),4,NCVX(3),2,NCVX(4),4)
      CALL DATA4(XZONE(1),0.5,XZONE(2),0.9,XZONE(3),0.6,XZONE(4),1.)
      CALL INTA6(NZY,5,NCVY(1),2,NCVY(2),2,NCVY(3),2,NCVY(4),4,
     1NCVY(5),2)
      CALL DATA5(YZONE(1),0.4,YZONE(2),0.4,YZONE(3),0.6,YZONE(4),1.0,
     1YZONE(5),0.6)
      CALL ZGRID
      RETURN
C*-*-*-*-*-*-*-*-*-*-*-*-*-*-*-*-*-*-*-*-*-*-*-*-*-*-*-*-*-*-*-*-*-*-*
      ENTRY BEGIN
      TITLE(1)='    TEMPERATURE '
      CALL INTA4(KSOLVE(1),1,KPRINT(1),1,KPLOT(1),1,LAST,10)
      CALL DATA6(TW1,400.,TW2,500.,TINF,300.,AQ,50.,BQ,2.E-8,AK,12.)
      DO 100 J=1,M1
      DO 100 I=1,L1
        T(I,J)=TW2
  100 CONTINUE
      RETURN
C*-*-*-*-*-*-*-*-*-*-*-*-*-*-*-*-*-*-*-*-*-*-*-*-*-*-*-*-*-*-*-*-*-*-*
      ENTRY OUTPUT
      HTFLY=0.
      DO 200 I=2,L2
        HTFLY=HTFLY+XCV(I)*FLUXJ1(I,1)
  200 CONTINUE
      DO 210 IUNIT=IU1,IU2
        IF(ITER.EQ.0) WRITE(IUNIT,220)
```

```
 220      FORMAT(2X,'ITER',3X,'T(2,12)',3X,'T(5,4)',3X,'T(10,12)'
    1    ,4X,'HEAT FLOW (BOTTOM FACE)')
          WRITE(IUNIT,230) ITER,T(2,12),T(5,4),T(10,12),HTFLY
 230      FORMAT(2X,I2,2X,1P3E10.2,9X,1PE11.3)
 210 CONTINUE
      IF(ITER.EQ.LAST) THEN
          CALL PRINT
C......
COME HERE TO FILL IBLOCK(I,J) BEFORE CALLING PLOT
          DO 240 J=2,M2
          DO 240 I=2,L2
             IF(X(I).GT.0.5.AND.X(I).LT.1.4.AND.Y(J).GT.0.8.AND.
    1          Y(J).LT.2.4) IBLOCK(I,J)=1
             IF(X(I).GT.2..AND.Y(J).GT.0.4.AND.Y(J).LT.1.4) IBLOCK(I,J)=1
 240      CONTINUE
          CALL PLOT
C......
      ENDIF
      RETURN
C* *-*-*-*-*-*-*-*-*-*-*-*-*-*-*-*-*-*-*-*-*-*-*-*-*-*-*-*-*-*-*
      ENTRY PHI
      DO 300 J=2,M2
      DO 300 I=2,L2
         GAM(I,J)=AK
         IF(X(I).GT.0.5.AND.X(I).LT.1.4.AND.Y(J).GT.0.8.AND.
    1    Y(J).LT.2.4) GAM(I,J)=1.E12
         IF(X(I).GT.2.0.AND.Y(J).GT.0.4.AND.Y(J).LT.1.4) GAM(I,J)=1.E12
 300 CONTINUE
      SC(4,6)=BIG*TWI
      SP(4,6)=-BIG
COME HERE TO SPECIFY BOUNDARY CONDITIONS
      DO 310 I=2,L2
         KBCM1(I)=2
         KBCJ1(I)=2
         FLXCJ1(I)=AQ*TINF+BQ*TINF**4+3.*BQ*T(I,1)**4
         FLXPJ1(I)=-(AQ+4.*BQ*T(I,1)**3)
 310 CONTINUE
      DO 320 J=2,M2
         KBCI1(J)=2
 320 CONTINUE
      RETURN
      END
```

cc

8.3-5   Results for Example 3

RESULTS OF CONDUCT FOR CARTESIAN COORDINATE SYSTEM
****

- -

STEADY CONDUCTION IN IRREGULAR GEOMETRY
- -

ITER	T(2,12)	T(5,4)	T(10,12)	HEAT FLOW (BOTTOM FACE)
0	5.00E+02	5.00E+02	5.00E+02	0.000E+00
1	4.22E+02	3.81E+02	4.66E+02	-7.193E+03
2	4.05E+02	3.73E+02	4.57E+02	-7.502E+03
3	4.02E+02	3.73E+02	4.56E+02	-7.486E+03
4	4.01E+02	3.73E+02	4.56E+02	-7.483E+03
5	4.01E+02	3.73E+02	4.56E+02	-7.483E+03
6	4.01E+02	3.73E+02	4.56E+02	-7.483E+03
7	4.01E+02	3.73E+02	4.56E+02	-7.483E+03
8	4.01E+02	3.73E+02	4.56E+02	-7.482E+03
9	4.01E+02	3.73E+02	4.56E+02	7.482E+03
10	4.01E+02	3.73E+02	4.56E+02	7.482E+03

I =	1	2	3	4	5	6	7
X =	0.00E+00	1.25E-01	3.75E-01	6.13E-01	8.37E-01	1.06E+00	1.29E+00

I =	8	9	10	11	12	13	14
X =	1.55E+00	1.85E+00	2.13E+00	2.38E+00	2.63E+00	2.88E+00	3.00E+00

J =	1	2	3	4	5	6	7
Y =	0.00E+00	1.00E-01	3.00E-01	5.00E-01	7.00E-01	9.50E-01	1.25E+00

J =	8	9	10	11	12	13	14
Y =	1.53E+00	1.78E+00	2.03E+00	2.28E+00	2.55E+00	2.85E+00	3.00E+00

****** 　　　 TEMPERATURE 　　　 ******
- -

I =	1	2	3	4	5	6	7
J							
14	5.00E+02	4.02E+02	4.02E+02	4.03E+02	4.05E+02	4.09E+02	4.15E+02
13	4.02E+02	4.02E+02	4.02E+02	4.03E+02	4.05E+02	4.09E+02	4.14E+02
12	4.01E+02	4.01E+02	4.01E+02	4.02E+02	4.02E+02	4.04E+02	4.09E+02
11	4.00E+02	4.00E+02	4.00E+02	4.00E+02	4.00E+02	4.00E+02	4.00E+02
10	4.00E+02	4.00E+02	4.00E+02	4.00E+02	4.00E+02	4.00E+02	4.00E+02
9	3.99E+02	3.99E+02	4.00E+02	4.00E+02	4.00E+02	4.00E+02	4.00E+02
8	3.97E+02	3.98E+02	3.99E+02	4.00E+02	4.00E+02	4.00E+02	4.00E+02
7	3.94E+02	3.95E+02	3.98E+02	4.00E+02	4.00E+02	4.00E+02	4.00E+02
6	3.86E+02	3.87E+02	3.93E+02	4.00E+02	4.00E+02	4.00E+02	4.00E+02
5	3.73E+02	3.74E+02	3.78E+02	3.88E+02	3.91E+02	3.93E+02	3.99E+02
4	3.60E+02	3.60E+02	3.64E+02	3.69E+02	3.73E+02	3.78E+02	3.86E+02
3	3.45E+02	3.45E+02	3.47E+02	3.50E+02	3.54E+02	3.59E+02	3.67E+02
2	3.28E+02	3.28E+02	3.30E+02	3.32E+02	3.34E+02	3.37E+02	3.43E+02
1	5.00E+02	3.20E+02	3.21E+02	3.22E+02	3.24E+02	3.26E+02	3.30E+02

I =	8	9	10	11	12	13	14
J							
14	4.25E+02	4.40E+02	4.55E+02	4.68E+02	4.81E+02	4.94E+02	5.00E+02
13	4.25E+02	4.40E+02	4.55E+02	4.69E+02	4.81E+02	4.94E+02	5.00E+02
12	4.21E+02	4.39E+02	4.56E+02	4.69E+02	4.82E+02	4.94E+02	5.00E+02
11	4.15E+02	4.39E+02	4.57E+02	4.71E+02	4.83E+02	4.95E+02	5.00E+02
10	4.15E+02	4.41E+02	4.61E+02	4.75E+02	4.86E+02	4.95E+02	5.00E+02
9	4.16E+02	4.46E+02	4.69E+02	4.81E+02	4.90E+02	4.97E+02	5.00E+02
8	4.19E+02	4.55E+02	4.84E+02	4.92E+02	4.96E+02	4.99E+02	5.00E+02
7	4.22E+02	4.69E+02	5.00E+02	5.00E+02	5.00E+02	5.00E+02	5.00E+02
6	4.22E+02	4.71E+02	5.00E+02	5.00E+02	5.00E+02	5.00E+02	5.00E+02
5	4.19E+02	4.66E+02	5.00E+02	5.00E+02	5.00E+02	5.00E+02	5.00E+02
4	4.06E+02	4.52E+02	5.00E+02	5.00E+02	5.00E+02	5.00E+02	5.00E+02
3	3.83E+02	4.16E+02	4.58E+02	4.67E+02	4.73E+02	4.87E+02	5.00E+02
2	3.53E+02	3.73E+02	3.95E+02	4.04E+02	4.16E+02	4.49E+02	5.00E+02
1	3.37E+02	3.51E+02	3.65E+02	3.72E+02	3.81E+02	4.08E+02	5.00E+02

8.3-6   Discussion of Results

Once again, our linearization practices can be seen to have produced a very rapid convergence. In the field printout of the temperature, we can identify the cut-out regions, where uniform values of 400 and 500 can be seen. The temperatures in the top-left corner are nearly equal to 400 because of the adiabatic boundary condition and the proximity to the cut-out region at 400. The temperature at the bottom boundary is close to the surrounding temperature of 300.

8.3-7   Final Remarks

The geometrical irregularities considered here were relatively simple. The cut-outs had a rectangular shape and their surfaces had a uniform temperature, which we could establish through a high conductivity. In the next example, we shall consider the treatment of a curved boundary and more complex boundary conditions.

8.4   Conduction in a Complex Cylindrical Geometry (Example 4)

8.4-1   Problem Description

We now consider the axisymmetric solid in Fig. 8.4, where the geometry and the boundary conditions are shown. The curved boundary on the right is a circular arc of radius 1. The numerical values of the various parameters are

$$T_{w1} = 200, \qquad T_{w2} = 100, \qquad T_{\infty} = 20 \qquad\qquad (8.7a)$$

$$k = 2, \qquad q_w = 1, \qquad h_e = 5 \qquad\qquad (8.7b)$$

$$S = 50 - 4\,T \qquad\qquad (8.7c)$$

8.4-2   Design of ADAPT

   GRID. Whereas the utilities such as EZGRID and ZGRID are very useful, we do not have to use them for every problem. Sometimes, it may be more appropriate to design the grid directly. This procedure is demonstrated in this problem.

   First, the axisymmetric geometry is specified by $MODE = 2$. The value of the inner radius $R(1)$ is then prescribed. A uniform grid is used in the y direction, with 9 control-volume widths so that the geometrical irregularities can be properly matched. In the x direction, a uniform grid is used up to $x = 1$. Beyond this point, the locations $XU(I)$ of the control-volume faces are calculated so as to fit the circular arc. The formula for $XU(I)$ is arranged such that the circular boundary passes through the centers of the vertical faces of the control volumes forming the blocked region. This can be seen from Fig. 8.5, where the control volumes are shown along with the geometry of the physical domain. For the coarse grid used

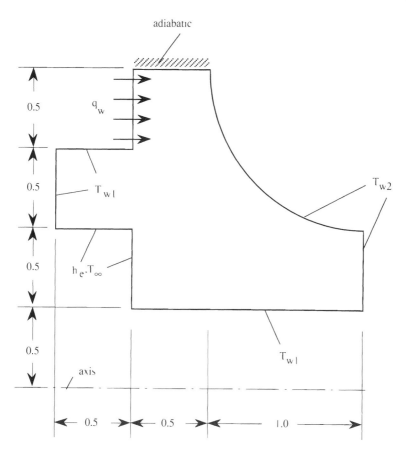

Fig. 8.4 Conduction in a complex cylindrical geometry

here, the stepped boundary of the blocked region represents a fairly good approximation to the curved boundary. Of course, the approximation can be improved by using finer grids.

The particular grid design used here is only one of many possibilities. We could have used a uniform grid in the x direction and calculated the corresponding YV(J) values to fit the circular arc. (This would, however, have caused problems in matching the geometrical irregularities on the left side of the domain.) It is also possible to allow grid nonuniformity in both x and y directions.

BEGIN. Here we specify the numerical values given in Eq. (8.7) for various parameters. The T(I,J) array is filled so as to provide T_{w1} and T_{w2} as known

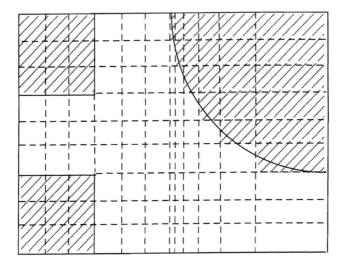

axis

Fig. 8.5 Design of control volumes for Example 4

boundary values on certain boundaries. and T_{w1} as the initial guess for all other points.

OUTPUT. For this problem. the output consists of the values of a few typical temperatures after every iteration and the field printout of $T(I,J)$ after the last iteration.

PHI. Here we use two types of practices to treat the boundary conditions at the geometrical irregularities. The uniform temperature on the circular boundary can be easily handled by setting the conductivity of the blocked region equal to a large number. For the cut-outs on the left side of the domain. we set the conductivity there equal to zero and then incorporate the given boundary conditions through additional source terms for the near-boundary control volumes. This practice was explained in Section 7.7-2.

While filling the $GAM(I,J)$ array, we first set it equal to k. and then overwrite it by zero in the cut-outs on the left. The specification of a large conductivity in the control volumes outside the circular arc is done in two stages. First. $GAM(I,J)$ in a square region enclosing the arc is equated to a large number; then, the GAM(I,J) values for the points lying below the arc are restored to k.

The temperature-dependent source term given in Eq. (8.7c) is specified through appropriate SC(I,J) and SP(I,J). To incorporate the boundary conditions on the surfaces of the cut-outs, *additional* source terms are added to the control volumes that receive heat flow from the cut-out region. The details of this treatment can be understood with reference to the description in Section 7.7-2. Please note that, while specifying the source terms for the near-boundary control volumes, we use SC(I,J) or SP(I,J) on the right-hand side of the Fortran statement as well. This is because we are providing *additional* source terms for these control volumes. The omission of SC(I,J) or SP(I,J) from the right-hand side would have the effect of removing the regular source term (given by Eq. (8.7c) in the problem description) for these control volumes.

Finally, the short adiabatic section at the top boundary is handled by setting KBC there equal to 2.

8.4-3 Additional Fortran Names

ABYV	area divided by volume
COND	thermal conductivity k, Eq. (8.7b)
CONS	ABYV/RES
DX,DY	control-volume widths in the x and y directions
GBYD	conductivity divided by distance
HE	heat transfer coefficient h_e, Eq. (8.7b)
QW	constant wall heat flux q_w, Eq. (8.7b)
RES	resistance to heat flow
T(I,J)	temperature T
TINF	surrounding temperature T_∞, Eq. (8.7a)
TW1,TW2	boundary temperatures T_{w1} and T_{w2}, Eq. (8.7a)
YM	y value of the center of control-volumes face

8.4-4 Listing of ADAPT for Example 4

```
CCCCCCCCCCCCCCCCCCCCCCCCCCCCCCCCCCCCCCCCCCCCCCCCCCCCCCCCCCCCCCCCCCCCCCCC
      SUBROUTINE ADAPT
C----------------------------------------------------------------------
C-----EXAMPLE 4 -- CONDUCTION IN A COMPLEX CYLINDRICAL GEOMETRY
C----------------------------------------------------------------------
$INCLUDE:'COMMON'
C**********************************************************************
      DIMENSION T(NI,NJ)
      EQUIVALENCE (F(1,1,1),T(1,1))
```

```
C*-*-*-*-*-*-*-*-*-*-*-*-*-*-*-*-*-*-*-*-*-*-*-*-*-*-*-*-*-*-*-*-*-*-*
      ENTRY GRID
      HEADER='CONDUCTION IN A COMPLEX CYLINDRICAL GEOMETRY'
      PRINTF='PRINT4'
      PLOTF='PLOT4'
      MODE=2
      R(1)=0.5
      L1=14
      M1=11
      YV(2)=0.
      DY=1.5/FLOAT(M1-2)
      DO 10 J=3,M1
         YV(J)=YV(J-1)+DY
   10 CONTINUE
      XU(2)=0.
      DX=1./6.
      DO 20 I=3,8
         XU(I)=XU(I-1)+DX
   20 CONTINUE
CONSTRUCT CONTROL-VOLUME FACES TO MATCH CIRCULAR BOUNDARY
      DO 30 I=9,13
         J=19-I
         YM=0.5*(YV(J)+YV(J-1))
         XU(I)=2.-SQRT(1.-(YM-1.5)**2.)
   30 CONTINUE
      XU(L1)=2.
      RETURN
C*-*-*-*-*-*-*-*-*-*-*-*-*-*-*-*-*-*-*-*-*-*-*-*-*-*-*-*-*-*-*-*-*-*-*
      ENTRY BEGIN
      TITLE(1)='   TEMPERATURE '
      CALL INTA4(KSOLVE(1),1,KPRINT(1),1,KPLOT(1),1,LAST,3)
      CALL DATA6(COND,2.,TW1,200.,TW2,100.,QW,1.,HE,5.,TINF,20.)
      DO 100 J=1,M1
      DO 100 I=1,L1
         T(I,J)=TW2
  100 CONTINUE
      DO 110 J=2,M2
         T(1,J)=TW1
  110 CONTINUE
      DO 120 I=2,L2
         T(I,1)=TW1
  120 CONTINUE
```

```
      RETURN
C*-*-*-*-*-*-*-*-*-*-*-*-*-*-*-*-*-*-*-*-*-*-*-*-*-*-*-*-*-*-*-*-*
      ENTRY OUTPUT
      DO 200 IUNIT=IU1,IU2
         IF(ITER.EQ.0) WRITE(IUNIT,210)
  210    FORMAT(2X,'ITER',2X,'T(4,5)',4X,'T(6,10)',3X,'T(10,4)',
     1        3X,'T(12,4)')
         WRITE(IUNIT,220) ITER,T(4,5),T(6,10),T(10,4),T(12,4)
  220    FORMAT(3X,I2,1P4E10.2)
  200 CONTINUE
      IF(ITER.EQ.LAST) THEN
         CALL PRINT
C······
COME HERE TO FILL IBLOCK(I,J) BEFORE CALLING PLOT
         DO 230 J=2,M2
         DO 230 I=2,L2
            IF(X(I).LT.0.5.AND.Y(J).GT.1.0) IBLOCK(I,J)=1
            IF(X(I).LT.0.5.AND.Y(J).LT.0.5) IBLOCK(I,J)=1
            IF(X(I).GT.1.0.AND.Y(J).GT.0.5) IBLOCK(I,J)=1
  230    CONTINUE
         DO 240 J=5,9
         DO 240 I=8,17-J
            IBLOCK(I,J)=0
  240    CONTINUE
         CALL PLOT
C······
      ENDIF
      RETURN
C*-*-*-*-*-*-*-*-*-*-*-*-*-*-*-*-*-*-*-*-*-*-*-*-*-*-*-*-*-*-*-*-*
      ENTRY PHI
      DO 300 J=2,M2
      DO 300 I=2,L2
         GAM(I,J)=COND
         IF(X(I).LT.0.5.AND.Y(J).GT.1.0) GAM(I,J)=0.
         IF(X(I).LT.0.5.AND.Y(J).LT.0.5) GAM(I,J)=0.
         IF(X(I).GT.1.0.AND.Y(J).GT.0.5) GAM(I,J)=BIG
         SC(I,J)=50.
         SP(I,J)=-4.
  300 CONTINUE
      DO 310 J=5,9
      DO 310 I=8,17-J
         GAM(I,J)=COND
```

```
  310 CONTINUE
COME HERE TO SPECIFY BOUNDARY CONDITIONS
       DO 320 J=8,M2
          SC(5,J)=SC(5,J)+QW/XCV(5)
  320 CONTINUE
       DO 330 J=2,4
          ABYV=1./XCV(5)
          RES=1./HE+0.5*XCV(5)/GAM(5,J)
          CONS=ABYV/RES
          SC(5,J)=SC(5,J)+CONS*TINF
          SP(5,J)=SP(5,J)-CONS
  330 CONTINUE
       DO 340 I=2,4
          GBYD=GAM(I,7)/(0.5*YCV(7))
          ABYV=RV(8)/YCVR(7)
          SC(I,7)=SC(I,7)+ABYV*GBYD*TW1
          SP(I,7)=SP(I,7)-ABYV*GBYD
          ABYV=RV(5)/YCVR(5)
          RES=1./HE+0.5*YCV(5)/GAM(I,5)
          CONS=ABYV/RES
          SC(I,5)=SC(I,5)+CONS*TINF
          SP(I,5)=SP(I,5)-CONS
  340 CONTINUE
       DO 350 I=2,L2
          KBCM1(I)=2
  350 CONTINUE
       RETURN
       END
CCCCCCCCCCCCCCCCCCCCCCCCCCCCCCCCCCCCCCCCCCCCCCCCCCCCCCCCCCCCCCCCCCCCCCCC
```

8.4-5   Results for Example 4

```
RESULTS OF CONDUCT FOR AXISYMMETRIC COORDINATE SYSTEM
******************************************************


- - - - - - - - - - - - - - - - - - - - - - - - - - - - - - - - - - - - - - - - - -
CONDUCTION IN A COMPLEX CYLINDRICAL GEOMETRY
- - - - - - - - - - - - - - - - - - - - - - - - - - - - - - - - - - - - - - - - - -


  ITER  T(4,5)    T(6,10)   T(10,4)   T(12,4)
```

```
0   1.00E+02   1.00E+02   1.00E+02   1.00E+02
1   1.16E+02   9.19E+01   1.15E+02   1.13E+02
2   1.16E+02   9.19E+01   1.15E+02   1.13E+02
3   1.16E+02   9.19E+01   1.15E+02   1.13E+02
```

```
I =     1        2        3        4        5        6        7
X = 0.00E+00 8.33E-02 2.50E-01 4.17E-01 5.83E-01 7.50E-01 9.17E-01
```

```
I =     8        9       10       11       12       13       14
X = 1.02E+00 1.06E+00 1.14E+00 1.26E+00 1.47E+00 1.80E+00 2.00E+00
```

```
J =     1        2        3        4        5        6        7
Y = 0.00E+00 8.33E-02 2.50E-01 4.17E-01 5.83E-01 7.50E-01 9.17E-01
```

```
J =     8        9       10       11
Y = 1.08E+00 1.25E+00 1.42E+00 1.50E+00
```

```
******        TEMPERATURE        ******
```

```
I =     1        2        3        4        5        6        7
J
11     1.00E+02 1.25E+01 1.25E+01 1.25E+01 9.04E+01 9.17E+01 9.58E+01
10     2.00E+02 1.25E+01 1.25E+01 1.25E+01 9.08E+01 9.19E+01 9.58E+01
 9     2.00E+02 1.25E+01 1.25E+01 1.25E+01 9.40E+01 9.38E+01 9.59E+01
 8     2.00E+02 1.25E+01 1.25E+01 1.25E+01 1.02E+02 9.81E+01 9.69E+01
 7     2.00E+02 1.93E+02 1.83E+02 1.68E+02 1.21E+02 1.05E+02 9.90E+01
 6     2.00E+02 1.83E+02 1.58E+02 1.39E+02 1.18E+02 1.06E+02 1.01E+02
 5     2.00E+02 1.68E+02 1.34E+02 1.16E+02 1.09E+02 1.06E+02 1.05E+02
 4     2.00E+02 1.25E+01 1.25E+01 1.25E+01 9.98E+01 1.10E+02 1.14E+02
 3     2.00E+02 1.25E+01 1.25E+01 1.25E+01 1.13E+02 1.28E+02 1.34E+02
 2     2.00E+02 1.25E+01 1.25E+01 1.25E+01 1.55E+02 1.68E+02 1.72E+02
 1     1.00E+02 2.00E+02 2.00E+02 2.00E+02 2.00E+02 2.00E+02 2.00E+02
```

```
I =     8        9       10       11       12       13       14
J
11     1.00E+02 1.00E+02 1.00E+02 1.00E+02 1.00E+02 1.00E+02 1.00E+02
10     1.00E+02 1.00E+02 1.00E+02 1.00E+02 1.00E+02 1.00E+02 1.00E+02
 9     9.94E+01 1.00E+02 1.00E+02 1.00E+02 1.00E+02 1.00E+02 1.00E+02
 8     9.84E+01 9.92E+01 1.00E+02 1.00E+02 1.00E+02 1.00E+02 1.00E+02
```

7	9.86E+01	9.87E+01	9.92E+01	1.00E+02	1.00E+02	1.00E+02	1.00E+02
6	1.00E+02	9.98E+01	9.96E+01	9.97E+01	1.00E+02	1.00E+02	1.00E+02
5	1.05E+02	1.04E+02	1.04E+02	1.04E+02	1.02E+02	1.00E+02	1.00E+02
4	1.15E+02	1.15E+02	1.15E+02	1.15E+02	1.13E+02	1.05E+02	1.00E+02
3	1.36E+02	1.36E+02	1.37E+02	1.37E+02	1.35E+02	1.23E+02	1.00E+02
2	1.72E+02	1.73E+02	1.73E+02	1.73E+02	1.72E+02	1.61E+02	1.00E+02
1	2.00E+02	2.00E+02	2.00E+02	2.00E+02	2.00E+02	2.00E+02	1.00E+02

8.4-6   Discussion of Results

For this linear problem, the solution can be seen to have converged in one iteration. In the field printout of the temperature, note the value of 100 established in the inactive region outside the curved boundary. The temperature values in the two cut-out regions on the left do not have any effect on our real calculation domain and should, therefore, be ignored. (Incidentally, these regions produce a uniform temperature value of 12.5. Since the conductivity is zero in these regions, temperature adjusts so as to make the source term there equal to zero. We can see from Eq. (8.7c) that a zero source term corresponds to a temperature of 12.5.)

8.4-7   Final Remarks

In the four example problems presented so far, all the possible complications in steady heat conduction problems have been illustrated. You should now be in a position to attempt many imaginative applications of CONDUCT. After these steady-state problems, we shall now consider two unsteady heat conduction situations. In the first one, the solid will undergo a time-dependent temperature change and eventually attain a steady state. In the second problem, there will be a periodic variation of temperature with time.

8.5   Unsteady Conduction with Heat Generation (Example 5)

8.5-1   Problem Description

In this problem, we shall calculate the transient temperature distribution in a solid as a result of suddenly imposed boundary conditions and heat generation. This is also our first example in the θr geometry.

For the semicircular solid shown in Fig. 8.6, a uniform initial temperature T_0 prevails throughout the solid. At time $t = 0$, a heat flux q_B is imposed on the flat boundaries as shown, the temperature of the inner surface at radius R_1 is raised to a value of T_i, the outer surface is exposed to a fluid at T_∞ (which provides a

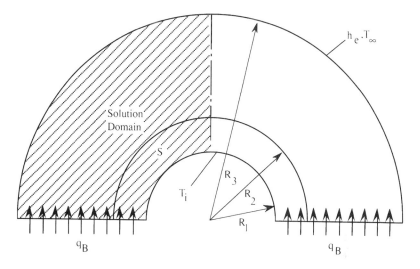

Fig. 8.6 Unsteady conduction with heat generation

convective heat transfer coefficient h_e), and heat generation at the rate S is started in the region between the radii R_1 and R_2. Since these conditions remain constant with time, the solid will eventually approach a steady state. Our objective is to calculate the transient temperature fields and the final steady-state temperature distribution.

We shall use the following numerical values

$$T_0 = 50, \qquad T_1 = 100, \qquad T_\infty = 20 \tag{8.8a}$$

$$R_1 = 0.5, \qquad R_2 = 0.75, \qquad R_3 = 1.5 \tag{8.8b}$$

$$q_B = 60, \qquad h_e = 5, \qquad S = 1000 \tag{8.8c}$$

$$k = 1, \qquad \rho c = 1 \tag{8.8d}$$

Because of the symmetry of the geometry and boundary conditions, we shall limit our calculation to the left half of the domain, as shown in Fig. 8.6.

8.5-2   Design of ADAPT

GRID. For the θr geometry ($\mathrm{MODE} = 3$), the X values in the computer program are to be interpreted as the angles θ in radians. Here we use ZGRID and create one zone in the θ direction and two zones in the r direction. The radius $R(1)$ of the inner surface must be specified if MODE equals 2 or 3.

BEGIN. For an unsteady problem. the iterations represent time steps. Here we decide to perform 35 time steps with a continuously increasing Δt. As we shall see later. a steady state will be reached after about 30 steps.

we assign numerical values to the various parameters given in Eq. (8.8). Here DT represents the initial value of Δt. which will be subsequently adjusted. The temperature array T(I,J) is filled with the initial value T_0. For steady-state problems. such initial values represent simply our starting *guesses*: for an unsteady situation. the values of T(I,J) specified in BEGIN for the internal grid points are the *known* initial temperatures at the start of the unsteady computation.

Finally. we provide the boundary temperature T_1 at the appropriate locations.

OUTPUT. In an unsteady problem. we obtain a new temperature field after each time step. Often. it is not convenient to call PRINT and get a full printout of each new field. Instead. we can print out a few useful values after each time step and get the final printout after a few selected time steps.

Here we arrange. at every time step. the output of four typical temperatures and the total heat flows at the inner and outer surfaces of the solid. In the calculation of these heat flow rates. the quantities XCV(I)*RV(2) and XCV(I)*RV(M1) represent the areas (per unit depth) of the control-volume faces on the inner and outer boundaries respectively.

In many time-dependent problems. the temperatures change with time very rapidly in the beginning and rather slowly later on. Therefore. it is desirable to use very small time steps to resolve the initial changes and then gradually increase the time step Δt. One convenient way of arranging this is through a statement such as

```
DT=DT*1.2
```

where the time step is increased every time by 20 %. We can start with a very small value of DT. get reasonably small values of Δt during the initial stage. and still eventually reach sufficiently large Δt values appropriate for the final approach to the steady-state solution.

In this problem. we call PRINT only after the final time step and get the field printout of the steady-state temperatures. If you wish to get a number of intermediate field printouts. you will find it convenient to ser KPGR = 0 after the first call to PRINT: this will ensure that the same grid coordinates are not printed out with every temperature field.

PHI. An unsteady problem mainly requires two extra actions. They are the specification of DT and the filling of the array ALAM(I,J). Therefore. in the PHI routine. we specify the ALAM(I,J) array in addition to providing GAM(I,J) and SC(I,J).

The specification of the boundary conditions includes the convective condition on the outer surface, the prescribed heat flux on the flat surface, and the zero-flux condition on the vertical centerline. These details are given in terms of the appropriate values of KBC, FLXC, and FLXP.

8.5-3   Additional Fortran Names

COND	thermal conductivity k. Eq. (8.8d)
HE	heat transfer coefficient h_e. Eq. (8.8c)
PI	the constant π
QB	constant boundary heat flux q_B. Eq. (8.8c)
QBTM	total heat flow through the bottom boundary
QTOP	total heat flow through the top boundary
RHOCP	heat capacity ρc. Eq. (8.8d)
SOURCE	heat generation rate S. Eq. (8.8c)
T(I,J)	temperature T
TI	inner surface temperature T_i, Eq. (8.8a)
TINF	freestream temperature T_∞. Eq. (8.8a)
TZERO	temperature T_0 at time t = 0. Eq. (8.8a)

8.5-4   Listing of ADAPT for Example 5

```
CCCCCCCCCCCCCCCCCCCCCCCCCCCCCCCCCCCCCCCCCCCCCCCCCCCCCCCCCCCCCCCCCCCCCCCCCC
      SUBROUTINE ADAPT
C..........................................................................
C.....EXAMPLE 5 -- UNSTEADY CONDUCTION WITH HEAT GENERATION
C..........................................................................
$INCLUDE:'COMMON'
C*************************************************************************
      DIMENSION T(NI,NJ)
      EQUIVALENCE (F(1,1,1),T(1,1))
C*.*.*.*.*.*.*.*.*.*.* *.*.*.*.*.*.*.* *.*.*.*.*.*.*.*.*.*.*.*.*.*
      ENTRY GRID
      HEADER='UNSTEADY CONDUCTION WITH HEAT GENERATION'
      PRINTF='PRINT5'
      PLOTF='PLOT5'
      MODE=3
      PI=3.14159
      CALL INTA5(NZX,1,NCVX(1),10,NZY,2,NCVY(1),3,NCVY(2),9)
      CALL DATA4(XZONE(1),0.5*PI,POWRX(1),1.2,YZONE(1),0.25,
     1    YZONE(2),0.75)
      CALL ZGRID
```

```
      R(1)=0.5
      RETURN
C*-*-*-*-*-*-*-*-*-*-*-*-*-*-*-*-*-*-*-*-*-*-*-*-*-*-*-*-*-*-*-*-*
      ENTRY BEGIN
      TITLE(1)='   TEMPERATURE '
      CALL INTA4(KSOLVE(1),1,KPRINT(1),1,KPLOT(1),1,LAST,35)
      CALL DATA6(COND,1.,RHOCP,1.,SOURCE,1000.,HE,5.,QB,60.,DT,0.001)
      CALL DATA3(TINF,20.,TZERO,50.,TI,100.)
      DO 100 J=1,M1
      DO 100 I=1,L1
         T(I,J)=TZERO
  100 CONTINUE
      DO 110 I=2,L2
         T(I,1)=TI
  110 CONTINUE
      RETURN
C*-*-*-*-*-*-*-*-*-*-*-*-*-*-*-*-*-*-*-*-*-*-*-*-*-*-*-*-*-*-*-*-*
      ENTRY OUTPUT
         QBTM=0.
         QTOP=0.
         DO 200 I=2,L2
            QBTM=QBTM+XCV(I)*RV(2)*FLUXJ1(I,1)
            QTOP=QTOP+XCV(I)*RV(M1)*FLUXM1(I,1)
  200    CONTINUE
      DO 210 IUNIT=IU1,IU2
         IF(ITER.EQ.0) WRITE(IUNIT,220)
  220    FORMAT(1X,'ITER',3X,'TIME',4X,'T(4,4)',3X,'T(10,4)',
     1   2X,'T(7,10)',1X,'T(10,10)',4X,'QBTM',6X,'QTOP')
         WRITE(IUNIT,230) ITER,TIME,T(4,4),T(10,4),
     1   T(7,10),T(10,10),QBTM,QTOP
  230    FORMAT(2X,I2,1X,1P5E9.2,1P2E10.2)
  210 CONTINUE
      DT=DT*1.2
      IF(ITER.EQ.LAST) THEN
         CALL PRINT
         CALL PLOT
      ENDIF
      RETURN
C*-*-*-*-*-*-*-*-*-*-*-*-*-*-*-*-*-*-*-*-*-*-*-*-*-*-*-*-*-*-*-*-*
      ENTRY PHI
      DO 300 J=2,M2
      DO 300 I=2,L2
```

```
          ALAM(I,J)=RHOCP
          GAM(I,J)=COND
          IF(Y(J).LT.0.25) SC(I,J)=SOURCE
  300 CONTINUE
COME HERE TO SPECIFY BOUNDARY CONDITIONS
        DO 310 J=2,M2
          KBCI1(J)=2
          KBCL1(J)=2
          FLXCI1(J)=QB
  310 CONTINUE
        DO 320 I=2,L2
          KBCM1(I)=2
          FLXCM1(I)=HE*TINF
          FLXPM1(I)=-HE
  320 CONTINUE
        RETURN
        END
CCCCCCCCCCCCCCCCCCCCCCCCCCCCCCCCCCCCCCCCCCCCCCCCCCCCCCCCCCCCCCCCCCCCCCCCCC
```

8.5-5   Results for Example 5

```
RESULTS OF CONDUCT FOR POLAR COORDINATE SYSTEM
**************************************************


.  .  .  .  .  .  .  .  .  .  .  .  .  .  .  .  .  .  .  .  .  .  .  .  .  .  .  .  .  .  .  .  .  .  .  .  .  .  .  .

UNSTEADY CONDUCTION WITH HEAT GENERATION
.  .  .  .  .  .  .  .  .  .  .  .  .  .  .  .  .  .  .  .  .  .  .  .  .  .  .  .  .  .  .  .  .  .  .  .  .  .  .  .
```

ITER	TIME	T(4,4)	T(10,4)	T(7,10)	T(10,10)	QBTM	QTOP
0	0.00E+00	5.00E+01	5.00E+01	5.00E+01	5.00E+01	0.00E+00	0.00E+00
1	1.20E-03	5.13E+01	5.12E+01	5.00E+01	5.00E+01	8.71E+02	2.88E+02
2	2.64E-03	5.29E+01	5.29E+01	5.00E+01	5.00E+01	6.02E+02	2.71E+02
3	4.37E-03	5.52E+01	5.51E+01	4.99E+01	4.99E+01	4.21E+02	2.55E+02
4	6.44E-03	5.80E+01	5.77E+01	4.98E+01	4.98E+01	3.00E+02	2.40E+02
5	8.93E-03	6.13E+01	6.09E+01	4.96E+01	4.96E+01	2.18E+02	2.26E+02
6	1.19E-02	6.50E+01	6.43E+01	4.94E+01	4.94E+01	1.60E+02	2.13E+02
7	1.55E-02	6.90E+01	6.79E+01	4.90E+01	4.90E+01	1.15E+02	2.00E+02
8	1.98E-02	7.30E+01	7.16E+01	4.85E+01	4.85E+01	7.95E+01	1.88E+02
9	2.50E-02	7.71E+01	7.53E+01	4.79E+01	4.79E+01	4.93E+01	1.77E+02
10	3.12E-02	8.11E+01	7.88E+01	4.73E+01	4.72E+01	2.33E+01	1.67E+02

```
11   3.86E-02 8.50E+01 8.22E+01 4.67E+01 4.66E+01  4.30E-01 -1.57E+02
12   4.75E-02 8.86E+01 8.53E+01 4.62E+01 4.61E+01 -1.97E+01  1.49E+02
13   5.82E-02 9.20E+01 8.81E+01 4.59E+01 4.58E+01 -3.74E+01 -1.42E+02
14   7.10E-02 9.51E+01 9.06E+01 4.58E+01 4.56E+01 -5.29E+01 -1.37E+02
15   8.64E-02 9.79E+01 9.29E+01 4.60E+01 4.57E+01 -6.63E+01 -1.34E+02
16   1.05E-01 1.00E+02 9.48E+01 4.66E+01 4.61E+01 -7.77E+01 -1.34E+02
17   1.27E-01 1.03E+02 9.64E+01 4.73E+01 4.66E+01 -8.74E+01 -1.36E+02
18   1.54E-01 1.04E+02 9.78E+01 4.83E+01 4.74E+01 -9.56E+01 -1.40E+02
19   1.86E-01 1.06E+02 9.90E+01 4.94E+01 4.83E+01 -1.02E+02 -1.45E+02
20   2.24E-01 1.07E+02 1.00E+02 5.06E+01 4.92E+01 -1.08E+02 -1.51E+02
21   2.70E-01 1.09E+02 1.01E+02 5.18E+01 5.02E+01 -1.13E+02 -1.57E+02
22   3.25E-01 1.10E+02 1.02E+02 5.28E+01 5.10E+01 -1.17E+02  1.62E+02
23   3.91E-01 1.10E+02 1.02E+02 5.37E+01 5.18E+01 -1.20E+02 -1.67E+02
24   4.71E-01 1.11E+02 1.03E+02 5.44E+01 5.24E+01 -1.23E+02 -1.71E+02
25   5.66E-01 1.11E+02 1.03E+02 5.49E+01 5.29E+01 -1.24E+02 -1.74E+02
26   6.81E-01 1.12E+02 1.03E+02 5.53E+01 5.32E+01 -1.26E+02 -1.76E+02
27   8.18E-01 1.12E+02 1.03E+02 5.55E+01 5.34E+01 -1.26E+02 -1.77E+02
28   9.83E-01 1.12E+02 1.03E+02 5.56E+01 5.35E+01 -1.27E+02 -1.78E+02
29   1.18E+00 1.12E+02 1.03E+02 5.57E+01 5.36E+01 -1.27E+02 -1.78E+02
30   1.42E+00 1.12E+02 1.03E+02 5.57E+01 5.36E+01 -1.27E+02 -1.78E+02
31   1.70E+00 1.12E+02 1.03E+02 5.58E+01 5.36E+01 -1.27E+02 -1.78E+02
32   2.04E+00 1.12E+02 1.03E+02 5.58E+01 5.36E+01 -1.27E+02 -1.78E+02
33   2.46E+00 1.12E+02 1.03E+02 5.58E+01 5.36E+01 -1.27E+02 -1.78E+02
34   2.95E+00 1.12E+02 1.03E+02 5.58E+01 5.36E+01 -1.27E+02 -1.78E+02
35   3.54E+00 1.12E+02 1.03E+02 5.58E+01 5.36E+01 -1.27E+02 -1.78E+02


I =    1        2        3        4        5        6        7
TH = 0.00E+00 4.96E-02 1.63E-01 2.99E-01 4.47E-01 6.03E-01 7.67E-01

I =    8        9       10       11       12
TH = 9.37E-01 1.11E+00 1.29E+00 1.48E+00 1.57E+00


J =    1        2        3        4        5        6        7
Y = 0.00E+00 4.17E-02 1.25E-01 2.08E-01 2.92E-01 3.75E-01 4.58E-01

J =    8        9       10       11       12       13       14
Y = 5.42E-01 6.25E-01 7.08E-01 7.92E-01 8.75E-01 9.58E-01 1.00E+00

******      TEMPERATURE      ******
      . . . . . . . . . . . . . . . . . . . . .
```

I =	1	2	3	4	5	6	7
J							
14	5.00E+01	4.41E+01	3.99E+01	3.72E+01	3.56E+01	3.45E+01	3.38E+01
13	5.28E+01	4.90E+01	4.40E+01	4.09E+01	3.88E+01	3.75E+01	3.67E+01
12	6.18E+01	5.81E+01	5.23E+01	4.82E+01	4.56E+01	4.38E+01	4.27E+01
11	7.02E+01	6.66E+01	6.04E+01	5.58E+01	5.26E+01	5.05E+01	4.91E+01
10	7.81E+01	7.47E+01	6.84E+01	6.34E+01	5.98E+01	5.74E+01	5.58E+01
9	8.56E+01	8.24E+01	7.63E+01	7.11E+01	6.73E+01	6.46E+01	6.28E+01
8	9.30E+01	9.00E+01	8.42E+01	7.90E+01	7.50E+01	7.22E+01	7.03E+01
7	1.00E+02	9.74E+01	9.20E+01	8.70E+01	8.30E+01	8.02E+01	7.82E+01
6	1.07E+02	1.05E+02	9.98E+01	9.51E+01	9.13E+01	8.86E+01	8.67E+01
5	1.14E+02	1.12E+02	1.08E+02	1.03E+02	1.00E+02	9.75E+01	9.57E+01
4	1.22E+02	1.19E+02	1.15E+02	1.12E+02	1.09E+02	1.07E+02	1.06E+02
3	1.21E+02	1.19E+02	1.16E+02	1.13E+02	1.11E+02	1.10E+02	1.09E+02
2	1.11E+02	1.10E+02	1.08E+02	1.07E+02	1.06E+02	1.05E+02	1.05E+02
1	5.00E+01	1.00E+02	1.00E+02	1.00E+02	1.00E+02	1.00E+02	1.00E+02

I =	8	9	10	11	12
J					
14	3.33E+01	3.30E+01	3.29E+01	3.28E+01	5.00E+01
13	3.61E+01	3.58E+01	3.56E+01	3.55E+01	3.55E+01
12	4.20E+01	4.15E+01	4.12E+01	4.11E+01	4.11E+01
11	4.82E+01	4.76E+01	4.72E+01	4.71E+01	4.71E+01
10	5.47E+01	5.40E+01	5.36E+01	5.34E+01	5.34E+01
9	6.16E+01	6.09E+01	6.04E+01	6.02E+01	6.02E+01
8	6.90E+01	6.82E+01	6.77E+01	6.75E+01	6.75E+01
7	7.69E+01	7.60E+01	7.55E+01	7.53E+01	7.53E+01
6	8.54E+01	8.45E+01	8.40E+01	8.38E+01	8.38E+01
5	9.45E+01	9.37E+01	9.33E+01	9.31E+01	9.30E+01
4	1.05E+02	1.04E+02	1.03E+02	1.03E+02	1.03E+02
3	1.08E+02	1.08E+02	1.07E+02	1.07E+02	1.07E+02
2	1.05E+02	1.05E+02	1.05E+02	1.04E+02	1.04E+02
1	1.00E+02	1.00E+02	1.00E+02	1.00E+02	5.00E+01

8.5-6   Discussion of Results

The solution of this problem shows a number of interesting trends in the variation of local temperatures. Among the four typical temperatures printed out. $T(4,4)$ and $T(10,4)$ are close to the inner surface. while $T(7,10)$ and $T(10,10)$ are in the outer region. Initially. the high temperature of the inner boundary. the start of the heat generation. and the provision of the heat flux q_B all contribute to increase the temperature of the inner region. Therefore, $T(4,4)$ and $T(10,4)$ can be seen to increase steadily. On the other hand, the outer boundary is exposed to a colder fluid at T_∞. As a result, $T(7,10)$ and $T(10,10)$ can be seen to gradually decrease for quite some time. However, eventually these temperatures too begin to increase as the high-temperature effects in the inner region exert their influence.

The heat flow QTOP from the outer surface is always negative indicating a heat *loss* to the surrounding cold fluid. At the inner surface, the heat flow QBTM is initially positive as heat enters the domain from the boundary at the high temperature T_i. Subsequently however, the internal heat generation causes the inside temperatures to rise above T_i and, as a result, QBTM becomes negative.

The final printout of the steady-state temperature distribution shows the expected effect of the various boundary conditions. The highest temperature in the domain is found at the heat flux boundary, while the lowest temperatures are along the outer boundary, which is in contact with the cold external fluid.

8.5-7 Final Remarks

This description of an adaptation of CONDUCT to an unsteady problem shows that the extra effort needed to perform unsteady computations is quite small. Yet, we are able to produce extensive and interesting information about the time-dependent temperature field. In this example, we did not include any geometrical irregularities; however, there is no reason why you cannot implement the techniques in the previous examples to analyze unsteady conduction in irregular domains. Here we used boundary conditions that remained constant in time. You can also experiment with time-varying boundary conditions. If the boundary temperatures or heat fluxes continuously vary in time, no final steady state is attained. For periodically varying boundary conditions, the entire temperature field eventually becomes periodic and exactly repeats itself cycle after cycle. Such a situation is considered in the next example.

8.6 Unsteady Conduction around an Underground Building (Example 6)

8.6-1 Problem Description

To illustrate periodic heat conduction, we shall now consider the variation of the temperature distribution in the ground caused by the seasonal changes in the air temperature. An additional influence is provided by an underground building, which is maintained at a constant temperature throughout the year.

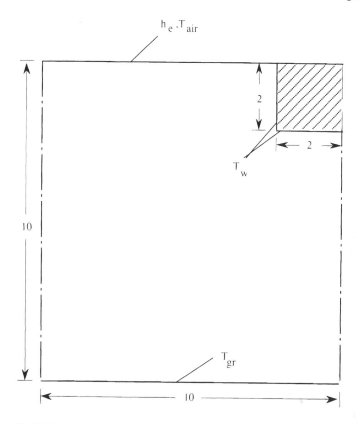

Fig.8.7 Unsteady conduction around an underground building

The physical situation and the boundary conditions are shown in Fig. 8.7. At the top surface, the ground exchanges heat by convection with the surrounding air at temperature T_{air}; this temperature varies during the year in a periodic manner. The building is shown shaded and is maintained at T_w. The bottom boundary, which is placed sufficiently away from the boundary, is at a known temperature T_{gr}. This deep-ground temperature is considered to remain constant throughout the year.

The left boundary of the domain is considered to be sufficiently far from the building; therefore, the heat flux across it is taken as zero. The right boundary represents a symmetry line through the building (the shaded area being only the left half of the building); therefore, it too is a zero-flux boundary.

We shall use the following variation of the air temperature.

$$T_{air} = 50 + 50 \sin (2\pi t/P) \tag{8.9}$$

where the period P is taken as 12. The other numerical values are given by

$$k = 2. \qquad \rho c = 3. \qquad h_e = 5 \qquad\qquad (8.10a)$$

$$T_w = 70. \qquad T_{gr} = 50 \qquad\qquad (8.10b)$$

Although we describe this problem in terms of a practical situation of heat conduction around an underground building. no particular attempt is made to specify the physical dimensions and material properties in a realistic manner in a certain system of units. Only the air temperatures can be thought of as roughly the winter-summer temperatures in degrees Fahrenheit. Further. the value P = 12 is intended to reflect the time t in months. However. no additional resemblance to a real situation is implied.

To start the unsteady computation. we need an initial temperature distribution. If we assume the entire ground to be initially at a uniform temperature and perform computations for a number of periods (i.e.. years). we shall find that. after the first few periods. the entire solution becomes periodic and exactly repeats itself period after period. It is this "steady periodic" state that is of interest in such problems.

8.6-2 Design of ADAPT

GRID. The geometry shown in Fig. 8.7 lends itself to the use of ZGRID. for which we use two zones each in x and y directions. Since the temperature in the ground will vary steeply in the vicinity of the building surface. it is desirable to provide a fine grid there. This is done by setting POWRX(1) and POWRY(1) equal to -1.5: the negative value gives a fine grid at the *end* of the zone.

BEGIN. Here we choose 120 as the total number of time steps. The value of DT is set equal to 0.5 (one-half month). Thus. in a given period P = 12. we perform 24 time steps. and the 120 steps correspond to 5 periods (years). The numerical data in Eqs. (8.9) and (8.10) is specified in BEGIN. This is followed by the setting of T(I,J) equal to T_{gr}. except over a portion of the top boundary where the boundary temperature is given as T_w.

OUTPUT. If the field printout of the temperature were to be arranged at a number of instants during the cycle. it will lead to a voluminous output. It is inconvenient to include so much printout in this book: however. you may wish to modify the ADAPT routine to get more detailed output.

For the purpose of this book. we shall not arrange any field printout at all. Rather. we shall include in the output the values of the temperatures at five grid points on a vertical line. This will give us some idea of the temperature profile in the ground at different stages in the cycle. Also, instead of providing this output after every time step, we shall try to save space by arranging the output after every third time step.

To judge the periodic behavior of the solution, we arrange to print a separate line (a string of ***) after a complete period, i.e., after 24 time steps.

PHI. Here the specification of ALAM(I,J) and GAM(I,J) is straightforward. For the region occupied by the building, a very large value of conductivity is used, so that this entire region would acquire the temperature T_w specified on its top boundary.

The left and right boundaries are specified as zero-flux boundaries through the appropriate KBC values. To provide the boundary condition for the portion of the top boundary that is exposed to the surrounding air, the time-varying value of T_{air} is first calculated. Please note here that the value of time used in this calculation is TIME+DT, and not just TIME. We are providing boundary conditions for the temperature field at the end of the time step, which corresponds to TIME+DT. (By the way, please do not write here: TIME=TIME+DT. The variable TIME is incremented internally by the invariant part of CONDUCT. Your redefinition of TIME would interfere with that calculation.)

Once the value of T_{air} is obtained, the convective boundary condition over the exposed part of the top boundary is given through the appropriate values of KBC, FLXC, and FLXP.

8.6-3   Additional Fortran Names

ARGU	argument in the expression for the air temperature, $2\pi t/P$
COND	thermal conductivity k, Eq. (8.10a)
HE	heat transfer coefficient h_e, Eq. (8.10a)
P	period of variation of the air temperature
PI	the constant π
RHOCP	heat capacity ρc
T(I,J)	temperature T
TAIR	air temperature T_{air}, Eq. (8.9)
TGR	ground temperature T_{gr}, Eq. (8.10b)
TW	wall temperature T_w, Eq. (8.10b)

8.6-4   Listing of ADAPT for Example 6

```
ccccccccccccccccccccccccccccccccccccccccccccccccccccccccccccccccccccc
      SUBROUTINE ADAPT
C..................................................................
C.....EXAMPLE 6 .. UNSTEADY CONDUCTION AROUND AN UNDERGROUND BUILDING
```

```
C· · · · · · · · · · · · · · · · · · · · · · · · · · · · · · · · · · · · · · · · · · ·
$INCLUDE:'COMMON'
C*********************************************************************
      DIMENSION T(NI,NJ)
      EQUIVALENCE (F(1,1,1),T(1,1))
C*·*·*·*·*·*·*·*-*·*·*·*·*·*-*·*·*·*·*·*·*·*·*-*·*·*·*·*·*·*-*·*-*·*
      ENTRY GRID
      HEADER='UNSTEADY CONDUCTION AROUND AN UNDERGROUND BUILDING'
      PRINTF='PRINT6'
      PI=3.14159
      CALL INTA3(NZX,2,NCVX(1),9,NCVX(2),3)
      CALL DATA2(XZONE(1),8.,XZONE(2),2.)
      CALL INTA3(NZY,2,NCVY(1),9,NCVY(2),3)
      CALL DATA2(YZONE(1),8.,YZONE(2),2.)
      CALL DATA2(POWRX(1),-1.5,POWRY(1),-1.5)
      CALL ZGRID
      RETURN
C*·*·*·*·*·*·*·*·*·*·*·*·*·*·*·*·*·*·*·*·*·*·*·*·*·*·*·*·*·*·*·*·*·*
      ENTRY BEGIN
      TITLE(1)='  TEMPERATURE '
      CALL INTA2(KSOLVE(1),1,LAST,120)
      CALL DATA7(COND,2.,RHOCP,3.,HE,5.,P,12.,DT,0.5,TGR,50.,TW,70.)
      DO 100 J=1,M1
      DO 100 I=1,L1
         T(I,J)=TGR
  100 CONTINUE
      DO 110 I=2,L2
         IF(X(I).GT.8.) T(I,M1)=TW
  110 CONTINUE
      RETURN
C*·*·*·*·*·*·*·*·*·*·*·*·*·*·*·*·*·*·*·*·*·*·*·*·*·*·*·*·*·*·*·*·*
      ENTRY OUTPUT
      DO 200 IUNIT=IU1,IU2
         IF(ITER.EQ.0) WRITE(IUNIT,210)
  210    FORMAT(1X,'ITER',3X,'TIME',5X,'T(2,14)',3X,'T(2,13)',3X,
     1       'T(2,12)',3X,'T(2,11)',3X,'T(2,10)')
         IF(MOD(ITER,3).EQ.0) THEN
            WRITE(IUNIT,240) ITER,TIME,T(2,14),T(2,13),T(2,12),T(2,11),
     1                       T(2,10)
  240       FORMAT(1X,I3,1P6E10.2)
            IF(MOD(ITER,24).EQ.0) WRITE(IUNIT,250)
  250       FORMAT(1X,69('*'))
```

```
        ENDIF
  200 CONTINUE
      RETURN
C*-*-.*-.*-.*-*-.*-.*-.*-.*-.*-.*-.*-.*-.*-.*-.*-.*-.*-.*-.*-.*-.*-.*-.*-.*-.*-.*-.*-.*-.*
      ENTRY PHI
      DO 300 J=2,M2
      DO 300 I=2,L2
         ALAM(I,J)=RHOCP
         GAM(I,J)=COND
         IF(X(I).GT.8.0.AND.Y(J).GT.8.0) GAM(I,J)=BIG
  300 CONTINUE
COME HERE TO SPECIFY BOUNDARY CONDITIONS
      DO 310 J=2,M2
         KBCI1(J)=2
         KBCL1(J)=2
  310 CONTINUE
      ARGU=2.*PI*(TIME+DT)/P
      TAIR=50.+50.*SIN(ARGU)
      DO 320 I=2,L2
         IF(X(I).LT.8.) THEN
            KBCM1(I)=2
            FLXCM1(I)=HE*TAIR
            FLXPM1(I)=-HE
         ENDIF
  320 CONTINUE
      RETURN
      END
CCCCCCCCCCCCCCCCCCCCCCCCCCCCCCCCCCCCCCCCCCCCCCCCCCCCCCCCCCCCCCCCCCCCCCCCCCCCCC
```

8.6-5 Results for Example 6

```
RESULTS OF CONDUCT FOR CARTESIAN COORDINATE SYSTEM
**************************************************


. . . . . . . . . . . . . . . . . . . . . . . . . . . . . . . . . . . . . . . . . . . . . .
UNSTEADY CONDUCTION AROUND AN UNDERGROUND BUILDING
. . . . . . . . . . . . . . . . . . . . . . . . . . . . . . . . . . . . . . . . . .


ITER  TIME     T(2,14)   T(2,13)   T(2,12)   T(2,11)   T(2,10)
  0  0.00E+00  5.00E+01  5.00E+01  5.00E+01  5.00E+01  5.00E+01
```

```
*********************************************************************
  3   1.50E+00   7.47E+01   6.68E+01   5.77E+01   5.34E+01   5.19E+01
  6   3.00E+00   8.96E+01   8.15E+01   6.88E+01   6.06E+01   5.70E+01
  9   4.50E+00   8.28E+01   8.04E+01   7.29E+01   6.58E+01   6.17E+01
 12   6.00E+00   5.77E+01   6.29E+01   6.61E+01   6.46E+01   6.24E+01
 15   7.50E+00   2.85E+01   3.88E+01   5.14E+01   5.70E+01   5.80E+01
 18   9.00E+00   1.24E+01   2.20E+01   3.72E+01   4.68E+01   5.06E+01
 21   1.05E+01   1.85E+01   2.21E+01   3.14E+01   3.99E+01   4.43E+01
 24   1.20E+01   4.34E+01   3.90E+01   3.74E+01   3.99E+01   4.26E+01
*********************************************************************
 27   1.35E+01   7.23E+01   6.28E+01   5.14E+01   4.69E+01   4.63E+01
 30   1.50E+01   8.84E+01   7.94E+01   6.53E+01   5.65E+01   5.32E+01
 33   1.65E+01   8.21E+01   7.91E+01   7.08E+01   6.32E+01   5.91E+01
 36   1.80E+01   5.72E+01   6.21E+01   6.47E+01   6.29E+01   6.06E+01
 39   1.95E+01   2.83E+01   3.83E+01   5.05E+01   5.58E+01   5.68E+01
 42   2.10E+01   1.22E+01   2.16E+01   3.66E+01   4.61E+01   4.98E+01
 45   2.25E+01   1.84E+01   2.19E+01   3.10E+01   3.94E+01   4.38E+01
 48   2.40E+01   4.33E+01   3.89E+01   3.71E+01   3.96E+01   4.23E+01
*********************************************************************
 51   2.55E+01   7.23E+01   6.27E+01   5.13E+01   4.67E+01   4.61E+01
 54   2.70E+01   8.84E+01   7.93E+01   6.52E+01   5.64E+01   5.31E+01
 57   2.85E+01   8.21E+01   7.91E+01   7.07E+01   6.31E+01   5.91E+01
 60   3.00E+01   5.72E+01   6.21E+01   6.47E+01   6.29E+01   6.07E+01
 63   3.15E+01   2.83E+01   3.83E+01   5.05E+01   5.58E+01   5.68E+01
 66   3.30E+01   1.22E+01   2.17E+01   3.66E+01   4.61E+01   4.99E+01
 69   3.45E+01   1.84E+01   2.19E+01   3.11E+01   3.94E+01   4.39E+01
 72   3.60E+01   4.33E+01   3.89E+01   3.72E+01   3.97E+01   4.23E+01
*********************************************************************
 75   3.75E+01   7.23E+01   6.27E+01   5.13E+01   4.67E+01   4.62E+01
 78   3.90E+01   8.84E+01   7.93E+01   6.52E+01   5.65E+01   5.31E+01
 81   4.05E+01   8.21E+01   7.91E+01   7.08E+01   6.32E+01   5.92E+01
 84   4.20E+01   5.72E+01   6.21E+01   6.47E+01   6.30E+01   6.07E+01
 87   4.35E+01   2.83E+01   3.83E+01   5.06E+01   5.59E+01   5.69E+01
 90   4.50E+01   1.22E+01   2.17E+01   3.67E+01   4.62E+01   5.00E+01
 93   4.65E+01   1.84E+01   2.19E+01   3.11E+01   3.95E+01   4.39E+01
 96   4.80E+01   4.33E+01   3.89E+01   3.72E+01   3.97E+01   4.24E+01
*********************************************************************
 99   4.95E+01   7.23E+01   6.27E+01   5.13E+01   4.68E+01   4.62E+01
102   5.10E+01   8.84E+01   7.94E+01   6.53E+01   5.65E+01   5.32E+01
105   5.25E+01   8.21E+01   7.91E+01   7.08E+01   6.32E+01   5.92E+01
108   5.40E+01   5.73E+01   6.21E+01   6.47E+01   6.30E+01   6.08E+01
111   5.55E+01   2.83E+01   3.83E+01   5.06E+01   5.59E+01   5.70E+01
```

```
114   5.70E+01   1.22E+01   2.17E+01   3.67E+01   4.62E+01   5.00E+01
117   5.85E+01   1.84E+01   2.19E+01   3.11E+01   3.95E+01   4.40E+01
120   6.00E+01   4.33E+01   3.89E+01   3.72E+01   3.97E+01   4.24E+01
**********************************************************************
```

8.6-6   Discussion of Results

The output for the five cycles shows that the temperature field has become almost completely periodic by the fifth cycle. In the periodic temperature distribution, some effect of the thermal inertia of the ground can be seen. At $ITER = 99$, the surface temperature of the ground $T(2,14)$ is quite high; but, at deeper locations temperatures less than 50 can be found. This can be regarded as the "memory" of the previous winter retained by the ground. Similarly, in an early winter month corresponding to $ITER = 111$, whereas the surface temperature $T(2,14)$ is low, temperatures higher than 50 can be found at other points along the same vertical line. This shows that, although the temperature at each location varies periodically with time, the thermal inertia creates a phase lag between different locations.

8.6-7   Final Remarks

The ability to compute the steady periodic phenomena is useful in a number of applications such as heat conduction in the walls of an internal combustion engine, the daily heat gain or loss of a building, and heat transfer in regenerators. Interesting physical insights can be obtained from a periodic temperature field, which often shows that the periodic variation of the boundary temperature causes a temperature wave along a space coordinate into the solid. Further, the amplitude of this wave decays exponentially along the coordinate.

The six examples presented in this chapter should enable you to use CONDUCT for a large variety of steady and unsteady problems in heat conduction. You should use this ability to solve complex and interesting problems and, at the same time, to enrich your understanding of the physical phenomena.

Heat conduction is only one class of problems that can be solved by CONDUCT. Another important class is the fully developed flow and heat transfer in ducts. In the next chapter, we shall review the mathematical formulation for this class of problems, and some examples of adaptations of CONDUCT to duct flows will be presented in Chapter 10.

Problems

8.1 Initially, you may begin by using CONDUCT for *one-dimensional* problems. If you want to calculate the temperature variation in the x direction, specify a zero heat flux on the boundaries normal to the y direction. (Note that M1 should not be less than 4.) Solve the problem in Section 2.4-5 by using CONDUCT. For this problem, the answers given by CONDUCT may not exactly agree with the answers in Chapter 2 because the grid design in CONDUCT uses Practice B and not Practice A.

8.2 Solve the one-dimensional steady heat conduction in a hollow cylinder of inner radius 0.5 and outer radius 2. Take the temperatures at the inner and outer surfaces as 100 and 200 respectively. Show that the solution obtained is indeed one-dimensional. For this case, the exact temperature variation in the *radial* direction is logarithmic. Compare the numerical values of T at the grid points with those obtained from the exact solution. For a reasonable number of grid points in the radial direction, this agreement should be very good.

Calculate the heat fluxes at the inner and outer surfaces and compare them with those given by the exact solution.

8.3 Write a GRID subroutine for the problem shown in the figure. Use $L1 = 13$ and $M1 = 11$. Ensure that the control-volume faces coincide with various discontinuities.

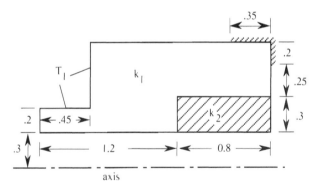

Problem 8.3

8.4 For a long solid circular cylinder the temperature distribution is to be calculated over its cross section. The conductivity equals 2.2 everywhere. The boundary conditions are given by: one-half of the circumference is adiabatic, while the remaining circumference is exposed to a fluid at a temperature of 500 with a heat transfer coefficient of 22. In the one-half cross section adjacent to the adiabatic boundary there is an internal heat generation of $S = 2000$; the source term is zero in the other half. Take $L1 = 12$, $M1 = 16$; assume a uniform grid. Write the PHI routine to specify the conductivity, source terms, and the boundary conditions.

8.5 For steady heat conduction in the plane solid of k = 5 shown in the figure. the boundary conditions are given by T_1 = 20. T_2 = 50. q = 250: the condition at the top surface is given by h = 2.9 and T_∞ = 5 and a radiative heat loss of $2.5\,(T_B^4 - T_\infty^4)$ where T_B is the boundary temperature. Assume a uniform grid with L1 = 11 and M1 = 8. Write the statements in PHI that are necessary to specify the heat flux boundary condition and the convection-radiation boundary condition.

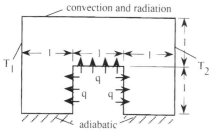

Problem 8.5

8.6 Calculate the temperature field for the situation described in Problem 8.5.

8.7 In a steady heat conduction problem for the solid shown. the boundary condition is that of given heat flux q = 50 entering the solid as shown. With L1 = 10 and M1 = 14, write the appropriate statements in PHI that are necessary to specify this boundary condition.

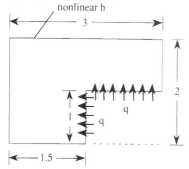

Problem 8.7

8.8 For Problem 8.7. the boundary condition at the top surface is specified in terms of a nonlinear convective heat transfer coefficient h and given outside fluid temperature T_∞ = 25. The heat transfer coefficient is given by the expression $h = 2.8\,(T(I,M1) - T_\infty)^{1/4}$. Assume that $T(I,M1)$ is always greater than T_∞. Write the appropriate statements in PHI to specify this boundary condition.

8.9 In a steady heat conduction problem in the axisymmetric solid shown. the boundary condition at boundary A is that of a convective heat transfer h = 5 and a fluid temperature $T_\infty = 25$. Assuming that the conductivities have been correctly specified in the array GAM(I , J). write the appropriate statements in PHI that are necessary to specify the boundary condition.

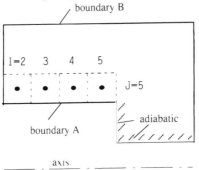

Problem 8.9

8.10 Modify the ADAPT routine for Example I in Section 8.1 to calculate the total heat flow rates at the four boundaries. Perform an overall heat balance and show that it is perfect within the conputer roundoff error.

Make further modifications to include a nonuniform heat source and a nonuniform conductivity. Observe how the temperature field changes. Continue to verify the overall heat balance.

8.11 In a steady heat conduction problem for the plane solid of conductivity $k = 10$. the boundary condition of heat flux $q = 50$ is applied on the boundary shown in the figure. Assuming that a uniform grid with $L1 = 11$ and $M1 = 10$ has been constructed. write the statements in PHI that are necessary to specify the heat flux boundary condition. Further. the boundary condition at the bottom surface is given by a convective heat coefficient $h = 3.8$ and a surrounding fluid temperature $T_\infty = 300$ *and* a radiative heat loss given by $2.34 \, (T (I , 1)^4 - T_\infty^4)$. You may assume that $T(I , 1)$ always remains greater than T_∞. Write appropriate statements required in PHI to specify this boundary condition.

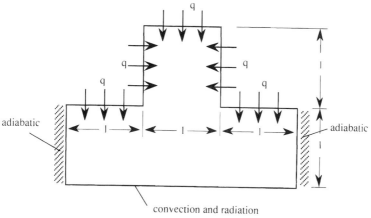

Problem 8.11

8.12 In a large concrete slab of thickness 3 units, reinforcing metal bars of 1×1 square cross section are embedded as shown. The top and bottom faces of the slab are maintained at temperatures of 0 and 100 respectively. The thermal conductivity of the metal is much greater than that of concrete. Compute the steady-state temperature distribution in the slab. Demonstrate that, as long as $k_{metal}/k_{concrete}$ is sufficiently large, different values of this ratio lead to practically the same solution. Print out the total heat flow through each face of the calculation domain and show that the overall heat balance is perfectly satisfied. On a drawing of the calculation domain and the grid used, approximately sketch the isotherms for

$$\frac{(T - T_{min})}{(T_{max} - T_{min})} = 0, 0.1, 0.2, ..., 0.9, 1.0$$

T = 0

3

1

T = 100

Problem 8.12

8.13 For the problem shown in the figure, write the GRID and PHI routines. The objective of the problem is to solve for the steady heat conduction with a uniform heat generation of $S = 300$ throughout the solid. Use $L1 = 11$ and $M1 = 8$.

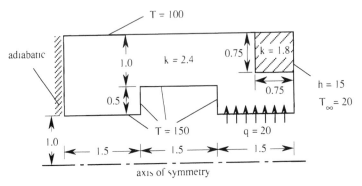

Problem 8.13

8.14 For the problem shown in the figure. write the GRID and PHI routines. The objective of the problem is to solve for the steady heat conduction with a uniform heat generation of $S = 300 - 0.2T$ throughout the solid. The solid has a rectangular hole in the center. the surface of the hole is maintained at a temperature $T_0 = 100$. The conductivity of the solid is 2.4. Use the smallest possible calculation domain. Use $L1 = 11$ and $M1 = 10$. Take $h = 7$. $T_\infty = 20$.

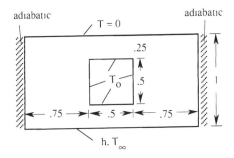

Problem 8.14

8.15 A thin metal plate of dimensions 7×4 has a thickness of 0.3. It is exposed on all its surfaces to a fluid at $T_\infty = 10$. The heat transfer coefficient between the fluid and the plate is 17. The shaded area represents an integrated circuit embedded in the plate. Heat generation in the circuit is at the rate of $S = 55$ per unit volume. The thermal conductivity of the metal plate is 5.5. while that of the shaded area is 1.2. The temperature variation over the thickness of the plate can be considered as negligible. Write the ADAPT routine and solve the steady temperature field in the plate. Also print out the total heat generation in the domain and the total heat loss from all the surfaces of the plate. Do you expect these quantities to be exactly equal?

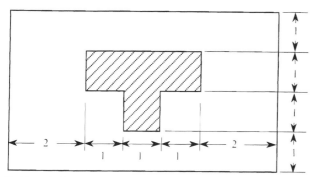

Problem 8.15

8.16 Solve the steady heat conduction problem shown in the figure. Use $L1 = 13$ and $M1 = 9$. Ensure that the control-volume faces coincide with the various discontinuities. The solid has a heat generation $S = 100 - 0.5T$ throughout. The conductivities and boundary conditions are shown in the figure.

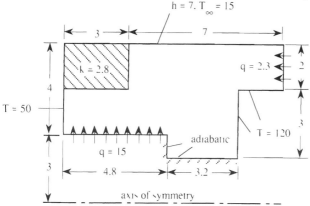

Problem 8.16

8.17 Calculate the steady state temperature distribution in the conduction situation shown in the figure. The geometry consists of a semi-circular solid with a 60° notch. The surface of the notch is in contact with a fluid at $T_\infty = 20$ and exchanges heat with the fluid with a heat transfer coefficient $h = 7.5$. The remaining circular boundary is adiabatic, while a uniform heat flux $q = 50$ enters the flat surface of the solid. The thermal conductivity k and specific heat c_p are 1.9 and 4.1 respectively. Calculate the total heat loss at the surface of the notch (which must finally equal the heat input through the flat surface).

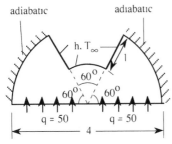

Problem 8.17

8.18 A concrete slab has I-shaped steel beams embedded in it for reinforcement. Take the conductivities of the concrete and steel as 1 and 100 respectively. The lower boundary is at a uniform temperature of 80, while the upper boundary loses heat to a surrounding fluid at temperature 20 with a heat transfer coefficient of 1.75. Using the smallest possible calculation domain, obtain the temperature distribution in the solid and the total heat loss from the upper boundary.

Compare this with the heat loss we would have if the whole domain were filled with concrete. (This will lead to a one-dimensional case and you should be able to obtain the heat loss from a simple formula. It is not necessary to run CONDUCT for this one-dimensional case.) You will notice that the region occupied by steel is nearly at a single uniform temperature. Why?

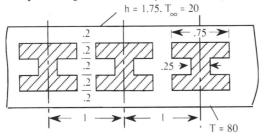

Problem 8.18

8.19 In Problem 8.18, introduce a volumetric internal heat generation given by $S = 2.5 - 0.01T$ in the region occupied by concrete. (This may represent a heat release caused by chemical reactions in the concrete.) Find the steady temperature distribution in the domain. Also evaluate:

    a) the total heat flow at the lower surface,

    b) the total heat flow at the upper surface, and

    c) the total amount of heat generated in the concrete.

Show that these three quantities *exactly* satisfy the overall heat balance.

8.20 A thin circular metal plate of diameter 8 and thickness 0.2 is exposed on all its surfaces to a fluid of temperature $T_\infty = 20$. The convective heat transfer coefficient between the fluid and plate surface is 25. The shaded areas represent integrated circuits embedded in the plate. The circuits generate heat at a rate of $S = 100$ per unit volume. The thermal conductivity of the metal plate and also the embedded circuits is 2.2. The temperature variation over the thickness of the metal plate can be regarded negligible. Prepare the ADAPT routine and obtain the steady state temperature distribution in the plate. Also calculate and print out (a) the total heat generation G in the domain. (b) the total heat loss Q from all the surfaces of the plate, and (c) some quantity obtained from G and Q such that it is zero (or very small) value would indicate perfect overall heat balance.

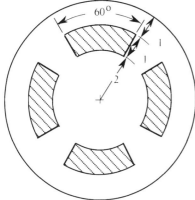

Problem 8.20

8.21 A solid cylinder of diameter 3 and uniform conductivity $k = 2$ has a heat generation $S = 500$ taking place only in the inner core of diameter 1. One half of the outer surface is insulated while the other half loses heat to a surrounding fluid at $T_\infty = 15$. The heat transfer coefficient (which may result from natural convection on the outer surface) is dependent on the local temperature T_s of the outer surface such that $h = 2.5(T_s - T_\infty)^{1/4}$. Prepare an ADAPT and calculate the steady-state temperature distribution. Arrange to printout, in addition to the temperature field, the local heat flux at all the grid points on the outer surface. Also provide a print out of the overall heat balance. Use proper linearization in handling the nonlinear boundary condition at the outer surface.

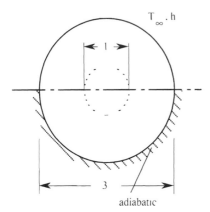

Problem 8.21

8.22 A circular rod of diameter 2 has a semicircular fin of outer radius 3.5 and thickness 0.3 attached to it. The surface of the rod is maintained at a constant temperature of 250. The fin loses heat to the surrounding fluid at $T_\infty = 27$ with a convective heat transfer coefficient of 12. This heat loss from the fin occurs at the flat edges 1 and 2, the semicircular edge 3 and the top and bottom surfaces 4 and 5. The conductivity of the fin material is 3.7. The temperature variation over the fin thickness can be regarded as negligible. Prepare the ADAPT routine to obtain the steady-state temperature distribution in the fin. Arrange the printout of total heat loss by the fin.

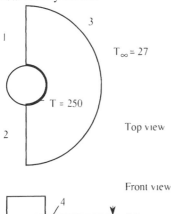

Problem 8.22

8.23 Consider a cylindrical fin of conductivity 2.5. diameter 5, and length 20. The base of the fin is at a temperature T_0. The end of the fin is adiabatic. The lateral surface of the fin loses heat by convection to a surrounding fluid at T_∞ with a heat transfer coefficient h. Use the following values: $T_0 = 100.$, $T_\infty = 20$, h = 0.35. Calculate the steady state temperature distribution in the fin. Find the total heat loss from the lateral surface of the fin. Also find the total heat flow into the fin at its base. Show that these quantities are *exactly* equal.

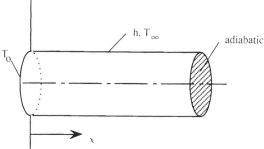

Problem 8.23

8.24 For the situation in Problem 8.23, consider that the fin was initially at a temperature of 100 throughout. With the same boundary conditions, calculate the unsteady temperature distribution in the fin until it reaches the steady state.

8.25 See the article by Lienhard (1981) on the Yin-Yang bodies. Solve at least two steady heat conduction problems, one in MODE = 1 and the other in MODE = 3, by specifying any Yin-Yang figure of your choice. (As far as I can see, the Lienhard theorem does not apply to the MODE = 2 geometry.) Calculate the total heat transfer in each problem and thus verify the Lienhard theorem. In general, a coarse-grid solution may not show a good agreement with the theorem. Demonstrate that, as the grid is refined, the numerical solution agrees more closely with the theorem.

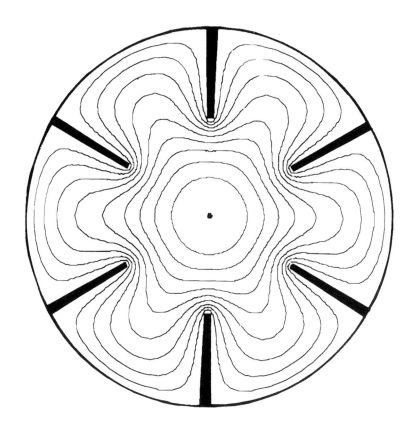

Contours of axial velocity in a tube with radial internal fins

Chapter 9

Flow and Heat Transfer in Ducts

As already mentioned, the velocity and temperature fields in a fully developed duct flow are governed by the general conduction-type equation. Eq. (3.6). Therefore, the program CONDUCT can be readily used for the analysis of fully developed flow and heat transfer in ducts. In this chapter, we shall discuss the main concepts and the governing equations for duct flows. In Chapter 10, you will find a number of adaptations of CONDUCT to duct flow problems.

9.1   General Characteristics of Duct Flows

Although ducts can have any geometry, let us consider, for convenience, a simple straight duct of rectangular cross section as shown in Fig. 9.1. The flow is primarily in the z direction (known as the axial or streamwise coordinate). We shall refer to x and y as the cross-stream coordinates. Out of the three velocity components u, v, and w, the streamwise velocity w is usually much larger than the cross-stream velocities u and v. The duct flow is driven by the pressure gradient $\partial p / \partial z$, which is usually negative. The pressure is nearly uniform over a cross section and varies mainly in the z direction.

Because the gradients of velocity and temperature in the z direction are usually small compared to the gradients in the x or y directions, it is customary to neglect the viscous stresses caused by $\partial w / \partial z$ and the heat conduction due to $\partial T / \partial z$. It is also common to replace the pressure gradient $\partial p / \partial z$ in the streamwise momentum equation by $d \bar{p} / dz$, where $\bar{p}$ represents the mean pressure over a cross section.

The flow in a duct can be steady or unsteady. We shall restrict attention to only steady flows.

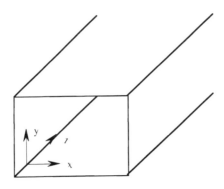

Fig. 9.1 Flow through a rectangular duct

9.2   Developing and Fully Developed Duct Flows

Let us consider the flow shown in Fig. 9.2, where the fluid enters a duct with a uniform velocity profile. As the fluid moves through the duct, the flow close to the duct walls slows down due to the friction at the walls. The flow velocity in the central core remains uniform in the cross-stream direction but increases in the streamwise direction. There is, in general, a migration of the fluid from the near-wall region to the central core. The boundary layers on the duct walls gradually grow and eventually merge with one another. At this stage, the central core vanishes and nonzero velocity gradients prevail over the entire cross section. By now, the adjustment of the velocity profile is complete, and the flow field for all downstream locations becomes independent of the streamwise coordinate z. Also, the pressure gradient d$\bar{p}$/dz becomes a constant.

This region of the duct flow, in which the velocity components do not depend on the streamwise location, is known as the *hydrodynamically fully developed* region. The region on the upstream side, in which the velocity distribution adjusts to the duct geometry and the wall friction, is known as the *developing* region or the *entrance* region.

The flow in a rectangular duct is, in general, a three-dimensional problem. However, the flow in the fully developed region presents only a two-dimensional problem since variations in the z direction are absent. (For two-dimensional duct flows in geometries like a circular tube or a parallel-plate channel, the fully developed flow becomes a one-dimensional problem.) Further, we can solve for the fully developed region directly without first obtaining the entrance-region solution. Thus, by restricting our attention to the fully developed flow, we effectively reduce the dimensionality of the problem and benefit from the resulting computational simplification.

In many practical applications, the length of the duct is very large in comparison with the cross-sectional dimensions. The length of the entrance region then represents only a small fraction of the total duct length. In such cases, it is

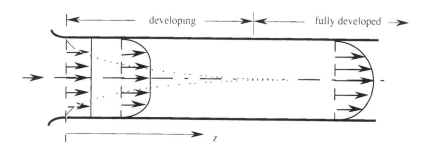

Fig. 9.2 Developing and fully developed flow

reasonable to design the entire duct on the basis of the fully developed behavior. The analysis of fully developed flows is. therefore. of considerable practical utility.

Let us consider what happens to the cross-stream velocity components in the fully developed region. For the situation shown in Fig. 9.2. the flow becomes purely axial in the fully developed region and the cross-stream velocities u and v equal to zero. There are. however. more complicated duct flows in which nonzero cross-stream velocities prevail even in the fully developed region. Examples of such behavior are found in curved ducts. rotating ducts. flows with buoyancy-induced secondary motion. and some turbulent flows in ducts of noncircular cross section. Depending on whether the cross-stream velocities are zero or nonzero, the fully developed duct flows can be classified as simple or complex.

A *complex fully developed duct flow* is characterized by:

$$u = u(x,y) \tag{9.1}$$

$$v = v(x,y) \tag{9.2}$$

$$w = w(x,y) \tag{9.3}$$

Note that the velocities u. v. and w are all independent of the streamwise coordinate z and that u and v can be nonzero as long as they do not depend on z.

A *simple fully developed duct flow* is defined by:

$$u = 0 \tag{9.4}$$

$$v = 0 \tag{9.5}$$

$$w = w(x,y) \tag{9.6}$$

A further characteristic of the simple fully developed duct flow is that the pressure is exactly uniform over a cross section and varies linearly in the streamwise direction. Thus,

$$p = p(z) \tag{9.7}$$

$$\frac{dp}{dz} = \text{constant} \tag{9.8}$$

As we shall see very shortly. only the *simple* fully developed duct flows are governed by conduction-type equations and therefore fall within the scope of CONDUCT. The *complex* fully developed duct flows do provide computational simplification by reducing the dimensionality; but the presence of the cross-stream velocities requires that the convection terms be included in the governing differential equations. Further, the calculation of these velocities must be done via the solution of the coupled momentum and continuity equations over the duct cross section—a task that is far too complicated to be included within the scope of this introductory book.

A *simple* fully developed duct flow usually occurs downstream of the entrance region in a straight duct of uniform cross section. For the flow to remain fully developed, the fluid properties such as viscosity and density (and thermal conductivity and specific heat for our later consideration of fully developed heat transfer) must normally remain constant. As to the geometry, the duct cross section can be represented in xy or θr coordinates (i.e., $MODE = 1$ or 3). The axisymmetric xr geometry ($MODE = 2$) has no role in analyzing duct flows. It can nicely represent a *curved* duct of rectangular cross section; but that would lead to a *complex* duct flow, which is outside the scope of CONDUCT.

9.3   Mathematical Formulation of the Velocity Field

9.3-1   The Governing Equation

For a steady, laminar, three-dimensional flow in a duct, the momentum equation in the z direction can be written as:

$$\rho u \frac{\partial w}{\partial x} + \rho v \frac{\partial w}{\partial y} + \rho w \frac{\partial w}{\partial z} = \frac{\partial}{\partial x} \left(\mu \frac{\partial w}{\partial x} \right) + \frac{\partial}{\partial y} \left(\mu \frac{\partial w}{\partial y} \right) + \frac{\partial}{\partial z} \left(\mu \frac{\partial w}{\partial z} \right) - \frac{\partial p}{\partial z} \qquad (9.9)$$

where the terms on the left-hand side represent the convection of momentum in the three directions, while the right-hand side stands for the viscous stress and pressure gradient. For the simple fully developed duct flow considered here, Eqs. (9.4)–(9.8) apply. As a result, the left-hand side of Eq. (9.9) vanishes, and the equation reduces to:

$$0 = \frac{\partial}{\partial x} \left(\mu \frac{\partial w}{\partial x} \right) + \frac{\partial}{\partial y} \left(\mu \frac{\partial w}{\partial y} \right) - \frac{dp}{dz} \qquad (9.10)$$

Physically, this equation implies a balance between the pressure force and the viscous stresses caused by the cross-stream gradients of w. It is interesting to observe that the fluid density does not appear in Eq. (9.10). Note also that dp/dz is simply a constant for the entire cross section. Equation (9.10) can be seen to conform to the steady-state form of Eq. (3.6) with the following choices:

$$\phi = w \qquad (9.11)$$

$$\Gamma = \mu \qquad (9.12)$$

$$S = -\frac{dp}{dz} \tag{9.13}$$

In other words, the computation of the axial velocity field is mathematically identical to solving a heat conduction problem with a uniform heat generation rate.

9.3-2 Dimensionless Form

To identify the parameters governing the velocity field, it is useful to cast Eq. (9.10) into a dimensionless form. For constant viscosity, a suitable form is:

$$0 = \frac{\partial}{\partial X}\left(\frac{\partial W}{\partial X}\right) + \frac{\partial}{\partial Y}\left(\frac{\partial W}{\partial Y}\right) + 1 \tag{9.14}$$

where the dimensionless coordinates X and Y are defined as:

$$X = \frac{x}{D} \tag{9.15}$$

$$Y = \frac{y}{D} \tag{9.16}$$

Here D is some characteristic length related to the duct cross section, such as the duct diameter or radius for a circular duct, one of the sides for a rectangular duct, or the hydraulic diameter (to be defined later in Eq. (9.18) in this chapter). The dimensionless velocity W is defined as :

$$W = \frac{\mu w}{-(dp/dz)\, D^2} \tag{9.17}$$

In a duct flow, the pressure p decreases with z. The quantity $(-dp/dz)$ is, therefore, a positive number.

Since no parameters appear in Eq. (9.14), its solution depends only on the boundary conditions. These are provided by the geometry of the duct and the requirement that the velocity w equals zero on the duct walls. The solution for W thus depends only on the shape of the duct (not its physical size) and is independent of the actual values of the viscosity, the pressure gradient, the Reynolds number, etc. This distribution of the dimensionless velocity W can be obtained only from the knowledge of the shape of the duct cross section. The dimensional velocity w can then be obtained from Eq. (9.17) by substituting the actual values of viscosity, pressure gradient, and characteristic length.

9.4 Presentation of the Overall Flow Characteristics

It is customary to characterize a duct flow by some special quantities and parameters. We shall look at some of them here.

Some useful definitions. The hydraulic diameter D_h of a duct cross section is defined as:

$$D_h = \frac{4\,A}{P} \tag{9.18}$$

where A is the cross-sectional area and P is the wetted perimeter (i.e.. the length of the perimeter over which the fluid is in contact with the duct walls). The number 4 in Eq. (9.18) is used so as to make the hydraulic diameter D_h for a circular tube equal to the actual diameter of the tube. Also. for a duct of square cross section, D_h equals the length of the side of the square.

The mean velocity $\overline{w}$ over the duct cross section is defined such that the volumetric flow through the duct is given by $\overline{w}A$. Thus.

$$\overline{w} = \frac{\iint w \, dx \, dy}{A} = \frac{\iint w \, dx \, dy}{\iint dx \, dy} \tag{9.19}$$

Here the double integrals are evaluated over the duct cross section. The Reynolds number Re of the duct flow is usually defined as:

$$Re = \frac{\rho \overline{w} D_h}{\mu} \tag{9.20}$$

where ρ and μ stand for the density and viscosity of the fluid.

A dimensionless measure of the pressure gradient is the friction factor f, which is defined in three or four different ways. The definition that we shall use in this book is:

$$f = \frac{-\,(dp/dz)\,D_h}{(1/2)\rho \overline{w}^{\,2}} \tag{9.21}$$

Because many other definitions exist, you should be careful in comparing the values of f obtained from Eq. (9.21) with the values available from other sources. Other common definitions give the f values that are 0.5 or 0.25 times the f values given by Eq. (9.21).

Implications for the fully developed flow. By combining Eqs. (9.17), (9.20), (9.21), we can derive that:

$$fRe = \frac{2\,(D_h/D)^2}{\overline{W}} \tag{9.22}$$

where $\overline{W}$ is related to $\overline{w}$ according to Eq. (9.17). Alternatively. after W has been obtained as a function of X and Y. $\overline{W}$ can be calculated from a formula similar to Eq. (9.19). Since $\overline{W}$ simply depends on the duct shape. and so does the ratio D_h/D. we can conclude from Eq. (9.22) that for a fully developed laminar flow. the product fRe for a duct of given shape is a constant. The value of fRe for the circular pipe is 64: for a parallel-plate channel. fRe equals 96.

Although W is the appropriate dimensionless velocity for Eq. (9.14). another useful dimensionless quantity is $w/\overline{w}$. Again. for a duct of given shape. there is a universal distribution of $w/\overline{w}$.

Significance of overall parameters. The computational task of obtaining the velocity field in a duct ends when Eq. (9.10) or (9.14) is solved. The calculation of $\overline{W}$ or fRe simply represents the post-processing of the results. You should understand that such post-processing is not essential to the primary task of obtaining the velocity field. It is done simply to present the overall results in a form that is customary and familiar to practicing designers. Even in the established practice. different definitions of the friction factor and related quantities are quite common. No particular definition should be regarded as more correct or superior to others. As long as proper consistency is maintained between the definition and the computed value. any form of presenting the overall results is valid.

9.5 Fully Developed Heat Transfer

A preliminary examination. Just as the velocity field in a duct exhibits the developing and fully developed regions. the temperature field undergoes similar transformation. Figure 9.3 shows the temperature profiles at various cross sections in a duct. The fluid enters the duct with a uniform inlet temperature. Subsequently. thermal boundary layers form and grow on the duct walls. Ultimately. the temperature variation in the cross section is fully established and ceases to change with the coordinate z. This region where the temperature $T = T(x,y)$ and $\partial T/\partial z = 0$

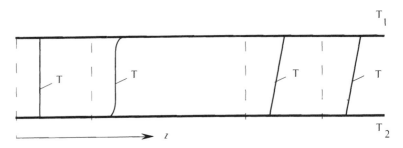

Fig. 9.3 Devloping and fully developed temperature field

may be considered as a thermally fully developed region; however, as explained in the next paragraph, such a definition would restrict the analysis of fully developed heat transfer to rather uninteresting cases.

In the situation shown in Fig. 9.3, the fluid receives heat from the top wall and loses it to the bottom wall. In the fully developed region, the fluid itself does not experience any net gain or loss of heat. (That is why the temperature T does not change in the z direction.) The motion of the fluid has absolutely no influence on the temperature distribution in the fully developed region. The problem reduces to a simple steady heat conduction problem in the cross section without any heat sources. Although such a problem can easily be solved by using CONDUCT, it does not represent a practical duct flow situation, in which the goal is to heat or cool the fluid.

If the duct fluid is to undergo a net heat loss or gain, the temperature T must change with the coordinate z. Even then, it is possible to define a particular kind of fully developed behavior for the temperature field.

Thermally fully developed region. We shall consider the temperature field fully developed when, although T continues to depend on x, y, and z, some dimensionless temperature Θ becomes independent of z. Thus, for a thermally fully developed region,

$$\Theta = \Theta\,(x,y) \tag{9.23}$$

This means that the *shape* of the temperature profile is the same at different streamwise locations. Alternatively, the thermally fully developed region is considered to be characterized by a constant heat transfer coefficient, a quantity that we shall define in Section 9.6-2.

Since the temperature field is influenced by the velocity field, an important prerequisite for a thermally fully developed region is that the velocity field be fully developed. Further, a certain degree of regularity is necessary in the thermal boundary conditions so that the resulting temperature field has an invariant shape. Some common thermal boundary conditions for the thermally developed region are discussed in Sections 9.6-3 to 9.6-6.

9.6   Mathematical Formulation of the Temperature Field

9.6-1   Differential Equation

The energy equation written for a steady, low-speed, duct flow without viscous dissipation is:

$$\rho c_p \left(u\,\frac{\partial T}{\partial x} + v\,\frac{\partial T}{\partial y} + w\,\frac{\partial T}{\partial z} \right) = \frac{\partial}{\partial x}\left(k\,\frac{\partial T}{\partial x} \right) + \frac{\partial}{\partial y}\left(k\,\frac{\partial T}{\partial y} \right) + \frac{\partial}{\partial z}\left(k\,\frac{\partial T}{\partial z} \right) \tag{9.24}$$

where c_p is the constant-pressure specific heat and k is the thermal conductivity of the fluid. The left-hand side of Eq. (9.24) represents the convection of enthalpy in the duct. while the terms on the right-hand side describe the heat conduction in the fluid. Since the conduction in the streamwise direction z is usually very small in comparison with the cross-stream conduction. it is common to omit the last term in Eq. (9.24). Further. for *simple* fully developed duct flows. the cross-stream velocities u and v are equal to zero. Therefore. Eq. (9.24) reduces to

$$\rho c_p w \frac{\partial T}{\partial z} = \frac{\partial}{\partial x} \left(k \frac{\partial T}{\partial x} \right) + \frac{\partial}{\partial y} \left(k \frac{\partial T}{\partial y} \right) \tag{9.25}$$

This conforms to the steady-state form of the general differential equation. Eq. (3.6). with the following choices:

$$\phi = T \tag{9.26}$$

$$\Gamma = k \tag{9.27}$$

$$S = -\rho c_p w \frac{\partial T}{\partial z} \tag{9.28}$$

Thus. the analysis of heat transfer in a duct can be treated as a conduction-like problem if the streamwise convection is formally considered as a source term. Of course. the problem can be solved only if the source term can be specified. This means that the gradient $\partial T/\partial z$ must be known or obtainable from related information. These conditions are met when the temperature profiles at different z locations exhibit some kind of similarity. Incidentally. the distribution of w in Eq. (9.28) is supposed to be known from the solution of the fully developed velocity field.

9.6-2 Some Useful Definitions

Let q_w stand for the local heat flux at the duct wall. This may. in general. be nonuniform over the heated perimeter of the duct. Further. let Q_w denote the total heat flow rate per unit axial length of the duct. It then follows that

$$Q_w = \int q_w \, ds \tag{9.29}$$

where s is the curvilinear distance along the perimeter of the duct cross section and the integral is taken over the entire heated perimeter.

The mean temperature of the fluid in a given cross section is usually taken as the so-called bulk temperature; it is defined as:

$$T_b = \frac{\iint \rho\, c_p w\, T\, dx\, dy}{\iint \rho\, c_p w\, dx\, dy} \tag{9.30}$$

The integrals here are taken over the duct cross section. For uniform values of ρ and c_p, Eq. (9.30) simplifies to

$$T_b = \frac{\iint wT\, dx\, dy}{\overline{w}\, A} \tag{9.31}$$

where $\overline{w}$ is the mean velocity and A is the cross-sectional area of the duct. A useful consequence of the definition of T_b is that the total rate of enthalpy flow through the duct cross section is given simply by $(\rho\overline{w}A)\, c_p T_b$.

The local heat transfer coefficient h at a point on the duct wall is defined as :

$$h = \frac{q_w}{T_w - T_b} \tag{9.32}$$

where T_w is the local wall temperature. It is also possible to base the heat transfer coefficient on an average wall temperature or any other representative wall temperature. Further, an average heat transfer coefficient $\overline{h}$ can be obtained as an average of the local values of h or based on the average heat flux q_w and the average wall temperature T_w. Whereas any of these definitions can be used, it is preferable to employ those definitions that make it easy to extract the physical quantities of interest.

The Nusselt number Nu is a dimensionless form of the heat transfer coefficient. It is defined as:

$$Nu = \frac{hD}{k} \tag{9.33}$$

where D is the characteristic dimension of the duct cross section and k is the thermal conductivity of the fluid. An average Nusselt number $\overline{Nu}$ is similarly based on $\overline{h}$. In interpreting the values of Nu or $\overline{Nu}$ reported in the literature, you should carefully examine the underlying definitions: otherwise, quite erroneous conclusions may be drawn. Again, remember that, for most part, there are no good or bad definitions: they are just different.

One way of characterizing the thermally fully developed region is to require that the heat transfer coefficient h or the Nusselt number Nu becomes independent of z. For laminar flow, the fully developed Nusselt number has a constant value that is independent of the Reynolds number and the Prandtl number: it depends only on the duct geometry and thermal boundary condition.

There are four commonly encountered thermal boundary conditions for which the thermally fully developed region exists. The mathematical characteristics of these boundary conditions are discussed in Sections 9.6-3 to 9.6-6.

9.6-3 Prescribed Local Heat Flux

If the distribution of the local heat flux q_w over the perimeter of the duct is specified and remains unchanged in the z direction, then in the thermally developed region, the temperatures at all locations in a cross section rise linearly with z at the same rate. Thus,

$$\frac{\partial T}{\partial z} = \frac{dT_w}{dz} = \frac{dT_b}{dz} = \text{constant} \tag{9.34}$$

In other words, the temperature distribution at one cross section can be obtained simply by adding a constant to the temperature field at another cross section.

An overall energy balance over the duct cross section requires that the increase in the enthalpy flow per unit axial length must be equal to the heat flow through the duct walls over that length. Thus,

$$Q_w = \rho \bar{w} A c_p \frac{dT_b}{dz} \tag{9.35}$$

where Q_w is the heat flow integrated over the duct perimeter, as defined by Eq. (9.29). In this manner the value of $\partial T/\partial z$ needed in Eq. (9.28) can now be obtained from Eq. (9.35) in terms of the known heat flow rate Q_w. Alternatively, if $\partial T/\partial z$ is known, Q_w can be calculated from Eq. (9.35).

To show that the dimensionless temperature distribution does not depend on any parameter such as the Reynolds number or the Prandtl number, we proceed as follows. The term on the left-hand side of Eq. (9.25) can now be written as:

$$\rho c_p w \frac{\partial T}{\partial z} = \left(\frac{w}{\bar{w}}\right) \rho c_p \bar{w} \frac{dT_b}{dz} \tag{9.36}$$

Using this equation, we can now cast Eq. (9.25) into a convenient dimensionless form as:

$$\frac{\partial^2 \Theta}{\partial X^2} + \frac{\partial^2 \Theta}{\partial Y^2} + \left(\frac{w}{\bar{w}}\right) = 0 \tag{9.37}$$

where the dimensionless cross-stream coordinates X and Y are given by Eqs. (9.15)–(9.16) as before, and the dimensionless temperature Θ is defined by:

$$\Theta = \frac{T_w - T}{(\overline{w}\, D^2/\alpha)\,(dT_b/dz)}$$ (9.38)

where α is the thermal diffusivity given by:

$$\alpha = \frac{k}{\rho c_p}$$ (9.39)

The temperature T_w is to be regarded as the wall temperature at some specific point on the duct perimeter.

Equation (9.37) conforms to the steady-state form of Eq. (3.6) with the source term given by $(w/\overline{w})$, which is known from the solution of the velocity field.

If an average Nusselt number $\overline{Nu}$ is defined as:

$$\overline{Nu} = \frac{D}{k\,(T_w - T_b)}\left(\frac{Q_w}{P}\right)$$ (9.40)

where P is the heated perimeter of the duct, then from Eqs. (9.35), (9.38) and (9.39) we can derive:

$$\overline{Nu} = \frac{1}{\Theta_b}\left(\frac{A}{PD}\right)$$ (9.41)

where Θ_b is dimensionless temperature corresponding to T_b. Since for a given duct shape and wall heat flux distribution, Eq. (9.38) has a universal solution, the Nusselt number as given by Eq. (9.41) is a constant that is independent of the Reynolds number and the Prandtl number. (The local Nusselt number will, in general, vary with position along the duct perimeter as a result of the duct geometry and the variation of the prescribed heat flux.)

Another useful dimensionless temperature based on T_w and T_b can be defined as $(T - T_w)/(T_b - T_w)$, which has some similarity with $w/\overline{w}$.

9.6-4   Axially Uniform Heat Flow and Peripherally Uniform Wall Temperature

If the duct wall has a high thermal conductivity, there will be peripheral conduction of heat so as to make the wall temperature uniform in a given cross section. In this case, the local variation of the heat flux q_w is not known, but a constant value of the heat flow Q_w per unit axial length is prescribed.

The analysis of this case is very similar to the case considered in the previous subsection. Equations (9.34) to (9.41) continue to apply except that now T_w is the uniform wall temperature.

9.6-5 Axially and Peripherally Uniform Wall Temperature

A thermally fully developed region of a somewhat different kind is formed when the duct wall has a constant wall temperature at all streamwise locations. In such a situation, the fluid continuously increases or decreases in temperature until it attains the temperature of the wall. In the cases considered in Sections 9.6-3 and 9.6-4, the temperature difference $(T_w - T_b)$ and the associated heat flow Q_w remain constant. In the case of constant wall temperature, the temperature difference and the heat flow decay exponentially in the z direction.

The thermally fully developed region is characterized by temperature profiles of similar shape: i.e., the ratio $(T_w - T)/(T_w - T_b)$ becomes independent of z. In other words, although the temperature difference $(T_w - T)$ decays with z, the difference $(T_w - T_b)$ decays in the same proportion. Since:

$$\frac{T_w - T}{T_w - T_b} = f(x,y) \tag{9.42}$$

it follows that:

$$\ln(T_w - T) - \ln(T_w - T_b) = \ln |f(x,y)| \tag{9.43}$$

Differentiating with respect to z, we get:

$$\frac{1}{T_w - T}\left(\frac{\partial T}{\partial z}\right) = \frac{1}{T_w - T_b}\left(\frac{dT_b}{dz}\right) \tag{9.44}$$

Thus,

$$\frac{\partial T}{\partial z} = \left(\frac{T_w - T}{T_w - T_b}\right)\frac{dT_b}{dz} \tag{9.45}$$

In this manner if dT_b/dz is known, the local $\partial T/\partial z$ can be related to it by Eq. (9.45). The relationship given by Eq. (9.35) between dT_b/dz and Q_w continues to be applicable here; therefore, if Q_w is known, dT_b/dz can be evaluated.

To solve Eq. (9.25), the source-term expression (9.28) can now be written as

$$S = -\rho c_p w \left(\frac{T_w - T}{T_w - T_b}\right)\frac{dT_b}{dz} \tag{9.46}$$

Here, because of the presence of T in the source term, we may be tempted to use the source-term linearization with:

$$S_C = -\frac{\rho c_p w T_w}{T_w - T_b}\frac{dT_b}{dz} \tag{9.47}$$

and

$$S_P = \frac{\rho c_p w}{T_w - T_b}\frac{dT_b}{dz} \tag{9.48}$$

However, this produces a positive S_P, since when $(T_w - T_b) > 0$, $dT_b/dz > 0$. As explained earlier, the use of a positive S_P is considered undesirable for a satisfactory convergence of the iteration process. Therefore, the recommended treatment of the source term in this case is:

$$S_C = -\rho c_p w \left(\frac{T_w - T}{T_w - T_b} \right) \frac{dT_b}{dz}$$

and
$$(9.49)$$

$$S_P = 0 \qquad\qquad (9.50)$$

Of course, S_C is recalculated at each iteration from the values of T and T_b available from the previous iteration. Fortunately, this iterative process converges very quickly. The reason is that the ratio $(T_w-T)/(T_w-T_b)$ depends only on the shape of the temperature distribution and not on the magnitude of T. Therefore, a computation started even from an arbitrary guess for T converges quite rapidly.

An appropriate dimensionless form of Eq. (9.25) can be written as:

$$\frac{\partial^2 \Theta}{\partial X^2} + \frac{\partial^2 \Theta}{\partial Y^2} + \frac{w}{\overline{w}} \left(\frac{\Theta}{\Theta_b} \right) = 0 \qquad\qquad (9.51)$$

where Θ is still defined by Eq. (9.38). We should note that, since $(T_w - T)$ and (dT_b/dz) both decay in the z direction, the dimensionless temperature Θ remains independent of z. The boundary condition for Eq. (9.51) is supplied by the requirement that Θ equals zero at the duct wall.

The Nusselt number $\overline{Nu}$ for the constant-wall-temperature case is still given by Eq. (9.41). It then follows that the Nusselt number depends only on the shape of the duct cross section and is independent of the other parameters.

9.6-6 Uniform External Heat Transfer Coefficient

An extension of the constant-wall-temperature boundary condition occurs when the duct passes through an environment at a uniform temperature T_∞ and exchanges heat with the external fluid such that:

$$q_w = h_e(T_\infty - T_w) \qquad\qquad (9.52)$$

where h_e is the constant heat transfer coefficient prescribed at the external surface of the duct. The formulation for this case is very similar to the formulation in Section 9.6-5. The only difference is that the temperature T_∞ should be used in place of T_w. (The temperature T_w does not remain constant; it exponentially approaches the temperature T_∞.) Thus, the dimensionless temperature Θ is defined as:

$$\Theta = \frac{T_\infty - T}{(\overline{w} D^2/\alpha)(dT_b/dz)} \tag{9.53}$$

which obeys Eq. (9.51).

The thermal boundary condition specified in terms of an external convective heat transfer coefficient and a constant temperature of the surrounding fluid can be considered a more general boundary condition, from which more restrictive boundary conditions can be derived. For example, if the external heat transfer coefficient is very large, then the wall temperature T_w becomes very nearly equal to T_∞ and we get the boundary condition of constant wall temperature. For a very small heat transfer coefficient, the difference $(T_\infty - T_w)$ becomes much greater than the temperature difference within the duct; then the constant heat flux boundary condition is approached. A more detailed discussion of this behavior can be found in Sparrow and Patankar (1977).

9.6-7   More Complex Boundary Conditions

We can construct more complex thermal boundary conditions by requiring that the stated boundary condition (of prescribed heat flux or wall temperature) applies over only a portion of the duct perimeter, the remaining portion being adiabatic. The formulations worked out so far continue to be applicable when such inactive parts (having zero heat flux) are present along the duct perimeter.

However, if we have a prescribed (nonzero) heat flux over some portion of the duct perimeter and a prescribed wall temperature over the rest, the concept of the thermally fully developed region must be examined again. Such boundary conditions would actually lead to the rather uninteresting physical situations as described in the beginning of Section 9.5. In the fully developed condition, the prescribed heat flow will enter the duct fluid through a part of the boundary and leave the duct through the boundary regions where the wall temperature is specified. The temperature field in the duct will not change with the axial coordinate z. Thus, the fluid will experience no net gain or loss of heat. Computationally, the problem reduces to a pure conduction problem, with the fluid motion playing no part whatsoever.

9.7   Introduction to Duct Flow Adaptations

Using the background from this chapter, we shall apply CONDUCT to four duct flow problems in Chapter 10. Additionally, application to some advanced duct flows will be illustrated in Chapter 11. Here we shall consider some of the common features of these adaptations.

Fluid properties and physical parameters. For a given duct shape and thermal boundary condition, our objective would be to find the dimensionless solution, which is independent of the values of viscosity, conductivity, pressure gradient, heat flux, etc. Therefore, we can give any arbitrary values to all these physical quantities. Our practice would be to use simple numbers such as $\mu = 1$, $k = 1$, etc. However, it can be demonstrated that the *dimensionless* solution does not change if these values are changed to some other numbers such as $\mu = 27.9$, $k = 0.458$, etc.

If a boundary temperature T_w is given, its value too can be set arbitrarily. However, it is better to set this value as zero, rather than nonzero. The reason is that we may not know the magnitude of the temperature difference that will be produced over the duct cross section. If we set $T_w = 100$, we may get all interior temperatures as 99.997, 99.998, etc. and lose accuracy in calculating the temperature differences. In absence of any information about the temperature drop, setting $T_w = 0$ would be the best choice.

Sequence of computation. For the duct flow problems in Chapter 10, we shall use $NF = 1$ for the velocity w and $NF = 2$ for the temperature T. We shall first solve for the velocity w, and then use the resulting w field for the source term in the temperature equation. Since the velocity equation is linear, it can be solved in one iteration. Still according to our practice for linear problems, we shall perform three iterations for w. During these three iterations, only KSOLVE(1) will be nonzero. At the end of three iterations, we shall set KSOLVE(1) = 0 and KSOLVE(2) = 1. In this manner the temperature equation will be switched on. For the boundary conditions in Sections 9.6-3 and 9.6-4, the temperature equation is linear; therefore, only three iterations will be performed. If we use the boundary conditions in Sections 9.6-5 and 9.6-6, the nonlinear nature of the temperature equation would require a number of iterations until it converges.

When the boundary heat flux is specified, the heat flow rate Q_w and the gradient dT_b/dz are related by Eq. (9.35); their values cannot be arbitrarily specified. If you set Q_w and dT_b/dz independently, overall heat balance will not be satisfied and you will not get a converged solution.

Overall results. In duct flow problems, it is customary to calculate the values of fRe and Nu. Although we shall include these in our example problems, you need not consider the evaluation of these quantities as essential to duct flow computation. Also, we shall use a certain definition of the Nusselt number in each problem; but other definitions would be alright too.

The calculation of these quantities is based on $\overline{w}$ and T_b, for which we use Eqs. (9.19) and (9.31). If w and T are stored as $W(I,J)$ and $T(I,J)$, the calculation of $\overline{w}$ and T_b will be done as follows.

```
       ASUM=0.
       WSUM=0.
       TSUM=0.
       DO 200 J=2,M2
       DO 200 I=2,L2
       AR=XCV(I)*YCVR(J)
       ASUM=ASUM+AR
       WSUM=WSUM+W(I,J)*AR
       TSUM=TSUM=TSUM+W(I,J)*T(I,J)*AR
   200 CONTINUE
       WBAR=WSUM/ASUM
       TB=TSUM/(WSUM+SMALL)
```

Here ASUM is the cross-sectional area A, WBAR is $\overline{w}$, and TB is T_b. The quantity WSUM gives the volumetric flow rate through the duct. AR is the part of the cross-sectional area occupied by one control volume. The expression used for AR is correct for MODE = 1 and 3. (MODE = 2 is not used for simple fully developed duct flows.) In calculating TB, we add the quantity SMALL in the denominator to avoid a possible division by zero.

Representation of fins. The duct geometries will often have fins and other solid obstacles. These will be treated by the general method outlined in Section 7.7. When the fins are attached to the outer boundary, the velocity field can be handled by the use of a large value of $GAM(I,J)$. When a solid object forms an island, the velocity at (at least) one grid point on the object must be set equal to zero through large S_C and S_P, in addition to making $GAM(I,J)$ large in the solid.

Conjugate heat transfer in the fin and the duct fluid can be calculated quite easily. All that is necessary is to set the values of $GAM(I,J)$ in the solid and fluid regions equal to the respective thermal conductivities of the two regions.

In such situations, if we need to obtain the temperature at the surface of the fin, Eq. (2.81) should be used.

9.8 Final Remarks

We have seen that the restriction to the fully developed flow enables us to reduce the dimensionality of the problem and facilitates the computational task. The simple fully developed flows are governed by conduction-type equations. The solution of the velocity field involves the use of a constant pressure gradient as the source term. For certain thermal boundary conditions, a thermally fully developed region exists, in which the temperature profiles exhibit some kind of similarity. The streamwise convection term can then be treated as a source term, which depends on the velocity distribution in the duct cross section.

The pressure drop and heat flow information obtained from the analysis of fully developed duct flows can be directly used in the design of long ducts, in which a large proportion of the duct length can be assumed to be under the fully developed conditions.

For additional information on flow and heat transfer in ducts, you may refer to Kays and Crawford (1980) and Shah and London (1978).

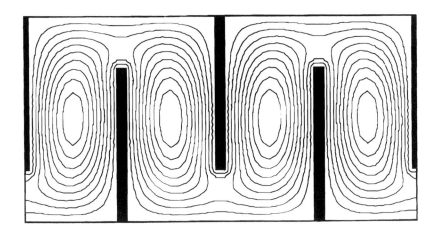

Fluid temperature contours in an axial flow through a staggered fin array

Chapter 10

Adaptation Examples: Duct Flow and Associated Heat Transfer

An interesting and important class of the applications of CONDUCT is the fully developed flow and heat transfer in ducts. In this chapter, we implement the theory developed in Chapter 9 to analyze different duct geometries and thermal boundary conditions.

10.1 Rectangular Duct with One Heated Wall (Example 7)

10.1-1 Problem Description

Here we consider the rectangular duct shown in Fig. 10.1. The thermal boundary conditions consist of the uniform heat flux q_w at the bottom wall and a zero heat flux at the three remaining walls of the duct. Because of symmetry, we shall use the left half of the duct as our calculation domain. (For the velocity field, we could have used the symmetry about the horizontal centerline also and limited our calculation to one-fourth of the duct cross section. However, the given thermal boundary condition would not make the temperature field symmetrical about the horizontal centerline.)

Our task is to obtain the velocity and temperature distributions over the duct cross section. We shall also calculate the product fRe, an overall Nusselt number, and local Nusselt numbers along the heated wall.

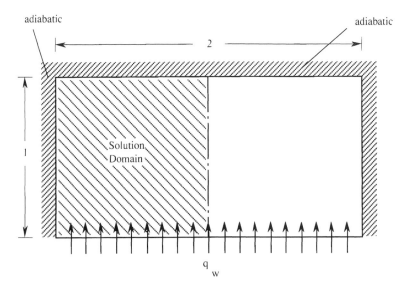

Fig. 10.1 Rectangular duct with one heated wall

10.1-2   Design of ADAPT

GRID. Here we construct a nonuniform grid such that the spacing is fine near the duct walls. In the x direction, this can be done through only one zone with a fine spacing at the left end (by setting POWRX(1) = 1.2). In the y direction, we define two zones; if we set the values of POWRY(1) and POWRY(2) as 1.2 and −1.2 respectively, we can get a fine grid near the top and bottom walls.

BEGIN. Here we use NF = 1 and 2 for the *dimensional* velocity and temperature respectively, and NF = 3 and 4 for the corresponding *dimensionless* variables. The problem is linear for both w and T; we perform the first 3 iterations for the solution of w and the next 3 iterations for T. Therefore, initially, only KSOLVE(1) is set equal to 1.

The fluid properties, pressure gradient, and heat flux are given arbitrary values. As to the starting values of w and T, the zero default values are considered acceptable. They give the desired velocities at the wall boundaries and serve as satisfactory guess for the temperature.

OUTPUT. After the first three iterations, we wish to switch off the solution of w and switch on the solution of T. This is arranged at the beginning of OUTPUT. Then we obtain overall quantities such as $\overline{w}$, T_b, fRe, Nu, etc. For the definition of the overall Nusselt number, an average temperature of the heated wall is used. This quantity TWAV is calculated by a line integration of the local wall temperature over the heated wall. The hydraulic diameter is used as the length scale for defining the Nusselt number.

After every iteration, typical values of w and T, and the values of fRe and Nu are printed out. The final printout consists of the values of local Nusselt numbers along the bottom wall and the two-dimensional fields of dimensionless w and T. These dimensionless values are calculated and stored as F(I , J , 3) and F(I , J , 4) before calling PRINT.

PHI. Here, since we have two dependent variables w and T, we fill the arrays GAM(I , J) and SC(I , J) separately for the two variables. For the velocity w, the diffusion coefficient Γ equals the viscosity μ, and the source term equals −dp/dz. For the temperature T, we set Γ equal to the fluid conductivity k, while the source term is given by

$$S_C = -\rho\, c_p\, w\, \frac{\partial T}{\partial z} \tag{10.1}$$

where ∂T/∂z is related to the prescribed heat flux q_w by Eqs. (9.29) and (9.35).

It is very important to maintain consistency between the prescribed heat fluxes and the corresponding ∂T/∂z. The total heat input at the boundaries must exactly cancel the integrated source term over the whole domain. Otherwise, no steady-state solution can be reached.

In specifying the boundary conditions, we set $KBCL1(J) = 2$ to represent the right boundary as a symmetry boundary (zero flux) for both w and T. No further boundary-condition information is needed for w since the velocity is known (as zero) on the three remaining boundaries. For these boundaries, the known-flux conditions are specified for the temperature only, via appropriate values of KBC and FLXC.

10.1-3   Additional Fortran Names

AMU	viscosity μ
ANU	average Nusselt number Nu
ANULOC	local Nusselt number along the bottom wall
AR	area of control volume, dA
ASUM	cross-sectional area
COND	thermal conductivity k
CP	specific heat c_p
DEN	density ρ
DH	hydraulic diameter D_h
DPDZ	axial pressure gradient dp/dz
DTDZ	axial temperature gradient $\partial T/\partial z$
FRE	the product fRe
QW	wall heat flux q_w
RE	Reynolds number
RHOCP	heat capacity ρc_p
T(I,J)	temperature T
TB	bulk temperature T_b
TSUM	$\int w \, T \, dA$
TWAV	average temperature of the heated wall
W(I,J)	axial velocity w
WBAR	cross-sectional mean velocity, $\overline{w}$
WP	wetted perimeter
WSUM	$\int w \, dA$

10.1-4   Listing of ADAPT for Example 7

```
CCCCCCCCCCCCCCCCCCCCCCCCCCCCCCCCCCCCCCCCCCCCCCCCCCCCCCCCCCCCCCCCCCCCCCC
      SUBROUTINE ADAPT
C..................................................................
C------EXAMPLE 7 -- RECTANGULAR DUCT WITH ONE HEATED WALL
C..................................................................
```

```
$INCLUDE:'COMMON'
C***********************************************************************
      DIMENSION W(NI,NJ),T(NI,NJ)
      EQUIVALENCE (F(1,1,1),W(1,1)),(F(1,1,2),T(1,1))
C*.*.*-*-*.*.*.*.*.*.*-*.*.*.*.*.*.*.*.*.*-*-*.*.*.*.*.*-*.*.*.*.*
      ENTRY GRID
      HEADER='RECTANGULAR DUCT WITH ONE HEATED WALL'
      PRINTF='PRINT7'
      PLOTF='PLOT7'
      CALL INTA2(NZX,1,NCVX(1),5)
      CALL DATA2(XZONE(1),1.,POWRX(1),1.2)
      CALL INTA3(NZY,2,NCVY(1),5,NCVY(2),5)
      CALL DATA4(YZONE(1),0.5,YZONE(2),0.5,POWRY(1),1.2,POWRY(2),-1.2)
      CALL ZGRID
      RETURN
C*.*.*.*-*.*.*.*.*.*-*.*.*.*.*.*.*.*.*.*-*.*-*-*-*.*.*.*.*-*.*.*.*.*.*
      ENTRY BEGIN
      TITLE(1)=' AXIAL VELOCITY W '
      TITLE(2)='    TEMPERATURE    '
      TITLE(3)='      W/WBAR       '
      TITLE(4)='(T-TWAV)/(TB-TWAV)'
      CALL INTA4(KSOLVE(1),1,KPLOT(3),1,KPLOT(4),1,LAST,6)
      DO 100 N=1,4
  100 KPRINT(N)=1
      CALL DATA6(AMU,1.,COND,1.,CP,1.,DEN,1.,DPDZ,-1.,QW,1.)
      RHOCP=DEN*CP
C
C-- SINCE THE ZERO DEFAULT VALUES OF W(I,J) AND T(I,J) ARE SATISFACTORY,
C   THESE ARRAYS ARE NOT FILLED HERE.
      RETURN
C*.*.*.*.*.*.*.*.*.*.*.*.*.*.*.*.*.*.*.*.*.*.*.*.*.*.*.*.*.*.*.*.*.*.*
      ENTRY OUTPUT
      IF(ITER.EQ.3) THEN
         KSOLVE(1)=0
         KSOLVE(2)=1
      ENDIF
      ASUM=0.
      WSUM=0.
      TSUM=0.
      DO 200 J=2,M2
      DO 200 I=2,L2
         AR=XCV(I)*YCV(J)
```

```
            ASUM=ASUM+AR
            WSUM=WSUM+W(I,J)*AR
            TSUM=TSUM+W(I,J)*T(I,J)*AR
  200 CONTINUE
        WBAR=WSUM/ASUM
        TB=TSUM/(WSUM+SMALL)
        WP=2.*X(L1)+Y(M1)
        DH=4.*ASUM/WP
        RE=DH*WBAR*DEN/AMU
        FRE=-2.*DPDZ*DH/(DEN*WBAR**2+SMALL)*RE
        TWAV=0.
        DO 210 I=2,L2
            TWAV=TWAV+XCV(I)*T(I,1)
  210 CONTINUE
        TWAV=TWAV/X(L1)
        ANU=QW*DH/(COND*(TWAV-TB)+SMALL)
        DO 220 IUNIT=IU1,IU2
            IF(ITER.EQ.0) WRITE(IUNIT,230)
  230       FORMAT(2X,'ITER',4X,'W(5,8)',4X,'W(3,7)',4X,'T(5,8)',
     1      4X,'T(3,7)',5X,'FRE',8X,'NU')
            WRITE(IUNIT,240) ITER,W(5,8),W(3,7),T(5,8),T(3,7),FRE,ANU
  240       FORMAT(3X,I2,2X,1P6E10.2)
  220 CONTINUE
        IF(ITER.EQ.LAST) THEN
COME HERE TO CALCULATE LOCAL NU ON BOTTOM WALL
            DO 250 I=2,L2
                ANULOC=QW*DH/(COND*(T(I,1)-TB)+SMALL)
                DO 260 IUNIT=IU1,IU2
                    IF(I.EQ.2) WRITE(IUNIT,270)
  270               FORMAT(//,' I',8X,'X(I)',7X,'LOCAL NU (BOTTOM WALL)')
                    WRITE(IUNIT,280) I,X(I),ANULOC
  280               FORMAT(1X,I2,5X,1PE9.2,11X,1PE9.2)
  260           CONTINUE
  250       CONTINUE
            DO 290 J=1,M1
            DO 290 I=1,L1
                F(I,J,3)=W(I,J)/WBAR
                F(I,J,4)=(T(I,J)-TWAV)/(TB-TWAV)
  290       CONTINUE
            CALL PRINT
            CALL PLOT
        ENDIF
```

```
       RETURN
C*-*-*-*-*-*-*-*-*-*-*-*-*-*-*-*-*-*-*-*-*-*-*-*-*-*-*-*-*-*-*-*-*-*-*
       ENTRY PHI
       IF(NF.EQ.1) THEN
          DO 300 J=2,M2
          DO 300 I=2,L2
             GAM(I,J)=AMU
             SC(I,J)=-DPDZ
  300     CONTINUE
       ENDIF
       IF(NF.EQ.2) THEN
          DTDZ=QW*X(L1)/(WSUM*RHOCP)
          DO 320 J=2,M2
          DO 320 I=2,L2
             GAM(I,J)=COND
             SC(I,J)=-RHOCP*DTDZ*W(I,J)
  320     CONTINUE
       ENDIF
COME HERE TO SPECIFY BOUNDARY CONDITIONS
          DO 310 J=2,M2
             KBCL1(J)=2
  310     CONTINUE
       IF(NF.EQ.2) THEN
          DO 330 J=2,M2
             KBCI1(J)=2
  330     CONTINUE
          DO 340 I=2,L2
             KBCJ1(I)=2
             FLXCJ1(I)=QW
             KBCM1(I)=2
  340     CONTINUE
       ENDIF
       RETURN
       END
CCCCCCCCCCCCCCCCCCCCCCCCCCCCCCCCCCCCCCCCCCCCCCCCCCCCCCCCCCCCCCCCCCCCCCC
```

10.1-5   Results for Example 7

```
RESULTS OF CONDUCT FOR CARTESIAN COORDINATE SYSTEM
****************************************************
```

. . . . - - - - - . . . .

RECTANGULAR DUCT WITH ONE HEATED WALL

. . . .

ITER	W(5,8)	W(3,7)	T(5,8)	T(3,7)	FRE	NU
0	0.00E+00	0.00E+00	0.00E+00	0.00E+00	0.00E+00	1.33E+20
1	9.46E-02	6.40E-02	0.00E+00	0.00E+00	6.11E+01	1.33E+20
2	9.46E-02	6.40E-02	0.00E+00	0.00E+00	6.11E+01	1.33E+20
3	9.46E-02	6.40E-02	0.00E+00	0.00E+00	6.11E+01	1.33E+20
4	9.46E-02	6.40E-02	4.41E-02	1.43E-01	6.11E+01	3.49E+00
5	9.46E-02	6.40E-02	4.41E-02	1.43E-01	6.11E+01	3.49E+00
6	9.46E-02	6.40E-02	4.41E-02	1.43E-01	6.11E+01	3.49E+00

I	X(I)	LOCAL NU (BOTTOM WALL)
2	7.25E-02	3.14E+00
3	2.39E-01	3.24E+00
4	4.37E-01	3.45E+00
5	6.53E-01	3.68E+00
6	8.83E-01	3.84E+00

I =	1	2	3	4	5	6	7
X =	0.00E+00	7.25E-02	2.39E-01	4.37E-01	6.53E-01	8.83E-01	1.00E+00

J =	1	2	3	4	5	6	7
Y =	0.00E+00	3.62E-02	1.19E-01	2.19E-01	3.27E-01	4.41E-01	5.59E-01

J =	8	9	10	11	12
Y =	6.73E-01	7.81E-01	8.81E-01	9.64E-01	1.00E+00

****** AXIAL VELOCITY W ******

. . . .

I =	1	2	3	4	5	6	7
J							
12	0.00E+00	0.00E+00	0.00E+00	0.00E+00	0.00E+00	0.00E+00	0.00E+00
11	0.00E+00	4.77E-03	1.03E-02	1.36E-02	1.53E-02	1.61E-02	1.62E-02
10	0.00E+00	1.25E-02	2.99E-02	4.05E-02	4.61E-02	4.85E-02	4.88E-02
9	0.00E+00	1.84E-02	4.65E-02	6.44E-02	7.41E-02	7.82E-02	7.88E-02

```
8     0.00E+00 2.22E 02 5.81E 02 8.17E 02 9.46E 02 1.00E 01 1.01E 01
7     0.00E+00 2.41E 02 6.40E 02 9.09E 02 1.06E 01 1.12E 01 1.13E 01
6     0.00E+00 2.41E 02 6.40E 02 9.09E 02 1.06E 01 1.12E 01 1.13E 01
5     0.00E+00 2.22E 02 5.81E 02 8.17E 02 9.46E 02 1.00E 01 1.01E 01
4     0.00E+00 1.84E 02 4.65E 02 6.44E 02 7.41E 02 7.82E 02 7.88E 02
3     0.00E+00 1.25E 02 2.99E 02 4.05E 02 4.61E 02 4.85E 02 4.88E 02
2     0.00E+00 4.77E 03 1.03E 02 1.36E 02 1.53E 02 1.61E 02 1.62E 02
1     0.00E+00 0.00E+00 0.00E+00 0.00E+00 0.00E+00 0.00E+00 0.00E+00
```

******   TEMPERATURE   ******

```
I =     1       2       3       4       5       6       7
J
12    0.00E+00 8.18E 02 6.89E 02 4.47E 02 2.01E 02 4.59E 03 0.00E+00
11    8.37E 02 8.23E 02 6.92E 02 4.49E 02 2.01E 02 4.58E 03 2.59E 03
10    8.76E 02 8.60E 02 7.17E 02 4.60E 02 2.04E 02 4.51E 03 2.46E 03
9     9.99E 02 9.81E 02 8.15E 02 5.35E 02 2.62E 02 9.43E 03 7.29E 03
8     1.24E 01 1.22E 01 1.04E 01 7.32E 02 4.41E 02 2.65E 02 2.42E 02
7     1.65E 01 1.63E 01 1.43E 01 1.11E 01 8.06E 02 6.24E 02 6.00E 02
6     2.24E 01 2.22E 01 2.02E 01 1.70E 01 1.39E 01 1.21E 01 1.19E 01
5     2.98E 01 2.96E 01 2.77E 01 2.47E 01 2.17E 01 2.00E 01 1.97E 01
4     3.81E 01 3.79E 01 3.63E 01 3.35E 01 3.07E 01 2.91E 01 2.89E 01
3     4.68E 01 4.67E 01 4.52E 01 4.27E 01 4.01E 01 3.85E 01 3.83E 01
2     5.47E 01 5.46E 01 5.33E 01 5.09E 01 4.84E 01 4.68E 01 4.66E 01
1     0.00E+00 5.82E 01 5.69E 01 5.45E 01 5.20E 01 5.05E 01 0.00E+00
```

******   W/WBAR   ******

```
I =     1       2       3       4       5       6       7
J
12    0.00E+00 0.00E+00 0.00E+00 0.00E+00 0.00E+00 0.00E+00 0.00E+00
11    0.00E+00 8.20E 02 1.78E 01 2.34E 01 2.63E 01 2.76E 01 2.78E 01
10    0.00E+00 2.16E 01 5.15E 01 6.96E 01 7.93E 01 8.34E 01 8.39E 01
9     0.00E+00 3.16E 01 8.00E 01 1.11E+00 1.27E+00 1.35E+00 1.35E+00
8     0.00E+00 3.82E 01 9.98E 01 1.40E+00 1.63E+00 1.72E+00 1.74E+00
7     0.00E+00 4.15E 01 1.10E+00 1.56E+00 1.82E+00 1.93E+00 1.94E+00
6     0.00E+00 4.15E 01 1.10E+00 1.56E+00 1.82E+00 1.93E+00 1.94E+00
5     0.00E+00 3.82E 01 9.98E 01 1.40E+00 1.63E+00 1.72E+00 1.74E+00
4     0.00E+00 3.16E 01 8.00E 01 1.11E+00 1.27E+00 1.35E+00 1.35E+00
3     0.00E+00 2.16E 01 5.15E 01 6.96E 01 7.93E 01 8.34E 01 8.39E 01
```

```
2    0.00E+00 8.20E 02 1.78E 01 2.34E 01 2.63E 01 2.76E 01 2.78E 01
1    0.00E+00 0.00E+00 0.00E+00 0.00E+00 0.00E+00 0.00E+00 0.00E+00
```

```
******   (T TWAV)/(TB TWAV)   ******
         . . . . . . . . . . . . . . . . . . .
```

```
I =    1       2       3       4       5       6       7
J
12   1.41E+00 1.20E+00 1.23E+00 1.30E+00 1.36E+00 1.40E+00 1.41E+00
11   1.19E+00 1.20E+00 1.23E+00 1.30E+00 1.36E+00 1.40E+00 1.41E+00
10   1.18E+00 1.19E+00 1.23E+00 1.29E+00 1.36E+00 1.40E+00 1.41E+00
 9   1.15E+00 1.16E+00 1.20E+00 1.27E+00 1.34E+00 1.39E+00 1.39E+00
 8   1.09E+00 1.09E+00 1.14E+00 1.22E+00 1.30E+00 1.34E+00 1.35E+00
 7   9.80E 01 9.86E 01 1.04E+00 1.12E+00 1.20E+00 1.25E+00 1.26E+00
 6   8.27E 01 8.32E 01 8.84E 01 9.69E 01 1.05E+00 1.10E+00 1.10E+00
 5   6.33E 01 6.39E 01 6.87E 01 7.68E 01 8.44E 01 8.90E 01 8.96E 01
 4   4.15E 01 4.20E 01 4.63E 01 5.37E 01 6.08E 01 6.52E 01 6.57E 01
 3   1.88E 01 1.92E 01 2.29E 01 2.96E 01 3.63E 01 4.05E 01 4.10E 01
 2   2.02E 02 1.65E 02 1.78E 02 8.14E 02 1.46E 01 1.87E 01 1.92E 01
 1   1.41E+00 1.10E 01 7.64E 02 1.31E 02 5.14E 02 9.20E 02 1.41E+00
```

10.1-6 Discussion of Results

As expected the solutions of the *linear* equations for both w and T have converged in one iteration each. For the first three iterations, the printed values of the Nusselt number should be ignored since the temperature equation is solved only from the fourth iteration onwards.

The value of fRe from the exact solution for a duct of aspect ratio of 0.5 is 62.19. Even for the coarse grid employed, our computed value of 61.1 agrees with this exact solution within 2%.

Along the heated wall, the local Nusselt number can be seen to increase from left to right. This is to be expected since the fluid moves with a smaller velocity near the corner than around the middle of the heated wall.

The field printout of the velocity shows the expected symmetry between the top and bottom halves of the domain. The maximum value of $w/\bar{w}$ is 1.94, which agrees very well with the exact value of 1.99.

The negative numbers in the printout of the dimensionless temperature field requires some explanation. Since the heat flux along the heated wall is uniform, the corresponding wall temperature is not uniform. For the definition of the dimensionless temperature, we use an average wall temperature TWAV. The points where the local temperature is greater than the average wall temperature give a negative value of the dimensionless temperature.

10.1-7   Final Remarks

This adaptation to a simple rectangular duct has illustrated a number of basic features involved in the computation of duct flows. If the shape of the duct was square, then the calculation of the velocity field here and the calculation of the temperature field in Example 1 (Chapter 8) would be mathematically identical problems. It is interesting to know that the seemingly different problems of flow in a duct and the temperature field in a solid with heat generation have identical mathematical characters.

In the subsequent examples in this chapter, we shall consider more complex geometries and boundary conditions and compute conjugate heat transfer as well.

10.2   Circular Tube with Radial Internal Fins (Example 8)

10.2-1   Problem Description

It is common to put fins on the duct walls to increase the surface area and thus augment heat transfer. The presence of the fins also leads to a higher pressure drop for the same flow rate. In this example, we consider the finned circular tube shown in Fig. 10.2. For the calculation of the temperature fields, we need to account for the conduction in the solid material of the fin along with the convection in the duct fluid. Such a combined conduction-convection situation is called conjugate heat transfer. You will see that, with the method incorporated in CONDUCT, the treatment of a conjugate problems requires little extra effort and is almost automatic.

The finned tube shown in Fig. 10.2 has six fins that are uniformly spaced. With the lengths and angles indicated in the figure, the geometry of the fins can be specified as

height ratio $\qquad \dfrac{H}{R} = 0.3$ $\qquad\qquad\qquad$ (10.2a)

thickness ratio $\qquad \dfrac{\alpha_t}{\alpha} = 1.5$ $\qquad\qquad\qquad$ (10.2b)

The thermal boundary condition consists of a peripherally uniform temperature T_w along the outer wall of the tube; this temperature is considered to vary linearly in the axial direction. Thus, the tube receives a uniform heat input per unit axial length. Under this boundary condition, all the temperatures in the duct cross section vary linearly with z at the same rate. In other words, $\partial T/\partial z$ is a constant.

To complete the problem specification, we need the ratio of the conductivities of the solid and fluid. Here the ratio will be taken as

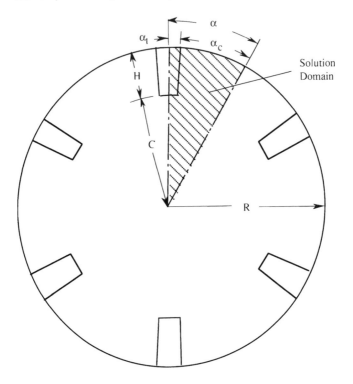

Fig. 10.2 Circular tube with radial internal fins

$$\frac{k_{fin}}{k_{fluid}} = 3.5 \tag{10.2c}$$

The objective of the calculation is to determine the fields of axial velocity and temperature and hence find the values of the product fRe and the Nusselt number. Because of symmetry, the solution will be confined to the shaded domain shown in Fig. 10.2: this region is bounded by the centerline passing through the thickness of one fin and the line of symmetry between two adjacent fins.

10.2-2   Design of ADAPT

GRID. To get the geometrical quantities, we set the radius of the tube arbitrarily as RAD = 1. All the other required dimensions are then calculated in accordance with Eq. (10.2). We construct the grid by using ZGRID, with two zones in the θ direction and two zones in the r direction. The zone widths for the θ direction are α_t and α_c. In the r direction, the zone lengths are C and H. In addition to calling ZGRID, we specify the values of R(1) and MODE.

BEGIN. Unlike in Example 7. here we do not use separate arrays for the dimensional and dimensionless versions of the velocity field and the temperature field. We employ the arrays $W(I,J)$ and $T(I,J)$ (which are equivalent to $F(I,J,NF)$ for $NF = 1$ and 2 respectively) to store the *dimensional* velocity and temperature; however, just before the final printout is produced, we convert the same arrays to *dimensionless* velocity and temperature. To indicate the true meanings of these arrays at the printout stage, the dimensionless descriptions are used in assigning $TITLE(1)$ and $TITLE(2)$.

Quantities such as the fluid properties. the pressure and temperature gradients, and the wall temperature are given arbitrary values. As mentioned earlier, you may wish to verify that the values of these quantities do not affect the *dimensionless* solution of the problem.

The field of $W(I,J)$ is initially set equal to zero; this provides the correct boundary values on the solid boundaries of the domain. The temperatures $T(I,J)$ are made equal to T_w. which gives the correct boundary value on the outer surface of the domain.

OUTPUT. Here, much of the structure is very similar to the treatment in Example 7. The switching from the velocity solution to the temperature solution and the summations over the cross section to calculate $\overline{w}$ and T_b are standard items. In the summation loop, the product $XCV(I)*YCVR(J)$ represents, for the polar coordinate system, the cross-sectional area occupied by one control volume. Since the mean velocity $\overline{w}$ should be calculated over the true flow area, the area contribution of the control volumes in the fin cross section is set equal to zero.

For the calculation of an average Nusselt number, we need to decide whether the average heat flux will be based on the actual heat transfer area (including the extra area provided by the fins) or the nominal area corresponding to the original area of the unfinned tube. If we would like the Nusselt number to be a measure of the total heat transfer by the finned tube, the use of the nominal root area will be more desirable. Then, for different geometries of the fins, we could compare the Nusselt numbers to judge their heat transfer performance. On the other hand, the use of the actual heat transfer area would lead to a lower Nusselt number since the average heat flux per unit of the *actual* area may well be small.

For these considerations, we define the heated perimeter HP different from the wetted perimeter WP. We take HP to be simply the outer area of our calculation domain. The total heat flow rate per unit axial length is calculated from the value of $\partial T/\partial z$ and then it is converted to the average heat flux by dividing by HP. In defining the overall Nusselt number, we use the tube diameter as the length scale.

The printout after each iteration includes the values of some typical velocities and temperatures, the product fRe. and the Nusselt number. In the final output, the variation of the local Nusselt number along the outer wall of the tube is printed out. In this calculation, the local heat flux $FLUXM1(I,2)$ is used, where I denotes the location along the outer boundary and 2 is the value of NF for temperature as the dependent variable. (The quantity FLUXM1(I,1) stands for the viscous stress at the boundary experienced by the axial velocity.)

Before calling PRINT for the field printout, we convert W(I,J) and T(I,J) to their dimensionless counterparts. As we shall see later, our treatment of the velocity in the solid material of the fin gives very small (but nonzero) values of the velocity there. To facilitate the identification of the fin region in the printout, we replace all small values of W(I,J) by zero. After this preparation, PRINT is called to produce the field printout.

PHI. In the calculation of the velocity field, the solid region occupied by the fin will be treated as a region of very high viscosity. As a result, the value of w throughout this region will be uniform and equal to the zero value at the boundary. In this manner, our solution will produce zero values of velocity in the fin region. Of course, the computed values of w in the fin region will not be exactly zero. Their smallness will depend on the largeness of the viscosity value used in the fin. If we set this fictitious value to be 1.E20, the "zero" velocities will probably be of the order of 1.E-20. In order to produce a more clean-looking printout, we have already arranged in OUTPUT the replacement of these small velocities by zeroes.

For NF = 1, we set GAM(I,J) equal to the fluid viscosity; if the grid point happens to lie in the fin region, this value is overwritten by a very large number. The source term SC(I,J) is made equal to $-dp/dz$.

The practice for the temperature (i.e., NF = 2) is very similar except that the value of GAM(I,J) in the fin region is not made large but set equal to the conductivity of the solid material. The expression used for the source term SC(I,J) is $-\rho c_p w(\partial T/\partial z)$.

For this problem, the boundary conditions for w are identical in nature to those for T. The values of both w and T are specified on the outer boundary of the domain. All the remaining boundaries can be considered as zero-flux boundaries for the variables. The KBC values are specified to reflect these considerations.

The applicability of identical boundary conditions for both w and T is somewhat uncommon. In most problems, there will be some difference in the boundary conditions for the two dependent variables. Therefore, care is needed in specifying the appropriate boundary conditions for the value of NF under consideration.

10.2-3 Additional Fortran Names

ALPHA	half of the angle between fins, Fig. 10.2
ALPHAC	α_c, Fig. 10.2
ALPHAT	α_t, Fig. 10.2
AMU	viscosity μ

ANU	average Nusselt number Nu
ANULOC	local Nusselt number along the tube wall
AR	area of control volume, dA
ASUM	cross-sectional area
C	clearance C, Fig. 10.2
CNDFIN	thermal conductivity of the fin, k_{fin}, Eq. (10.2c)
COND	thermal conductivity of the fluid, k_{fluid}, Eq. (10.2c)
CP	specific heat c_p
DEN	density ρ
DH	hydraulic diameter D_h
DIA	diameter of tube
DPDZ	axial pressure gradient dp/dz
DTDZ	axial temperature gradient $\partial T/\partial z$
FRE	the product fRe
H	fin height H
HP	heated perimeter
HRATIO	ratio H/R of fin height to radius of tube, Eq. (10.2a)
NFINS	number of fins
PI	the constant π
QW	average wall heat flux q_w
RAD	radius of tube, R
RE	Reynolds number
RHOCP	heat capacity ρc_p
T(I,J)	temperature T
TB	bulk temperature T_b
TRATIO	ratio α_t/α, Eq. (10.2b)
TSUM	$\int_w T\, dA$
TW	wall temperature T_w
W(I,J)	axial velocity w
WBAR	cross-sectional mean velocity $\overline{w}$
WP	wetted perimeter
WSUM	$\int_w dA$

10.2-4 Listing of ADAPT for Example 8

```
CCCCCCCCCCCCCCCCCCCCCCCCCCCCCCCCCCCCCCCCCCCCCCCCCCCCCCCCCCCCCCCCCCCCCCCCCC
      SUBROUTINE ADAPT
C - - - - - - - - - - - - - - - - - - - - - - - - - - - - - - - - - - - - - - - - - - - - - - - - - - -
C - - - - - EXAMPLE 8 - - CIRCULAR TUBE WITH RADIAL INTERNAL FINS
C - - - - - - - - - - - - - - - - - - - - - - - - - - - - - - - - - - - - - - - - - - - - - - - - - - -
$INCLUDE:'COMMON'
C*********************************************************************
```

```
      DIMENSION W(NI,NJ),T(NI,NJ)
      EQUIVALENCE (F(1,1,1),W(1,1)),(F(1,1,2),T(1,1))
C*-*-*-*-*-*-*-*-*-*-*-*-*-*-*-*-*-*-*-*-*-*-*-*-*-*-*-*-*-*-*-*-*-*-*
      ENTRY GRID
      HEADER='CIRCULAR TUBE WITH RADIAL INTERNAL FINS'
      PRINTF='PRINT8'
      PLOTF='PLOT8'
      NFINS=6
      CALL DATA2(HRATIO,0.3,TRATIO,0.15)
      RAD=1.
      DIA=2.*RAD
      H=RAD*HRATIO
      C=RAD·H
      PI=3.14159
      ALPHA=PI/FLOAT(NFINS)
      ALPHAT=ALPHA*TRATIO
      ALPHAC=ALPHA·ALPHAT
      CALL INTA3(NZX,2,NCVX(1),3,NCVX(2),7)
      CALL INTA3(NZY,2,NCVY(1),8,NCVY(2),4)
      CALL DATA4(XZONE(1),ALPHAT,XZONE(2),ALPHAC,YZONE(1),C,
     1      YZONE(2),H)
      CALL ZGRID
      R(1)=0.
      MODE=3
      RETURN
C*-*-*-*-*-*-*-*-*-*-*-*-*-*-*-*-*-*-*-*-*-*-*-*-*-*-*-*-*-*-*-*-*-*-*
      ENTRY BEGIN
      TITLE(1)='     W/WBAR     '
      TITLE(2)='  (T-TW)/(TB-TW) '
      CALL INTA6(KSOLVE(1),1,KPRINT(1),1,KPRINT(2),1,KPLOT(1),1,
     1           KPLOT(2),1,LAST,6)
      CALL DATA7(AMU,1.,DEN,1.,CP,1.,COND,1.,DPDZ,-1.,DTDZ,1.,TW,0.)
      RHOCP=DEN*CP
      CNDFIN=COND*3.5
      DO 100 J=1,M1
      DO 100 I=1,L1
         W(I,J)=0.
         T(I,J)=TW
  100 CONTINUE
      RETURN
C*-*-*-*-*-*-*-*-*-*-*-*-*-*-*-*-*-*-*-*-*-*-*-*-*-*-*-*-*-*-*-*-*-*-*
      ENTRY OUTPUT
```

```
     IF(ITER.EQ.3) THEN
        KSOLVE(1)=0
        KSOLVE(2)=1
     ENDIF
     ASUM=0.
     WSUM=0.
     TSUM=0.
     DO 200 J=2,M2
     DO 200 I=2,L2
        AR=XCV(I)*YCVR(J)
        IF(X(I).LT.ALPHAT.AND.Y(J).GT.C) AR=0.
        ASUM=ASUM+AR
        WSUM=WSUM+W(I,J)*AR
        TSUM=TSUM+W(I,J)*T(I,J)*AR
 200 CONTINUE
     WBAR=WSUM/ASUM
     TB=TSUM/(WSUM+SMALL)
     WP=ALPHAC*RAD+H+ALPHAT*C
     DH=4.*ASUM/WP
     RE=DH*WBAR*DEN/AMU
     FRE=-2.*DPDZ*DH/(DEN*WBAR**2+SMALL)*RE
     HP=ALPHA*RAD
     QW=DTDZ*RHOCP*WSUM/HP
     ANU=QW*DIA/(COND*(TW-TB)+SMALL)
     DO 210 IUNIT=IU1,IU2
        IF(!TER.EQ.0) WRITE(IUNIT,220)
 220    FORMAT(1X,'ITER',2X,'W(8,8)',4X,'W(11,4)',3X,'T(8,8)'
    1   ,4X,'T(11,4)',4X,'FRE',8X,'NU')
        WRITE(IUNIT,230) ITER,W(8,8),W(11,4),T(8,8),T(11,4),FRE,ANU
 230    FORMAT(2X,I2,1P6E10.2)
 210 CONTINUE
     IF(ITER.EQ.LAST) THEN
        DO 240 IUNIT=IU1,IU2
           WRITE(IUNIT,250)
 250       FORMAT(//,'   I',8X,'TH(I)',6X,'LOCAL NU (TUBE WALL)')
 240    CONTINUE
        DO 260 I=2,L2
           ANULOC=FLUXMI(I,2)*DIA/(COND*(TW-TB))
           DO 270 IUNIT=IU1,IU2
           WRITE(IUNIT,280) I,X(I),ANULOC
 280       FORMAT(1X,I2,5X,1PE9.2,10X,1PE9.2)
 270       CONTINUE
```

```
  260     CONTINUE
          DO 290 J=1,M1
          DO 290 I=1,L1
             W(I,J)=W(I,J)/WBAR
             IF(W(I,J).LT.1.E-10) W(I,J)=0.0
             T(I,J)=(T(I,J)-TW)/(TB-TW)
  290     CONTINUE
          CALL PRINT
          DO 295 J=2,M2
          DO 295 I=2,L2
             IF(X(I).LT.ALPHAT.AND.Y(J).GT.C) IBLOCK(I,J)=1
  295     CONTINUE
          CALL PLOT
       ENDIF
       RETURN
C*-*-*-*-*-*-*-*-*-*-*-*-*-*-*-*-*-*-*-*-*-*-*-*-*-*-*-*-*-*-*-*-*
       ENTRY PHI
       IF(NF.EQ.1) THEN
          DO 300 J=2,M2
          DO 300 I=2,L2
             GAM(I,J)=AMU
             IF(X(I).LT.ALPHAT.AND.Y(J).GT.C) GAM(I,J)=BIG
             SC(I,J)=-DPDZ
  300     CONTINUE
       ENDIF
       IF(NF.EQ.2) THEN
          DO 310 J=2,M2
          DO 310 I=2,L2
             GAM(I,J)=COND
             IF(X(I).LT.ALPHAT.AND.Y(J).GT.C) GAM(I,J)=CNDFIN
             SC(I,J)=-RHOCP*DTDZ*W(I,J)
  310     CONTINUE
       ENDIF
COME HERE TO SPECIFY BOUNDARY CONDITIONS
       DO 320 J=2,M2
          KBCI1(J)=2
          KBCL1(J)=2
  320 CONTINUE
       DO 330 I=2,L2
          KBCJ1(I)=2
  330 CONTINUE
       RETURN
```

```
        END
CCCCCCCCCCCCCCCCCCCCCCCCCCCCCCCCCCCCCCCCCCCCCCCCCCCCCCCCCCCCCCCCCCCCCCCCCC
```

10.2-5   Results for Example 8

```
RESULTS OF CONDUCT FOR POLAR COORDINATE SYSTEM
**************************************************



CIRCULAR TUBE WITH RADIAL INTERNAL FINS

```

ITER	W(8,8)	W(11,4)	T(8,8)	T(11,4)	FRE	NU
0	0.00E+00	0.00E+00	0.00E+00	0.00E+00	0.00E+00	0.00E+00
1	9.23E-02	1.59E-01	0.00E+00	0.00E+00	4.15E+01	6.51E+18
2	9.23E-02	1.59E-01	0.00E+00	0.00E+00	4.15E+01	6.51E+18
3	9.23E-02	1.59E-01	0.00E+00	0.00E+00	4.15E+01	6.51E+18
4	9.23E-02	1.59E-01	-1.35E-02	-2.36E-02	4.15E+01	4.11E+00
5	9.23E-02	1.59E-01	-1.35E-02	-2.36E-02	4.15E+01	4.11E+00
6	9.23E-02	1.59E-01	-1.35E-02	-2.36E-02	4.15E+01	4.11E+00

I	TH(I)	LOCAL NU (TUBE WALL)
2	1.31E-02	9.00E+00
3	3.93E-02	9.03E+00
4	6.54E-02	9.09E+00
5	1.10E-01	2.71E+00
6	1.74E-01	2.94E+00
7	2.37E-01	3.15E+00
8	3.01E-01	3.32E+00
9	3.65E-01	3.45E+00
10	4.28E-01	3.53E+00
11	4.92E-01	3.58E+00

```
I =    1         2         3         4         5         6         7
TH = 0.00E+00 1.31E-02 3.93E-02 6.54E-02 1.10E-01 1.74E-01 2.37E-01

I =    8         9        10        11        12
TH = 3.01E-01 3.65E-01 4.28E-01 4.92E-01 5.24E-01
```

```
J =    1        2        3        4        5        6        7
Y = 0.00E+00 4.37E 02 1.31E 01 2.19E 01 3.06E 01 3.94E 01 4.81E 01

J =    8        9       10       11       12       13       14
Y = 5.69E 01 6.56E 01 7.37E 01 8.13E 01 8.88E 01 9.62E 01 1.00E+00
```

******   W/WBAR   ******

. . . .

```
I =    1        2        3        4        5        6        7
J
14  0.00E+00 0.00E+00 0.00E+00 0.00E+00 0.00E+00 0.00E+00 0.00E+00
13  0.00E+00 0.00E+00 0.00E+00 0.00E+00 3.54E 02 7.89E 02 1.09E 01
12  0.00E+00 0.00E+00 0.00E+00 0.00E+00 8.06E 02 2.01E 01 2.88E 01
11  0.00E+00 0.00E+00 0.00E+00 0.00E+00 1.22E 01 3.10E 01 4.47E 01
10  0.00E+00 0.00E+00 0.00E+00 0.00E+00 2.06E 01 4.61E 01 6.29E 01
 9  5.21E 01 5.23E 01 5.40E 01 5.76E 01 6.74E 01 8.06E 01 9.15E 01
 8  1.14E+00 1.14E+00 1.14E+00 1.15E+00 1.18E+00 1.22E+00 1.27E+00
 7  1.55E+00 1.55E+00 1.55E+00 1.56E+00 1.56E+00 1.58E+00 1.59E+00
 6  1.86E+00 1.86E+00 1.86E+00 1.86E+00 1.86E+00 1.86E+00 1.87E+00
 5  2.08E+00 2.08E+00 2.08E+00 2.08E+00 2.08E+00 2.09E+00 2.09E+00
 4  2.25E+00 2.25E+00 2.25E+00 2.25E+00 2.25E+00 2.25E+00 2.25E+00
 3  2.36E+00 2.36E+00 2.36E+00 2.36E+00 2.36E+00 2.36E+00 2.36E+00
 2  2.41E+00 2.41E+00 2.41E+00 2.41E+00 2.41E+00 2.41E+00 2.41E+00
 1  0.00E+00 2.42E+00 2.42E+00 2.42E+00 2.42E+00 2.42E+00 2.42E+00


I =    8        9       10       11       12
J
14  0.00E+00 0.00E+00 0.00E+00 0.00E+00 0.00E+00
13  1.29E 01 1.43E 01 1.52E 01 1.56E 01 1.56E 01
12  3.50E 01 3.91E 01 4.17E 01 4.29E 01 4.31E 01
11  5.43E 01 6.08E 01 6.47E 01 6.66E 01 6.69E 01
10  7.41E 01 8.16E 01 8.62E 01 8.84E 01 8.87E 01
 9  9.99E 01 1.06E+00 1.10E+00 1.12E+00 1.12E+00
 8  1.31E+00 1.34E+00 1.37E+00 1.38E+00 1.38E+00
 7  1.61E+00 1.63E+00 1.64E+00 1.64E+00 1.64E+00
 6  1.87E+00 1.88E+00 1.88E+00 1.88E+00 1.88E+00
 5  2.09E+00 2.09E+00 2.09E+00 2.09E+00 2.09E+00
 4  2.25E+00 2.25E+00 2.25E+00 2.25E+00 2.25E+00
 3  2.36E+00 2.36E+00 2.36E+00 2.36E+00 2.36E+00
```

```
2    2.41E+00 2.41E+00 2.41E+00 2.41E+00 2.41E+00
1    2.42E+00 2.42E+00 2.42E+00 2.42E+00 0.00E+00
```

******   (T TW)/(TB TW)   ******

. . . .

I =	1	2	3	4	5	6	7
J							
14	0.00E+00	0.00E+00	0.00E+00	0.00E+00	0.00E+00	0.00E+00	0.00E+00
13	4.90E 02	4.90E 02	4.92E 02	4.95E 02	5.17E 02	5.60E 02	6.00E 02
12	1.52E-01	1.52E 01	1.52E 01	1.53E 01	1.60E 01	1.74E 01	1.86E 01
11	2.58E-01	2.58E 01	2.59E-01	2.61E 01	2.75E 01	2.99E 01	3.19E-01
10	3.61E-01	3.61E 01	3.63E 01	3.68E 01	3.99E 01	4.37E 01	4.64E 01
9	5.64E 01	5.65E 01	5.68E 01	5.76E 01	5.95E 01	6.20E 01	6.41E 01
8	8.19E 01	8.19E 01	8.20E 01	8.22E 01	8.27E 01	8.36E 01	8.46E 01
7	1.03E+00	1.03E+00	1.03E+00	1.04E+00	1.04E+00	1.04E+00	1.04E+00
6	1.22E+00	1.22E+00	1.22E+00	1.22E+00	1.22E+00	1.22E+00	1.22E+00
5	1.37E+00	1.37E+00	1.37E+00	1.37E+00	1.37E+00	1.37E+00	1.37E+00
4	1.49E+00	1.49E+00	1.49E+00	1.49E+00	1.49E+00	1.49E+00	1.49E+00
3	1.57E+00	1.57E+00	1.57E+00	1.57E+00	1.57E+00	1.57E+00	1.57E+00
2	1.61E+00	1.61E+00	1.61E+00	1.61E+00	1.61E+00	1.61E+00	1.61E+00
1	0.00E+00	1.62E+00	1.62E+00	1.62E+00	1.62E+00	1.62E+00	1.62E+00

I =	8	9	10	11	12
J					
14	0.00E+00	0.00E+00	0.00E+00	0.00E+00	0.00E+00
13	6.32E 02	6.56E 02	6.72E 02	6.80E 02	6.81E 02
12	1.96E 01	2.03E 01	2.08E 01	2.10E-01	2.10E 01
11	3.35E 01	3.47E 01	3.54E 01	3.58E 01	3.58E 01
10	4.84E 01	4.98E 01	5.06E 01	5.11E 01	5.11E 01
9	6.57E 01	6.70E 01	6.78E 01	6.82E 01	6.82E 01
8	8.55E 01	8.62E 01	8.67E 01	8.70E 01	8.70E 01
7	1.05E+00	1.05E+00	1.05E+00	1.05E+00	1.05E+00
6	1.22E+00	1.22E+00	1.23E+00	1.23E+00	1.23E+00
5	1.37E+00	1.37E+00	1.37E+00	1.37E+00	1.37E+00
4	1.49E+00	1.49E+00	1.49E+00	1.49E+00	1.49E+00
3	1.57E+00	1.57E+00	1.57E+00	1.57E+00	1.57E+00
2	1.61E+00	1.61E+00	1.61E+00	1.61E+00	1.61E+00
1	1.62E+00	1.62E+00	1.62E+00	1.62E+00	0.00E+00

10.2-6   Discussion of Results

In this problem, the velocity and temperature equations are both linear. Therefore, the convergence is obtained in one iteration each.

The printout of the local Nusselt numbers on the outer surface shows very high values for $I = 2$, 3, and 4. These locations correspond to the root of the fin. Obviously, a large heat flux enters the fin material of reasonably high conductivity. The local Nusselt numbers along the rest of the outer wall show an increase in the direction away from the fin. This can be explained on the basis of the velocity field. The flow resistance offered by the fin surface reduces the velocity in the vicinity of the fin. A lack of strong flow reduces the local heat flux in that region.

In the field printout, the region occupied by the fin can be identified by the zero values of w. An overall effect of the fins on the velocity distribution is to reduce the velocity in the outer region and increase it in the inner region. Note that our predicted maximum value of $w/\overline{w}$ is 2.42: for a circular tube without fins, the corresponding value is 2.

The temperature field shows the variation of the temperature in the fin material along with the temperature distribution in the fluid. Even though the boundary conditions for w and T are similar, the calculated fields of dimensionless w and T are not identical. This is because of the differences in their source terms: the axial velocity has a uniform source over the whole cross section, while the temperature source term is proportional to the local velocity.

10.2-7   Final Remarks

A tube with internal fins represents a very interesting problem in many respects. Whereas the fins increase the surface area, they also slow down the fluid in the vicinity of the tube wall. If we use a large number of fins, they will—in the limit—simply reduce the effective diameter of the tube. Increased fin height gives more surface area, but the temperature drop along the fin may make the extra fin surface ineffective. This adaptation of CONDUCT enables you to examine these and related issues.

In the definition of the Nusselt number, we used the average heat flux based on the outer area of the tube. This enables us to compare the performance of different finned tubes. While defining fRe, however, we used the hydraulic diameter of the finned tube. If we had simply used the diameter of the outer tube, we could have judged the pressure drop performance of different finned tubes by their values of fRe.

Our computed value of fRe is 41.5. That this value is less than the value 64 for an unfinned tube does not mean that the finned tube requires less pressure drop. The seemingly small value of fRe results from our use of the hydraulic diameter in its definition. If we replace the hydraulic diameter by the outer diameter of the tube, we would get a value of 113.6 for fRe.

The thermal boundary conditions used in Examples 7 and 8 lead to a linear equation for temperature. In the next two examples, we shall encounter a nonlinear source term in their temperature equation.

10.3   Annular Sector Duct (Example 9)

10.3-1   Problem Description

The motivation for this problem comes from the double-pipe heat exchanger, which is made of two concentric pipes with one fluid flowing through the inner pipe and the other through the annular space between the two pipes. The heat exchange between the two fluids can occur only at the surface of the inner pipe, which has a rather small surface area. A remedy is to attach radial fins on the outside of the inner pipe thereby increasing the area available for heat transfer. If the fins extend all the way to the outer pipe, as shown in Fig. 10.3, the annular region gets subdivided into sectors. The computation of the flow and heat transfer in such an annular sector duct is the subject of this example.

Because of the symmetry, the solution domain will be the annular sector bounded by a fin on one side and the centerline between two adjacent fins on the other. This domain is shown shaded in Fig. 10.3. For the six fins shown, the angle α between the adjacent fins is 60 degrees. The radius ratio R_{in}/R_{out} is 0.2.

The thermal boundary conditions for this problem consist of an adiabatic outer wall and a constant known temperature over the inner pipe and over the fins. The fin thickness is negligible, yet their conductivity will be taken to be high enough to maintain a uniform temperature along them. The temperature for the inner pipe and the fins is not only uniform over a given cross section but also remains constant in the axial direction z. Such a condition may be attained in practice by a flow of a condensing fluid in the inner pipe.

10.3-2   Design of ADAPT

GRID. Since the geometry of the solution domain is very simple, EZGRID is used to prepare a uniform grid. Here, XL denotes the θ-direction extent of the domain in radians. The values of $MODE$ and $R(1)$ are also specified.

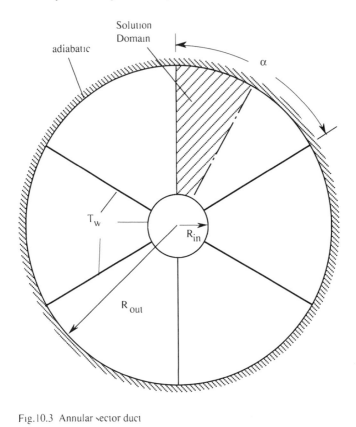

Fig.10.3 Annular sector duct

BEGIN. In this problem, although the velocity equation is linear, the temperature equation is nonlinear because of the particular boundary condition. Therefore, we allow the usual 3 iterations for velocity, but 7 iterations for temperature; thus, the value of LAST is set as 10. In addition to the usual values of the fluid properties, a value of dT_b/dz is specified. Choosing an arbitrary value for this quantity corresponds to the choice of a certain z location along the duct for our calculation. Of course, the dimensionless solution will not depend on the particular value used for dT_b/dz.

The arrays W(I,J) and T(I,J) are given initial values that are equal to their respective boundary values.

OUTPUT. Most of the design of the OUTPUT routine here is very similar to our previous two examples. For the overall Nusselt number, the heat transfer area used includes the surface area of the inner pipe and of the fin. The total heat flow rate per unit axial length is obtained from dT_b/dz. The fin height YL is used as the length scale to define the Nusselt number.

For the final printout, the local Nusselt numbers on the inner pipe and on the fin are calculated and printed out. The fields of $W(I,J)$ and $T(I,J)$ are converted into their dimensionless form before their field printout is arranged.

PHI. Here the specification of $GAM(I,J)$ and $SC(I,J)$ are straight-forward. To obtain the value of the local $\partial T/\partial z$ from dT_b/dz, we use Eq. (9.45). The quantity $TEMP$ stands for the ratio $(T - T_w)/(T_b - T_w)$. The fact that the values of $T(I,J)$ and TB are needed to calculate the source term $SC(I,J)$ indicates the nonlinear character of the temperature equation.

In the calculation of $TEMP$, the inclusion of $SMALL$ in the *denominator* is for avoiding a division by zero, which may arise at the beginning when T_b equals T_w. What is, however, the need for inclusion of $SMALL$ in the *numerator*? This is a very subtle point and requires some discussion. If we did not include $SMALL$ in the numerator, then initially $TEMP$ will be zero, and this will lead to a zero source term in the temperature equation. As a result, the solution for the temperature field would make all $T(I,J)$ values equal to the boundary temperature T_w. This will continue to keep $TEMP$ equal to zero in subsequent iterations, and we shall never get the desired solution for the temperature field. The inclusion of $SMALL$ in the numerator prevents the solution from getting stuck at the initial guess.

Of course, this complication arises from our particular practice of setting the initial guess. If we initially set the $T(I,J)$ values at the internal grid points equal to a number different from T_w, no difficulty would arise. Such a practice is illustrated in Example 10.

The boundary conditions for this problem consist of the zero-flux condition at the symmetry-line boundary and, for the temperature equation only, the zero-flux condition at the outer boundary. These are specified through the values of KBC at the appropriate boundaries.

10.3-3 Additional Fortran Names

ALPHA	angle between adjacent fins
AMU	viscosity μ
ANU	average Nusselt number Nu
ANUI	local Nusselt number along the inner wall
ANULFT	local Nusselt number along the left boundary
AR	area of control volume, dA
ASUM	cross-sectional area
COND	thermal conductivity k
CP	specific heat c_p
DEN	density ρ

DH	hydraulic diameter D_h
DPDZ	axial pressure gradient dp/dz
DTDZ	axial temperature gradient $\partial T/\partial z$
DTBDZ	axial bulk temperature gradient dT_b/dz
FRE	the product fRe
HP	heated perimeter
PI	the constant π
QW	wall heat flux q_w
RE	Reynolds number
RHOCP	heat capacity ρc_p
RIN	inner radius R_{in}
ROUT	outer radius R_{out}
T(I,J)	temperature T
TB	bulk temperature T_b
TEMP	temporary variable for $(T - T_w)/(T_b - T_w)$
TSUM	$\int_w T\, dA$
TW	wall temperature T_w
W(I,J)	axial velocity w
WBAR	cross-sectional mean velocity $\overline{w}$
WP	wetted perimeter
WSUM	$\int_w dA$

10.3-4   Listing of ADAPT for Example 9

```
CCCCCCCCCCCCCCCCCCCCCCCCCCCCCCCCCCCCCCCCCCCCCCCCCCCCCCCCCCCCCCCCCCCCCCCCCCCCC
      SUBROUTINE ADAPT
C.....................................................................
C....EXAMPLE 9 .. ANNULAR SECTOR DUCT
C.....................................................................
$INCLUDE:'COMMON'
C*******************************************************************************
      DIMENSION W(NI,NJ),T(NI,NJ)
      EQUIVALENCE (F(1,1,1),W(1,1)),(F(1,1,2),T(1,1))
C*.*.*.*.*.*.*.*.*.*.*.*.*.*.*.*.*.*.*.*.*.*.*.*.*.*.*.*.*.*.*.*.*
      ENTRY GRID
      HEADER='ANNULAR SECTOR DUCT'
      PRINTF='PRINT9'
      PLOTF='PLOT9'
      CALL DATA4(ROUT,1.,RIN,0.2,ALPHA,60.,PI,3.14159)
      XL=0.5*ALPHA*PI/180.
      CALL DATA2(R(1),RIN,YL,ROUT-RIN)
      CALL INTA3(NCVLX,5,NCVLY,10,MODE,3)
```

```
      CALL EZGRID
      RETURN
C*-*-*-*-*-*-*-*-*-*-*-*-*-*-*-*-*-*-*-*-*-*-*-*-*-*-*-*-*-*-*-*-*-*-*-*
      ENTRY BEGIN
      TITLE(1)='     W/WBAR     '
      TITLE(2)=' (T-TW)/(TB-TW)'
      CALL INTA6(KSOLVE(1),1,KPRINT(1),1,KPRINT(2),1,
     1           KPLOT(1),1,KPLOT(2),1,LAST,10)
      CALL DATA7(AMU,1.,COND,1.,CP,1.,DEN,1.,DPDZ,-1.,TW,0.,DTBDZ,1.)
      RHOCP=DEN*CP
      DO 100 J=1,M1
      DO 100 I=1,L1
         W(I,J)=0.
         T(I,J)=TW
  100 CONTINUE
      RETURN
C*-*-*-*-*-*-*-*-*-*-*-*-*-*-*-*-*-*-*-*-*-*-*-*-*-*-*-*-*-*-*-*-*-*-*-*
      ENTRY OUTPUT
      IF(ITER.EQ.3) THEN
         KSOLVE(1)=0
         KSOLVE(2)=1
      ENDIF
      ASUM=0.
      WSUM=0.
      TSUM=0.
      DO 200 J=2,M2
      DO 200 I=2,L2
         AR=XCV(I)*YCVR(J)
         ASUM=ASUM+AR
         WSUM=WSUM+W(I,J)*AR
         TSUM=TSUM+W(I,J)*T(I,J)*AR
  200 CONTINUE
      WBAR=WSUM/ASUM
      TB=TSUM/(WSUM+SMALL)
      WP=XL*(ROUT+RIN)+YL
      HP=XL*RIN+YL
      DH=4.*ASUM/WP
      RE=DH*WBAR*DEN/AMU
      FRE=-2.*DPDZ*DH/(DEN*WBAR**2+SMALL)*RE
      QW=DTBDZ*RHOCP*WSUM/HP
      ANU=QW*YL/(COND*(TW-TB)+SMALL)
      DO 210 IUNIT=IU1,IU2
```

```
          IF(ITER.EQ.0) WRITE(IUNIT,220)
  220     FORMAT(1X,'ITER',2X,'W(6,8)',4X,'W(4,11)',3X,'T(6,8)'
    1     ,4X,'T(4,11)',4X,'FRE',8X,'NU')
          WRITE(IUNIT,230) ITER,W(6,8),W(4,11),T(6,8),T(4,11),FRE,ANU
  230     FORMAT(2X,I2,1P7E10.2)
  210 CONTINUE
      IF(ITER.EQ.LAST) THEN
COME HERE TO CALCULATE LOCAL NU
          DO 240 I=2,L2
              ANUI=FLUXJI(I,2)*YL/(COND*(TW-TB)+SMALL)
              DO 250 IUNIT=IU1,IU2
                  IF(I.EQ.2) WRITE(IUNIT,260)
  260             FORMAT(//,' I',8X,'TH(I)',6X,'LOCAL NU (INNER WALL)')
                  WRITE(IUNIT,265) I,X(I),ANUI
  265             FORMAT(1X,I2,5X,1PE9.2,10X,1PE9.2)
  250         CONTINUE
  240     CONTINUE
          DO 270 J=M2,2,-1
              ANULFT=FLUXII(J,2)*YL/(COND*(TW-TB)+SMALL)
              DO 275 IUNIT=IU1,IU2
                  IF(J.EQ.M2) WRITE(IUNIT,280)
  280             FORMAT(//,' J',8X,'Y(J)',7X,'LOCAL NU (SIDE WALL)')
                  WRITE(IUNIT,285) J,Y(J),ANULFT
  285             FORMAT(1X,I2,5X,1PE9.2,10X,1PE9.2)
  275         CONTINUE
  270     CONTINUE
          DO 290 J=1,M1
          DO 290 I=1,L1
              W(I,J)=W(I,J)/WBAR
              T(I,J)=(T(I,J)-TW)/(TB-TW)
  290     CONTINUE
          CALL PRINT
          CALL PLOT
      ENDIF
      RETURN
C*-*-*-*-*-*-*-*-*-*-*-*-*-*-*-*-*-*-*-*-*-*-*-*-*-*-*-*-*-*-*-*-*-*-*-*
      ENTRY PHI
      IF(NF.EQ.1) THEN
          DO 300 J=2,M2
          DO 300 I=2,L2
              GAM(I,J)=AMU
              SC(I,J)=-DPDZ
```

```
300      CONTINUE
      ENDIF
      IF(NF.EQ.2) THEN
         DO 310 J=2,M2
         DO 310 I=2,L2
            GAM(I,J)=COND
            TEMP=(T(I,J)-TW+SMALL)/(TB-TW+SMALL)
            DTDZ=TEMP*DTBDZ
            SC(I,J)=-RHOCP*DTDZ*W(I,J)
310      CONTINUE
      ENDIF
COME HERE TO SPECIFY BOUNDARY CONDITIONS
         DO 320 J=2,M2
            KBCL1(J)=2
320      CONTINUE
      IF(NF.EQ.2) THEN
         DO 330 I=2,L2
            KBCM1(I)=2
330      CONTINUE
      ENDIF
      RETURN
      END
CCCCCCCCCCCCCCCCCCCCCCCCCCCCCCCCCCCCCCCCCCCCCCCCCCCCCCCCCCCCCCCCCCCCCCCCCCC
```

10.3-5   Results for Example 9

```
RESULTS OF CONDUCT FOR POLAR COORDINATE SYSTEM
**************************************************


..................................................................
ANNULAR SECTOR DUCT
..................................................................


ITER  W(6,8)    W(4,11)    T(6,8)     T(4,11)    FRE       NU
  0  0.00E+00  0.00E+00  0.00E+00  0.00E+00  0.00E+00  0.00E+00
  1  3.43E-02  7.16E-03  0.00E+00  0.00E+00  5.80E+01  3.80E+17
  2  3.43E-02  7.16E-03  0.00E+00  0.00E+00  5.80E+01  3.80E+17
  3  3.43E-02  7.16E-03  0.00E+00  0.00E+00  5.80E+01  3.80E+17
  4  3.43E-02  7.16E-03 -1.47E-03 -1.09E-03  5.80E+01  3.86E+00
  5  3.43E-02  7.16E-03 -1.84E-03 -1.33E-03  5.80E+01  3.27E+00
```

6	3.43E-02	7.16E-03	-1.89E-03	-1.38E-03	5.80E+01	3.21E+00
7	3.43E-02	7.16E-03	-1.90E-03	-1.39E-03	5.80E+01	3.20E+00
8	3.43E-02	7.16E-03	-1.90E-03	-1.39E-03	5.80E+01	3.19E+00
9	3.43E-02	7.16E-03	-1.90E-03	-1.39E-03	5.80E+01	3.19E+00
10	3.43E-02	7.16E-03	-1.90E-03	-1.39E-03	5.80E+01	3.19E+00

I	TH(I)	LOCAL NU (INNER WALL)
2	5.24E-02	2.13E-01
3	1.57E-01	6.24E-01
4	2.62E-01	9.75E-01
5	3.67E-01	1.23E+00
6	4.71E-01	1.37E+00

J	Y(J)	LOCAL NU (SIDE WALL)
11	7.60E-01	3.94E+00
10	6.80E-01	4.30E+00
9	6.00E-01	4.62E+00
8	5.20E-01	4.76E+00
7	4.40E-01	4.63E+00
6	3.60E-01	4.19E+00
5	2.80E-01	3.49E+00
4	2.00E-01	2.61E+00
3	1.20E-01	1.69E+00
2	4.00E-02	7.08E-01

I =	1	2	3	4	5	6	7
TH =	0.00E+00	5.24E-02	1.57E-01	2.62E-01	3.67E-01	4.71E-01	5.24E-01

J =	1	2	3	4	5	6	7
Y =	0.00E+00	4.00E-02	1.20E-01	2.00E-01	2.80E-01	3.60E-01	4.40E-01

J =	8	9	10	11	12
Y =	5.20E-01	6.00E-01	6.80E-01	7.60E-01	8.00E-01

******   W/WBAR   ******

. . . .

I =	1	2	3	4	5	6	7

```
 J
12      0.00E+00 0.00E+00 0.00E+00 0.00E+00 0.00E+00 0.00E+00 0.00E+00
11      0.00E+00 1.68E·01 3.32E·01 4.19E·01 4.67E·01 4.89E·01 4.92E·01
10      0.00E+00 3.50E·01 7.88E·01 1.04E+00 1.19E+00 1.25E+00 1.26E+00
 9      0.00E+00 4.31E·01 1.04E+00 1.42E+00 1.65E+00 1.75E+00 1.76E+00
 8      0.00E+00 4.54E·01 1.14E+00 1.60E+00 1.87E+00 2.01E+00 2.02E+00
 7      0.00E+00 4.37E·01 1.13E+00 1.60E+00 1.91E+00 2.05E+00 2.07E+00
 6      0.00E+00 3.92E·01 1.03E+00 1.49E+00 1.78E+00 1.93E+00 1.94E+00
 5      0.00E+00 3.28E·01 8.72E·01 1.27E+00 1.54E+00 1.67E+00 1.68E+00
 4      0.00E+00 2.51E·01 6.75E-01 9.93E·01 1.20E+00 1.31E+00 1.32E+00
 3      0.00E+00 1.68E·01 4.52E·01 6.66E·01 8.09E·01 8.81E·01 8.90E·01
 2      0.00E+00 7.29E·02 1.93E·01 2.81E·01 3.40E·01 3.69E·01 3.73E·01
 1      0.00E+00 0.00E+00 0.00E+00 0.00E+00 0.00E+00 0.00E+00 0.00E+00
```

```
******      (T·TW)/(TB·TW)      ******
            . . . . . . . . . . . . . . . . . .


I =     1       2       3       4       5       6       7
 J
12      0.00E+00 2.46E·01 7.32E·01 1.17E+00 1.50E+00 1.69E+00 0.00E+00
11      0.00E+00 2.46E-01 7.33E·01 1.17E+00 1.51E+00 1.69E+00 1.71E+00
10      0.00E+00 2.48E·01 7.41E·01 1.19E+00 1.54E+00 1.73E+00 1.76E+00
 9      0.00E+00 2.42E·01 7.28E·01 1.18E+00 1.53E+00 1.72E+00 1.74E+00
 8      0.00E+00 2.25E·01 6.75E-01 1.09E+00 1.42E+00 1.60E+00 1.62E+00
 7      0.00E+00 1.94E·01 5.80E·01 9.36E·01 1.21E+00 1.37E+00 1.38E+00
 6      0.00E+00 1.53E·01 4.56E·01 7.31E·01 9.43E·01 1.06E+00 1.07E+00
 5      0.00E+00 1.09E·01 3.23E·01 5.13E·01 6.58E·01 7.37E·01 7.47E·01
 4      0.00E+00 6.79E·02 2.00E·01 3.16E·01 4.03E-01 4.50E·01 4.55E·01
 3      0.00E+00 3.49E·02 1.03E·01 1.61E·01 2.04E-01 2.27E·01 2.30E·01
 2      0.00E+00 1.10E·02 3.22E·02 5.04E·02 6.37E·02 7.08E·02 7.17E·02
 1      0.00E+00 0.00E+00 0.00E+00 0.00E+00 0.00E+00 0.00E+00 0.00E+00
```

10.3-6 Discussion of Results

The nonlinearities of the temperature equation does not allow us to get a converged solution for T in one iteration; however, our iterative technique can be seen to have converged satisfactorily within 4 iterations.

The computed values of the product fRe is 58.0. This agrees extremely well with the exact value of 58.56 for this problem. Once again, the solution is quite accurate even for the relatively coarse grid.

The local Nusselt numbers on the inner wall increase as we go away from the fin. On the side wall (i.e., the fin), the maximum Nusselt number occurs somewhere in the middle. All these trends agree with the expectation that the local heat transfer rate will be high where there is a large flow velocity in the vicinity of the surface.

The field printout shows the expected distribution of dimensionless velocity and temperature. An annular duct is narrow at the inner radius and wide at the outer radius. Therefore, the maximum velocity in the cross section occurs at a radial location closer to the outer wall. This can be seen from our computed results. The largest value of the dimensionless temperature occurs, not on the outer adiabatic boundary, but at a smaller radius. This is the effect of the source term, which varies with the local flow velocity.

10.3-7   Final Remarks

The main new feature in this example is the constant-temperature boundary condition, which makes the temperature equation nonlinear. All other aspects of the problem should appear rather routine by now. A more general form of constant-T_w boundary condition occurs when an *external* heat transfer coefficient and a constant temperature of the surrounding fluid are prescribed. This is illustrated in the next example.

10.4 Staggered Fin Array (Example 10)

10.4-1   Problem Description

Figure 10.4 shows a set of parallel plates with fins attached to them in a staggered fashion. We wish to consider a fully developed flow in the direction normal to the plane of the diagram. The geometry of the duct is specified by

$$\frac{S}{H} = 1. \qquad \frac{H_{fin}}{H} = 0.75. \qquad \frac{t}{S} = 0.2 \qquad (10.3)$$

where S is the fin spacing. H is the distance between the parallel plates. H_{fin} is the fin height, and t is the fin thickness (and not time). These dimensions are shown in Fig. 10.4.

The thermal boundary conditions are given by the prescribed external heat transfer coefficients on the top and bottom plates and a constant temperature T_∞ of the surrounding fluid. The heat transfer coefficients are expressed through Biot numbers (Bi) for the two surfaces. The Biot number is similar in definition to the Nusselt number, except that the Biot number refers to *external* rather than internal heat transfer. The definition of Bi is

$$Bi = \frac{h_e H}{k} \qquad (10.4)$$

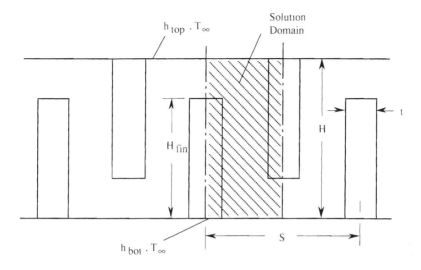

Fig.10.4 Staggered fin array

Where h_e is the external heat transfer coefficient. H is the length scale used in the definition. and k is the conductivity of the duct fluid. For the present problem, the Biot numbers for the bottom and top surfaces are given by

$$Bi_{bot} = 0.2, \qquad Bi_{top} = 5 \qquad\qquad\qquad (10.5)$$

Also. the ratio of conductivities. k_{fin}/k_{fluid} is specified as 1.8.

If the array is very long in the x direction. the vertical centerline through each fin represents a line of symmetry. Our solution domain is. therefore. chosen to be between two successive symmetry lines as shown in Fig. 10.4.

10.4-2   Design of ADAPT

GRID. For the geometry in this problem. the use of ZGRID is appropriate. Therefore. after prescribing the necessary geometrical information. we call ZGRID.

BEGIN. Here. the external heat transfer coefficients for the top and bottom boundaries are obtained from the given Biot numbers. The value of dT_b/dz is prescribed arbitrarily. as in the previous example.

In the assigning the initial values of $T(I,J)$. we could have used T_∞ as the value. However. this could lead to the same difficulty (of having the solution "stuck" at the initial guess) as we encountered in Example 9. and would require a similar resolution. Here we choose a different practice: we set the initial values of $T(I,J)$ equal to $T_\infty + 1$.

OUTPUT. Much of the design here follows the pattern of previous examples. In the calculation of the cross-sectional area, the region occupied by the thickness of the fins is omitted. An overall Nusselt number is based on the average heat flux calculated over the outer area of the top and bottom plates; the extra surface area provided by the fins is not used in determining the average heat flux. The driving temperature difference for the definition of the overall heat transfer coefficient is taken as the difference $(T_\infty - T_b)$. In addition to the overall Nusselt number, separate Nusselt numbers for the top and bottom surfaces are calculated from the average heat flux at these surfaces.

It should again be remembered that there are many valid ways of defining Nusselt numbers. Here we could have calculated some average wall temperature T_w and used $(T_w - T_b)$ as the temperature difference for defining an *internal* heat transfer coefficient. What definition you use depends on the purpose of the calculation, convenience, and personal taste. The practices included in the examples in this book are simply illustrations; they are certainly not the only practices. You should feel free to present the overall results of your duct flow computations in any manner you choose.

For the final printout, we use the procedure (employed earlier in Example 8) of replacing the very small values of velocity in the fin region by zeroes.

PHI. The main points about the specification in PHI are: for the fin region, we set the viscosity equal to a large number and the conductivity equal to the conductivity of the fin material; the local $\partial T / \partial z$ is obtained from the dT_b/dz according to Eq. (9.45); and, for the temperature equation, the convective boundary conditions are specified at the top and bottom surfaces.

10.4-3 Additional Fortran Names

AMU	viscosity μ
ANU	average Nusselt number Nu
ANUB	average Nusselt number along the bottom wall
ANUT	average Nusselt number along the top wall
AR	area of control volume, dA
ASUM	cross-sectional area
BIBOT	Biot number along the bottom wall, Eq. (10.5)
BITOP	Biot number along the top wall, Eq. (10.5)
C	clearance $H - H_{fin}$
CNDFIN	thermal conductivity of fins, k_{fin}
COND	thermal conductivity of fluid, k_{fluid}

CP	specific heat c_p
DEN	density ρ
DH	hydraulic diameter D_h
DPDZ	axial pressure gradient dp/dz
DTDZ	axial temperature gradient $\partial T/\partial z$
DTBDZ	axial bulk temperature gradient dT_b/dz
FRE	the product fRe
H	height H
HBOT	heat transfer coefficient h_{bot}
HFIN	fin height H_{fin}
HP	heated perimeter
HTOP	heat transfer coefficient h_{top}
PI	the constant π
QBOT	total heat flow through bottom wall
QTOP	total heat flow through top wall
QW	average wall heat flux q_w
RE	Reynolds number
RHOCP	heat capacity ρc_p
S	fin spacing S
SHALF	half of S
T(I,J)	temperature T
TB	bulk temperature T_b
TEMP	temporary variable for $(T - T_\infty)/(T_b - T_\infty)$
TH	thickness of fin, t
TINF	surrounding fluid temperature T_∞
TSUM	$\int_w T\, dA$
W(I,J)	axial velocity w
WBAR	cross-sectional mean velocity $\overline{w}$
WP	wetted perimeter
WSUM	$\int_w dA$
X1	half thickness of fin
X2	SHALF - X1

10.4-4 Listing of ADAPT for Example 10

```
CCCCCCCCCCCCCCCCCCCCCCCCCCCCCCCCCCCCCCCCCCCCCCCCCCCCCCCCCCCCCCCCCCCCCCCCC
      SUBROUTINE ADAPT
C - - - - - - - - - - - - - - - - - - - - - - - - - - - - - - - - - - - - - - - -
C - - - - EXAMPLE 10 - - STAGGERED FIN ARRAY
C - - - - - - - - - - - - - - - - - - - - - - - - - - - - - - - - - - - - - - - -
$INCLUDE:'COMMON'
C***********************************************************************
```

```
      DIMENSION W(NI,NJ),T(NI,NJ)
      EQUIVALENCE (F(1,1,1),W(1,1)),(F(1,1,2),T(1,1))
C* *.*.*.*.*.*.*.*.*.*.*.-*.*.*.*.*.*.*.*.*.*.*.*.*.*.*.*.*.*.*.*.*.*.*.*
      ENTRY GRID
      HEADER='STAGGERED FIN ARRAY'
      PRINTF='PRINT10'
      PLOTF='PLOT10'
      H=1.
      S=H
      SHALF=0.5*S
      TH=0.2*S
      X1=0.5*TH
      X2=SHALF-X1
      HFIN=0.75*H
      C=H-HFIN
      CALL INTA4(NZX,3,NCVX(1),2,NCVX(2),8,NCVX(3),2)
      CALL DATA3(XZONE(1),X1,XZONE(2),X2-X1,XZONE(3),X1)
      CALL INTA4(NZY,3,NCVY(1),3,NCVY(2),5,NCVY(3),3)
      CALL DATA3(YZONE(1),C,YZONE(2),HFIN-C,YZONE(3),C)
      CALL ZGRID
      RETURN
C*-*.*.*.*.*.*.*.*.*.*.*.*.*.*.*.*.*.*.*.-*.*.*.*.*.*.*.*.*.*.*.*.*.*.*.*
      ENTRY BEGIN
      TITLE(1)='     W/WBAR      '
      TITLE(2)='(T-TINF)/(TB-TINF)'
      CALL INTA6(KSOLVE(1),1,KPRINT(1),1,KPRINT(2),1,
     1           KPLOT(1),1,KPLOT(2),1,LAST,10)
      CALL DATA9(AMU,1.,COND,1.,CP,1.,DEN,1.,DPDZ,-1.,TINF,0.,
     1           DTBDZ,1.,BITOP,5.,BIBOT,0.2)
      HTOP=BITOP*COND/H
      HBOT=BIBOT*COND/H
      RHOCP=DEN*CP
      CNDFIN=1.8*COND
      DO 100 J=1,M1
      DO 100 I=1,L1
         W(I,J)=0.
         T(I,J)=TINF+1.
  100 CONTINUE
      RETURN
C*.*.*.*.*.*.*.*.*.-*.*.*.*.*.*.*.*.*.-*.*.*.*.*.*.*.*.-*.*.*.*.*.*.*.*
      ENTRY OUTPUT
      IF(ITER.EQ.3) THEN
```

```
          KSOLVE(1)=0
          KSOLVE(2)=1
      ENDIF
      ASUM=0.
      WSUM=0.
      TSUM=0.
      DO 200 J=2,M2
      DO 200 I=2,L2
          AR=XCV(I)*YCV(J)
          IF(X(I).LT.X1.AND.Y(J).LT.HFIN) AR=0.
          IF(X(I).GT.X2.AND.Y(J).GT.C) AR=0.
          ASUM=ASUM+AR
          WSUM=WSUM+W(I,J)*AR
          TSUM=TSUM+W(I,J)*T(I,J)*AR
  200 CONTINUE
      WBAR=WSUM/ASUM
      TB=TSUM/(WSUM+SMALL)
      WP=S+2.*HFIN
      DH=4.*ASUM/WP
      RE=DH*WBAR*DEN/AMU
      FRE=-2.*DPDZ*DH/(DEN*WBAR**2+SMALL)*RE
      HP=S
      QW=DTBDZ*RHOCP*WSUM/HP
      ANU=QW*H/(COND*(TINF-TB)+SMALL)
      QTOP=0.
      QBOT=0.
      DO 210 I=2,L2
          QTOP=QTOP+FLUXMI(1,2)*XCV(I)
          QBOT=QBOT+FLUXJI(1,2)*XCV(I)
  210 CONTINUE
      ANUT=QTOP*H/(SHALF*COND*(TINF-TB)+SMALL)
      ANUB=QBOT*H/(SHALF*COND*(TINF-TB)+SMALL)
      DO 220 IUNIT=IU1,IU2
          IF(ITER.EQ.0) WRITE(IUNIT,230)
  230     FORMAT(1X,'ITER',2X,'W(8,8)',5X,'T(8,8)',6X,
     1    'FRE',8X,'NU',7X,'NU(TOP)',3X,'NU(BOTTOM)')
          WRITE(IUNIT,240) ITER,W(8,8),T(8,8),FRE,ANU,ANUT,ANUB
  240     FORMAT(1X,I2,1P6E11.3)
  220 CONTINUE
      IF(ITER.EQ.LAST) THEN
          DO 250 J=1,M1
          DO 250 I=1,L1
```

```
            W(I,J)=W(I,J)/WBAR
            IF(W(I,J).LT.1.E-10) W(I,J)=0.
            T(I,J)=(T(I,J) TINF)/(TB TINF)
  250     CONTINUE
          CALL PRINT
C------
COME HERE TO FILL IBLOCK(I,J) BEFORE CALLING PLOT
          DO 260 J=2,M2
          DO 260 I=2,L2
            IF(X(I).LT.X1.AND.Y(J).LT.HFIN) IBLOCK(I,J)=1
            IF(X(I).GT.X2.AND.Y(J).GT.C) IBLOCK(I,J)=1
  260     CONTINUE
C------
          CALL PLOT
        ENDIF
        RETURN
C*-*-*-*-*-*-*-*-*-*-*-*-*-*-*-*-*-*-*-*-*-*-*-*-*-*-*-*-*-*-*-*-*
        ENTRY PHI
        IF(NF.EQ.1) THEN
          DO 300 J=2,M2
          DO 300 I=2,L2
            GAM(I,J)=AMU
            IF(X(I).LT.X1.AND.Y(J).LT.HFIN) GAM(I,J)=BIG
            IF(X(I).GT.X2.AND.Y(J).GT.C) GAM(I,J)=BIG
            SC(I,J)= DPDZ
  300     CONTINUE
        ENDIF
        IF(NF.EQ.2) THEN
          DO 310 J=2,M2
          DO 310 I=2,L2
            GAM(I,J)=COND
            IF(X(I).LT.X1.AND.Y(J).LT.HFIN) GAM(I,J)=CNDFIN
            IF(X(I).GT.X2.AND.Y(J).GT.C) GAM(I,J)=CNDFIN
            TEMP=(T(I,J) TINF)/(TB TINF+SMALL)
            DTDZ=TEMP*DTBDZ
            SC(I,J)=-RHOCP*DTDZ*W(I,J)
  310     CONTINUE
        ENDIF
COME HERE TO SPECIFY BOUNDARY CONDITIONS
        DO 320 J=2,M2
          KBC11(J)=2
          KBCL1(J)=2
```

```
320 CONTINUE
    IF(NF.EQ.2) THEN
        DO 330 I=2,L2
            KBCMI(I)=2
            KBCJI(I)=2
            FLXCMI(I)=HTOP*TINF
            FLXPMI(I)=-HTOP
            FLXCJI(I)=HBOT*TINF
            FLXPJI(I)=-HBOT
330     CONTINUE
    ENDIF
    RETURN
    END
CCCCCCCCCCCCCCCCCCCCCCCCCCCCCCCCCCCCCCCCCCCCCCCCCCCCCCCCCCCCCCCCCCCCCCCC
```

10.4-5 Results for Example 10

```
RESULTS OF CONDUCT FOR CARTESIAN COORDINATE SYSTEM
****************************************************


. . . . . . . . . . . . . . . . . . . . . . . . . . . . . . . . . . . . . . . . . .

STAGGERED FIN ARRAY

. . . . . . . . . . . . . . . . . . . . . . . . . . . . . . . . . . . . . . . . . .
```

ITER	W(8,8)	T(8,8)	FRE	NU	NU(TOP)	NU(BOTTOM)
0	0.000E+00	1.000E+00	0.000E+00	0.000E+00	0.000E+00	0.000E+00
1	1.153E-02	1.000E+00	8.069E+01	-2.720E-03	0.000E+00	0.000E+00
2	1.153E-02	1.000E+00	8.069E+01	-2.720E-03	0.000E+00	0.000E+00
3	1.153E-02	1.000E+00	8.069E+01	-2.720E-03	0.000E+00	0.000E+00
4	1.153E-02	-2.301E-03	8.069E+01	1.183E+00	2.125E+00	2.403E-01
5	1.153E-02	-2.393E-03	8.069E+01	1.124E+00	2.000E+00	2.472E-01
6	1.153E-02	-2.401E-03	8.069E+01	1.117E+00	1.987E+00	2.482E-01
7	1.153E-02	-2.402E-03	8.069E+01	1.117E+00	1.985E+00	2.483E-01
8	1.153E-02	-2.402E-03	8.069E+01	1.116E+00	1.985E+00	2.483E-01
9	1.153E-02	-2.402E-03	8.069E+01	1.116E+00	1.984E+00	2.483E-01
10	1.153E-02	-2.402E-03	8.069E+01	1.116E+00	1.984E+00	2.483E-01

```
I =    1       2       3       4       5       6       7
X = 0.00E+00 2.50E-02 7.50E-02 1.19E-01 1.56E-01 1.94E-01 2.31E-01
```

```
I =     8       9      10      11      12      13      14
X = 2.69E·01 3.06E·01 3.44E·01 3.81E·01 4.25E·01 4.75E·01 5.00E·01


J =     1       2       3       4       5       6       7
Y = 0.00E+00 4.17E·02 1.25E·01 2.08E·01 3.00E·01 4.00E-01 5.00·01

J =     8       9      10      11      12      13
Y = 6.00E·01 7.00E·01 7.92E·01 8.75E·01 9.58E·01 1.00E+00
```

******   W/WBAR   ******

```
I =     1       2       3       4       5       6       7
J
13   0.00E+00 0.00E+00 0.00E+00 0.00E+00 0.00E+00 0.00E+00 0.00E+00
12   6.49E·01 6.51E·01 6.63E·01 6.78E·01 6.87E·01 6.82E·01 6.58E·01
11   1.25E+00 1.26E+00 1.31E+00 1.38E+00 1.44E+00 1.45E+00 1.41E+00
10   8.15E·01 8.30E·01 9.52E·01 1.27E+00 1.48E+00 1.61E+00 1.62E+00
 9   0.00E+00 0.00E+00 0.00E+00 4.45E·01 1.04E+00 1.39E+00 1.55E+00
 8   0.00E+00 0.00E+00 0.00E+00 3.76E·01 9.39E·01 1.31E+00 1.49E+00
 7   0.00E+00 0.00E+00 0.00E+00 3.67E·01 9.20E·01 1.29E+00 1.47E+00
 6   0.00E+00 0.00E+00 0.00E+00 3.67E·01 9.21E·01 1.29E+00 1.48E+00
 5   0.00E+00 0.00E+00 0.00E+00 3.69E·01 9.27E·01 1.31E+00 1.52E+00
 4   0.00E+00 0.00E+00 0.00E+00 3.57E·01 8.98E·01 1.28E+00 1.52E+00
 3   0.00E+00 0.00E+00 0.00E+00 3.07E·01 7.59E·01 1.08E+00 1.29E+00
 2   0.00E+00 0.00E+00 0.00E+00 1.70E·01 3.83E·01 5.20E·01 6.07E·01
 1   0.00E+00 0.00E+00 0.00E+00 0.00E+00 0.00E+00 0.00E+00 0.00E+00


I =     8       9      10      11      12      13      14
J
13   0.00E+00 0.00E+00 0.00E+00 0.00E+00 0.00E+00 0.00E+00 0.00E+00
12   6.07E·01 5.20E·01 3.83E·01 1.70E·01 0.00E+00 0.00E+00 0.00E+00
11   1.29E+00 1.08E+00 7.59E·01 3.07E·01 0.00E+00 0.00E+00 0.00E+00
10   1.52E+00 1.28E+00 8.98E·01 3.57E·01 0.00E+00 0.00E+00 0.00E+00
 9   1.52E+00 1.31E+00 9.27E·01 3.69E·01 0.00E+00 0.00E+00 0.00E+00
 8   1.48E+00 1.29E+00 9.21E·01 3.67E·01 0.00E+00 0.00E+00 0.00E+00
 7   1.47E+00 1.29E+00 9.20E·01 3.67E·01 0.00E+00 0.00E+00 0.00E+00
 6   1.49E+00 1.31E+00 9.39E·01 3.76E·01 0.00E+00 0.00E+00 0.00E+00
 5   1.55E+00 1.39E+00 1.04E+00 4.45E·01 0.00E+00 0.00E+00 0.00E+00
 4   1.62E+00 1.61E+00 1.48E+00 1.27E+00 9.52E·01 8.30E·01 8.15E·01
```

```
3    1.41E+00 1.45E+00 1.44E+00 1.38E+00 1.31E+00 1.26E+00 1.25E+00
2    6.58E-01 6.82E-01 6.87E-01 6.78E-01 6.63E-01 6.51E-01 6.49E-01
1    0.00E+00 0.00E+00 0.00E+00 0.00E+00 0.00E+00 0.00E+00 0.00E+00
```

****** (T-TINF)/(TB-TINF) ******
. . . .

```
I =    1        2        3        4        5        6        7
J
13   ·4.10E+02 3.88E-01 3.87E-01 3.87E-01 3.86E-01 3.85E-01 3.85E-01
12    4.68E-01 4.68E-01 4.67E-01 4.66E-01 4.66E-01 4.65E-01 4.64E-01
11    6.24E-01 6.24E-01 6.23E-01 6.21E-01 6.20E-01 6.17E-01 6.15E-01
10    7.66E-01 7.66E-01 7.64E-01 7.58E-01 7.55E-01 7.53E-01 7.49E-01
9     8.77E-01 8.77E-01 8.77E-01 8.76E-01 8.76E-01 8.77E-01 8.75E-01
8     9.68E-01 9.68E-01 9.70E-01 9.76E-01 9.82E-01 9.87E-01 9.88E-01
7     1.05E+00 1.05E+00 1.05E+00 1.06E+00 1.07E+00 1.08E+00 1.08E+00
6     1.12E+00 1.12E+00 1.12E+00 1.13E+00 1.15E+00 1.16E+00 1.16E+00
5     1.17E+00 1.17E+00 1.17E+00 1.19E+00 1.20E+00 1.21E+00 1.22E+00
4     1.20E+00 1.20E+00 1.20E+00 1.22E+00 1.23E+00 1.25E+00 1.26E+00
3     1.21E+00 1.21E+00 1.21E+00 1.23E+00 1.24E+00 1.25E+00 1.27E+00
2     1.21E+00 1.21E+00 1.21E+00 1.22E+00 1.23E+00 1.24E+00 1.25E+00
1    ·4.10E+02 1.20E+00 1.21E+00 1.21E+00 1.22E+00 1.23E+00 1.24E+00


I =    8        9        10       11       12       13    ·   14
J
13    3.85E-01 3.86E-01 3.89E-01 3.96E-01 4.33E-01 4.38E-01 4.10E+02
12    4.63E-01 4.64E-01 4.67E-01 4.74E-01 4.86E-01 4.90E-01 4.90E-01
11    6.12E-01 6.09E-01 6.06E-01 6.03E-01 6.02E-01 6.02E-01 6.02E-01
10    7.44E-01 7.38E-01 7.31E-01 7.23E-01 7.17E-01 7.15E-01 7.14E-01
9     8.70E-01 8.63E-01 8.54E-01 8.43E-01 8.35E-01 8.31E-01 8.31E-01
8     9.86E-01 9.79E-01 9.69E-01 9.58E-01 9.49E-01 9.45E-01 9.44E-01
7     1.08E+00 1.08E+00 1.07E+00 1.06E+00 1.05E+00 1.04E+00 1.04E+00
6     1.16E+00 1.16E+00 1.15E+00 1.14E+00 1.13E+00 1.12E+00 1.12E+00
5     1.23E+00 1.22E+00 1.22E+00 1.20E+00 1.19E+00 1.19E+00 1.19E+00
4     1.26E+00 1.27E+00 1.26E+00 1.26E+00 1.25E+00 1.25E+00 1.25E+00
3     1.27E+00 1.28E+00 1.28E+00 1.28E+00 1.28E+00 1.28E+00 1.28E+00
2     1.26E+00 1.27E+00 1.27E+00 1.28E+00 1.28E+00 1.28E+00 1.28E+00
1     1.25E+00 1.26E+00 1.26E+00 1.27E+00 1.27E+00 1.27E+00·4.10E+02
```

10.4-6 Discussion of Results

The solution of the nonlinear temperature equation can be seen to have converged rapidly within a few iterations. The computed value of the overall Nusselt number lies, as expected, between the values of the separate Nusselt numbers for the top and bottom surfaces. We should also discuss the relationship of these Nusselt numbers to the Biot numbers specified as boundary conditions; however, we shall defer this to the end of this subsection.

In the field values of $w/\overline{w}$, the fin regions are easily noticeable due to the zero values there. For the given geometry of the staggered fin array, we expect a type of *inverted* symmetry in the velocity distribution. In other words, if we read the values of velocity along any vertical line in the left half of the domain going from bottom to top, then they should exactly correspond to the values along a symmetrically located line in the right half of the domain when read from top to bottom. Our computed solution does exhibit this inverted symmetry exactly. For example, the values of $w/\overline{w}$ along $I = 5$ for $J = 1, 2, 3....,$ etc. are identical to the values along $I = 10$ for $J = 13, 12, 11....,$ etc.

The temperature field would also have shown such symmetry if we had used equal Biot numbers on the top and bottom surfaces. However, because of the use of unequal Biot numbers, the values of the dimensionless temperature are much higher near the bottom boundary. To make this discussion more concrete, let us consider that the duct fluid is warmer than the surrounding fluid. For this situation, we would normally expect the highest temperature near the center of the duct and lower temperatures near the boundaries. However, because of the low Biot number at the bottom surface, our computed temperature distribution is significantly unsymmetrical: the temperatures in the lower half of the duct are appreciably greater than those in the upper half.

The temperature T_w at the bottom boundary, which should normally be less than T_b, is actually greater than T_b. If, for the bottom boundary, we calculate an *internal* heat transfer coefficient based on the temperature difference $(T_w - T_b)$, it will have a negative value. This simply means that, for such an unsymmetrical temperature field, the temperature difference $(T_w - T_b)$ is not the appropriate driving force for the heat transfer at the bottom boundary.

Let us return to the values of the separate Nusselt numbers calculated for the top and bottom boundaries. They are related to the respective Biot numbers at those boundaries. The only difference is that, for a given heat flux, a Biot number is based on the outside temperature difference $(T_\infty - T_w)$ as the driving temperature drop, while our Nusselt numbers are based on the overall temperature drop $(T_\infty - T_w)$. Normally, the magnitude of $(T_\infty - T_b)$ will be greater than that of $(T_\infty - T_w)$; therefore, the resulting Nusselt number will be less than the corresponding Biot number. This indeed is the case at the top boundary, where our computed Nusselt number of 1.984 is clearly less than Bi_{top}. However, at the bottom boundary, the magnitude of $(T_\infty - T_b)$ is smaller than that of $(T_\infty - T_w)$; as a result, the computed Nusselt number for the bottom boundary can be seen to be somewhat greater than the prescribed value of Bi_{bot}.

10.4-7 Final Remarks

As described in Section 9.6-6, the convective boundary condition employed here is a more general boundary condition from which the boundary conditions of constant wall temperature or constant heat flux can be derived. Also, the general boundary condition is more likely to be applicable in practical situations.

Through the four examples in this chapter, we have seen many features of the treatment of duct flows. They include the representation of fins, conjugate heat transfer, and a variety of thermal boundary conditions. With this background, you will now be in a position to apply CONDUCT to a wide variety of duct flow problems. While making imaginative use of the program, you should nevertheless be careful in setting only those problems for which a fully developed solution exists. One of the common mistakes is to apply thermal boundary conditions that either do not imply a fully developed solution or lead to an uninteresting situation like the one described in Section 9.5. Therefore, a good fundamental understanding of what you propose to compute is essential.

In the next chapter, we shall consider some advanced duct flows such as a nonNewtonian flow and a turbulent flow. We shall also apply CONDUCT to problems involving potential flow and flow through porous materials.

Problems

10.1 Use CONDUCT to solve the fully developed flow and heat transfer in a circular tube. For this geometry, the exact value of fRe is 64. For the uniform-heat-flux boundary condition, the Nusselt number is 4.364. When the wall temperature is axially and peripherally uniform, the Nusselt number becomes 3.657. Verify that the numerical solution approaches these values as you refine the grid.

10.2 Refer to the paper by Sparrow and Patankar (1977). Solve the problem for different values of the external heat transfer coefficient and compare with the results given in the paper.

10.3 The problem, shown in the figure, involves the fully developed flow and heat transfer in a circular tube with internal fins. There are six uniformly distributed fins. Other geometrical details are shown in the figure. Calculate the product fRe and $w/\overline{w}$. Use the hydraulic diameter to define Re. Also calculate the fully developed temperature distribution under the condition that heat is generated at the rate S per unit volume in the solid material forming the tube wall and fins. (This can normally be arranged by passing an electric current through the tube.) The outer surface of the tube is adiabatic. Calculate the Nusselt number based on the average heat flux and average temperature at the fluid-solid interface. The tube diameter may be used as the length scale for Nu. (Make sure that the value of $\partial T/\partial z$ is consistent with the value of S.)

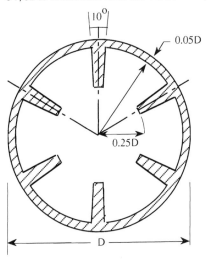

Problem 10.3

10.4 Design an ADAPT to calculate the fully developed flow and heat transfer in the shrouded fin array shown. The flow occurs only in the direction normal to the plane of the diagram. The shroud is adiabatic, while the base of the array transfers heat to the surrounding fluid at the temperature T_∞ with a constant heat transfer coefficient h_e. Take $h_e H/k = 3.5$, where k is the conductivity of the duct fluid. Also $k_{fin}/k_{fluid} = 2.8$. Arrange the printout of dimensionless axial velocity, dimensionless temperature, overall Nusselt number, and the product fRe.

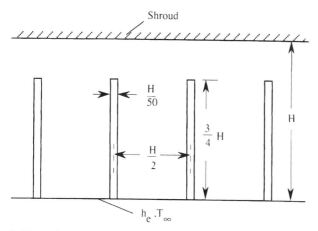

Problem 10.4

10.5 Refer to the paper by Sparrow, Baliga, and Patankar (1978), which deals with a shrouded fin array. Perform computations for the same cases and compare with the published results.

10.6 The figure shows the cross section of a rectangular duct with a partial thin partition in the middle. The partition offers no resistance to heat transfer. The thermal boundary conditions are given by a uniform heat flux of 100 at the side walls and a heat flux of 300 at the bottom wall. Write the ADAPT routine and solve the fully developed flow and heat transfer in the duct. Arrange the printout of $w/\overline{w}$ and a suitable dimensionless temperature field. Also printout fRe and Nu.

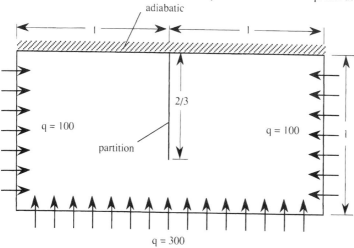

Problem 10.6

10.7 Compute the fully developed velocity and temperature field in a duct of semicircular cross section for the thermal boundary condition of a constant wall temperature for the flat wall and zero heat flux at the circular wall. Print out the distribution of $w/\bar{w}$ and $(T - T_w)/(T_b - T_w)$ over the duct cross section and the values of fRe and Nu. You may calculate Nu from the average heat flux on the heated wall and use the duct diameter as the characteristic dimension. Also print out the variation of the local Nusselt number along the heated wall. The exact value of the fRe (based on the hydraulic diameter) is 63.07.

10.8 A square duct has a rectangular blockage inside. The blockage is made of a material of very low conductivity. At the outer surface of the duct, the thermal boundary condition is specified by an external heat transfer coefficient $h_e = 2.5$ and an outside fluid temperature $T_\infty = 20$. Take the conductivity of the duct fluid as $k = 2$. The dimensions of the duct are shown in the figure. Print out the distributions of $w/\bar{w}$ and $(T - T_\infty)/(T_b - T_\infty)$ and the values of fRe and Nu.

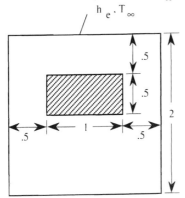

Problem 10.8

10.9 An electronic circuit board can be idealized as heat generating blocks placed over an adiabatic plate with another adiabatic plate acting as a shroud. The cooling fluid flows normal to the plane of the diagram. The conductivity of the solid material of the blocks is four times the conductivity of the fluid. There is a uniform heat generation in each block. Calculate the fully developed flow and heat transfer. Print out the distributions of $w/\bar{w}$ and $(T - T_w)/(T_b - T_w)$ and the values of fRe and Nu. Take T_w as the average temperature of the surface of the blocks.

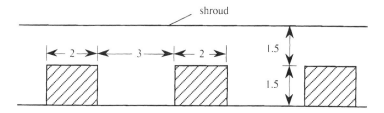

Problem 10.9

10.10 For the annular square duct shown, the inside surfaces are maintained at a constant temperature of T_∞, while the outer surfaces exchange heat with an external fluid at T_∞ and heat transfer coefficient h. The value of h is given by $hD/k = 5.8$ where k is the conductivity of the fluid. Write the ADAPT routine to calculate the fully developed flow and heat transfer in this duct. Print out the distributions of $w/\bar{w}$ and $(T - T_\infty)/(T_b - T_\infty)$ and the values of fRe and Nu. (You may define Nu in any reasonable manner and implement that definition in the program.)

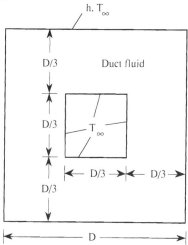

Problem 10.10

10.11 The figure shows an infinite array of rods of square (1×1) cross section. A fluid flows between the rods in the z direction normal to the diagram. The rod surfaces are maintained at a uniform temperature T_w in a given cross section, and this T_w rises linearly in the z direction. Write the ADAPT routine to calculate the velocity and temperature fields in the fully developed region. Also calculate fRe and Nu.

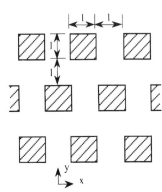

Problem 10.11

10.12 The figure shows a duct flow situation. The inner and outer duct diameters are d and D respectively, and D/d = 2. The inner cylinder of diameter d is made of solid material, while two different fluids flow in the annular space. The fluid in the lower half has a viscosity μ_1, that in the upper half has μ_2, and $\mu_2/\mu_1 = 2.5$. Both fluids have the same density ρ. Calculate the dimensionless velocity $w/\bar{w}$ and product fRe (where μ_1 is used in the definition of Re). The heat transfer situation is characterized by a heat generation rate S per unit volume in the upper half of the solid material forming the inner cylinder and a heat flux 0.25 Sd through the bottom 60° arc of the outer cylinder. The remainder of the outer cylinder is insulated. Also, the lower half of the inner cylinder is made of a zero-conductivity material. The two fluids and the heat-generating solid material all have the same conductivity k and the same specific heat c_p. Calculate the dimensionless temperature field and the Nusselt number Nu, where Nu is defined as h(D–d)/k, where h is based on the average temperature and the area of the heated surface. The heated surface consists of the 60° arc of the outer cylinder and the upper half of the inner cylinder.

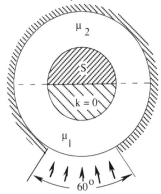

Problem 10.12

10.13 A duct flow situation is shown in the figure. Two different fluids of viscosity μ_1 and μ_2 flow in the regions shown. $\mu_2/\mu_1 = 4.5$. The right-wall temperature T_w remains constant in the axial direction. Both fluids have the same density ρ. Calculate the dimensionless velocity field $w/\bar{w}$ and the product fRe. You may use μ_1 in the definition of Re. The two fluids have the same conductivity k and the same specific heat c_p. Calculate the dimensionless temperature field and a Nusselt number based on the average heat flux over the right wall.

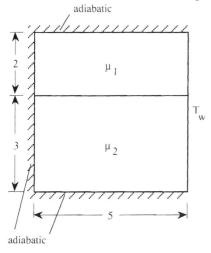

Problem 10.13

10.14 Consider the duct of semicircular cross section with a thin partition in the middle. The partition offers no resistance to heat transfer. The thermal boundary condition is given by a wall temperature T_w that remains constant in the axial direction and over the circular perimeter; the flat wall of the duct is insulated. Calculate the dimensionless velocity $w/\bar{w}$ and temperature. Also calculate the product fRe and the Nusselt number, based on the duct diameter.

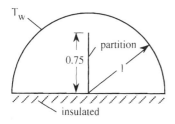

Problem 10.14

10.15 Fluid flows in the direction normal to the diagram shown. The top and bottom plates are adiabatic. The heat generating blocks are attached to them in a staggered fashion. The heat generation rate in the bottom blocks is twice the rate in the top blocks. The conductivity of the solid material is 2.5 times that of the fluid. Calculate the fully developed velocity and temperature fields. Arrange the printout of $w/\overline{w}$ and a suitable dimensionless temperature.

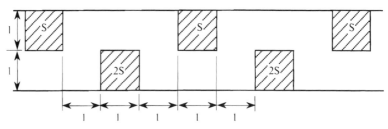

Problem 10.15

10.16 Write an ADAPT to calculate the fully developed flow and heat transfer in the duct shown. The flow is only in the z direction. The thermal boundary conditions as shown include some adiabatic surfaces, while the remaining surfaces transfer heat to a surrounding fluid at temperature T_∞ with a constant heat transfer coefficient h_e. Take $h_e H/k = 3.5$, where k is the conductivity of the fluid in the duct. Arrange the printout of dimensionless axial velocity, dimensionless temperature, fRe, and the overall Nusselt number (based on H, $T_\infty - T_b$, and the average heat flux over the nonadiabatic wall surface.)

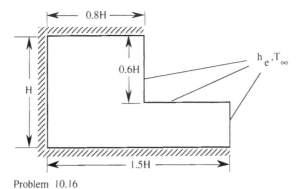

Problem 10.16

10.17 Consider the flow in an annulus formed by a solid rod of diameter 2 placed in an adiabatic pipe of diameter 5. The upper and lower halves of the annulus contain fluids of different viscosity, $\mu = 0.2$ and $\mu = 1$ as shown; both fluids have the same conductivities $k = 2$. The lower half of the solid rod has zero conductivity and no heat generation while the upper half has a conductivity of $k =$

5 and S = 50. Write the ADAPT routine and calculate the fully developed velocity and temperature fields. Arrange the printout of dimensionless velocity and dimensionless temperature. Also printout fRe and Nu.

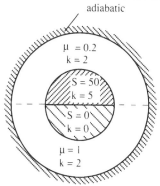

Problem 10.17

10.18 The figure shows an infinite array of rods of square cross section. A fluid flows between the rods in the z direction normal to the plane of the figure. The rods in alternate rows have a uniform heat generation rate of S while the remaining rods generate heat the a rate 2S. Take $k_{solid}/k_{fluid} = 2.1$. Write the ADAPT routine to calculate the fully developed velocity and temperature field.

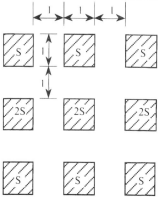

Problem 10.18

10.19 Refer to the paper by Sparrow, Patankar, and Shahrestani (1978). Compute the heat transfer situation situation for the variable external heat transfer coefficient and compare with the published results.

10.20 Refer to the book by Shah and London (1978). Find two interesting duct geometries for which fully developed results are given. Use CONDUCT for a range of parameter values for the chosen problems and compare the results.

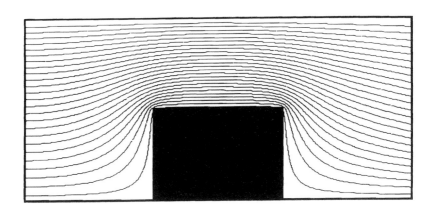

Potential flow around a block

Chapter 11

Additional Adaptations of CONDUCT

So far, we have seen how to apply CONDUCT to steady and unsteady heat conduction, various duct flows, and the associated heat transfer. In this chapter, we shall study the use of the program for more complex duct flows and for other analogous phenomena such as potential flow and flow in porous materials. Once again, the objective in presenting these examples is to stimulate your imagination for further uses of the program. Once you appreciate that, within some overall limits, the program can be used for a wide variety of applications, you will begin to explore the virtually endless opportunities that lie ahead.

Since detailed descriptions of nonNewtonian flow, turbulent flow, or flow in porous media are obviously outside the scope of this book, the relevant information given is relatively short. Nevertheless, there should be no difficulty in understanding the main concepts employed here.

11.1 NonNewtonian Flow in a Semicircular Duct (Example 11)

11.1-1 Problem Description

For a Newtonian flow in a duct, the viscous shear stress experienced by the axial flow is proportional to the gradient of the axial velocity. In a nonNewtonian flow, however, this relationship between the stress and the velocity gradient is more complex. For certain type of fluids, the stress can be taken as proportional to a certain power of the velocity gradient. These fluids are known as the power-law fluids.

If we try to fit the stress relationship for a power-law fluid into our standard framework of a Newtonian flow, we shall find that the resulting apparent viscosity depends on the velocity gradients. Thus, from a computational point of view, the solution of a nonNewtonian flow is simply a *nonlinear* problem, in which the viscosity is some function of the unknown velocity gradients.

We shall consider the fully developed nonNewtonian flow in the semicircular duct shown in Fig. 11.1. For a power-law fluid, the apparent viscosity μ is given by

$$\mu = K \left[\left(\frac{\partial w}{\partial r} \right)^2 + \frac{1}{r^2} \left(\frac{\partial w}{\partial \theta} \right)^2 \right]^{(n-1)/2} \tag{11.1}$$

where K is the consistency constant and n is the power-law index. The quantities K and n are the properties of the fluid. The value of n = 1 corresponds to a Newtonian flow. Here we shall take the value of n as 0.5 to illustrate a substantial departure from the Newtonian behavior.

For the definition of the Reynolds number, a reference viscosity is defined as

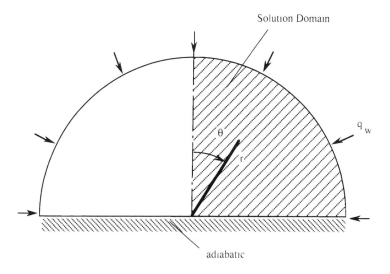

Fig. 11.1 NonNewtonian flow in a semicircular duct

$$\mu_{ref} = K \left(\frac{\overline{w}}{D_h} \right)^{(n-1)} \tag{11.2}$$

where D_h is the hydraulic diameter of the duct.

The thermal boundary condition is given by a uniform heat flux over the curved perimeter with the flat part of the perimeter remaining adiabatic. Because of symmetry. we shall perform our computation over only the right half of the duct.

11.1-2   Design of ADAPT

GRID. Since the geometry is rather simple. we construct a uniform grid by the use of EZGRID. For the θr coordinates. the value of MODE is given as 3 and $R(1)$ is set equal to zero.

BEGIN. In this problem. the velocity equation is nonlinear. but the temperature equation is linear. Therefore. we shall perform the first 15 iterations to get a converged solution for w and the final 3 iterations to obtain T. Thus. LAST is set equal to 18.

The usual fluid properties are specified in BEGIN. Here. RK is the consistency constant K in Eq. (11.1). and POWER is the power-law index n. The constant viscosity AMU is used. as will be explained later. only to get a starting solution. REGAM is an underrelaxation factor. also to be discussed later.

OUTPUT. For the calculation of the Reynolds number, we use the reference viscosity μ_{ref} defined in Eq. (11.2). This is given the Fortran name AMUR. For the curved perimeter with a uniform heat flux, an average wall temperature TWAV is calculated. Here, XCV(I) denotes the angular extent of each control-volume face. The remaining parts of OUTPUT should by now be very familiar to you.

PHI. Since the apparent viscosity μ as given by Eq. (11.1) depends on the velocity gradients, we shall not get any meaningful value of μ from the zero initial guess for W(I,J). To overcome this difficulty, we use the following strategy. We shall perform the first iteration for a Newtonian flow using the constant viscosity AMU. This will produce a reasonable velocity field that can be used to calculate the apparent viscosity μ from Eq. (11.1). Then we shall continue to update the values of μ after every iteration and proceed to the converged solution.

This procedure, however, has one major shortcoming. In the first iteration, the constant viscosity AMU will, under the action of the prescribed value of DPDZ, produce corresponding values of velocity w. The magnitude of μ that is obtained from Eq. (11.1) will depend on the magnitude of w. With these values of μ and the *same* value of DPDZ, the resulting magnitudes of w could be entirely different. These will in turn give very different μ values in the next iteration and consequently very different w values, and so on. The solution may require many iterations to attain convergence.

There is a way to overcome this shortcoming. After the first iteration, we calculate the values of μ by using the velocity field given by the Newtonian solution. Then we calculate an average value of μ. This may be quite different from the constant viscosity AMU we used in the first iteration. To get approximately the same magnitudes of w with the new field of μ, we now *adjust* the value of the pressure gradient DPDZ by

$$DPDZ=DPDZ*GAMAV/AMU \qquad\qquad (11.3)$$

where GAMAV is the average value of μ. This adjustment can, in principle, be performed after every iteration. However, its major impact would occur only after the first iteration. For this reason, we perform the adjustment only once.

To summarize, we calculate the Newtonian flow in the first iteration, get the corresponding values of the apparent viscosity μ, find their average, adjust the pressure gradient accordingly, and continue to perform the subsequent iterations by updating the values of μ but without changing the pressure gradient anymore.

Whereas the major swings in the values of w from iteration to iteration are eliminated by this procedure, we can further control the changes by underrelaxing w. This was explained in Section 5.7. Alternatively, we can underrelax the apparent viscosity μ according to the practice outlined in Eq. (5.65). Here we illustrate the underrelaxation of μ by using the underrelaxation factor REGAM.

Obtaining μ from Eq. (11.1) requires a calculation of the gradients of w. For a given control volume, we calculate the gradients in the θ direction from the interpolated values of w at the two faces that are normal to the θ direction. These interpolated values are denoted by WP and WM. Since we are using a uniform grid, the interpolation reduces to averages of the w values at the neighboring grid points, with a suitable adjustment for the near-boundary control volumes. A similar procedure is used for the gradients in the radial direction. If we had used a nonuniform grid, we would need a more elaborate interpolation. This will be illustrated in the context of the turbulent flow, Example 13, presented in Section 11.3.

Once the gradients are calculated, we use Eq. (11.1) to obtain μ. Finally, the GAM(I,J) array is filled with an appropriately underrelaxed value of μ. The calculation of μ is followed by the adjustment of DPDZ, which has been already described. This adjustment is performed only when ITER equals 1.

The remainder of the PHI routine should appear very familiar. It involves the specification of SC(I,J) for the velocity, both GAM(I,J) and SC(I,J) for the temperature, and appropriate boundary conditions for the two dependent variables.

11.1-3   Additional Fortran Names

AMU	Newtonian viscosity μ
AMUR	reference viscosity μ_{ref}, Eq. (11.2)
ANU	average Nusselt number Nu
AR	area of control volume, dA
ASUM	cross-sectional area
COND	thermal conductivity k
CP	specific heat c_p
DEN	density ρ
DH	hydraulic diameter D_h
DPDZ	axial pressure gradient dp/dz
DTDZ	axial temperature gradient ∂T/∂z
DWDX	$(1/r)(\partial w/\partial \theta)$
DWDY	∂w/∂r
FRE	the product fRe
GAMAV	average value of nonNewtonian viscosity μ
GAMT	temporary value for nonNewtonian viscosity μ
HP	heated perimeter
PI	the constant π
POWER	power-law index n, Eq. (11.1)
QW	wall heat flux q_w

RE	Reynolds number
REGAM	underrelaxation factor for nonNewtonian viscosity
REL	temporary variable for $1-$REGAM
RHOCP	heat capacity ρc_p
RK	consistency constant K, Eq. (11.1)
T(I,J)	temperature T
TB	bulk temperature T_b
TSUM	$\int_w T\, dA$
TWAV	average temperature of heated wall T_w
W(I,J)	axial velocity w
WBAR	cross-sectional mean velocity $\overline{w}$
WM	interface velocity
WP	wetted perimeter or interface velocity
WSUM	$\int_w dA$

11.1-4 Listing of ADAPT for Example 11

```
CCCCCCCCCCCCCCCCCCCCCCCCCCCCCCCCCCCCCCCCCCCCCCCCCCCCCCCCCCCCCCCCCCCCCCCCCC
      SUBROUTINE ADAPT
C.......................................................................
C-----EXAMPLE 11 -- NON-NEWTONIAN FLOW IN A SEMICIRCULAR DUCT
C.......................................................................
$INCLUDE:'COMMON'
C************************************************************************
      DIMENSION W(NI,NJ),T(NI,NJ)
      EQUIVALENCE (F(1,1,1),W(1,1)),(F(1,1,2),T(1,1))
C*.*.*.*.*.*.*.*.*.*.*.*.*.*.*.*.*.*.*.*.*.*.*.*.*.*.*.*.*.*.*.*.*.*.*.*.*
      ENTRY GRID
      HEADER='NON NEWTONIAN FLOW IN A SEMICIRCULAR DUCT'
      PRINTF='PRINT11'
      PLOTF='PLOT11'
      CALL DATA2(R(1),0.,PI,3.14159)
      CALL DATA2(XL,0.5*PI,YL,1.)
      CALL INTA3(MODE,3,NCVLX,10,NCVLY,16)
      CALL EZGRID
      RETURN
C*.*.*.*.*.*.*.*.*.*.*.*.*.*.*.*.*.*.*.*.*.*.*.*.*.*.*.*.*.*.*.*.*.*.*.*.*
      ENTRY BEGIN
      TITLE(1)='    W/WBAR      '
      TITLE(2)='(T-TWAV)/(TB-TWAV)'
      CALL INTA6(KSOLVE(1),1,KPRINT(1),1,KPRINT(2),1,
     1           KPLOT(1),1,KPLOT(2),1,LAST,18)
```

```
      CALL DATA9(DEN,1.,REGAM,0.8,RK,1.,POWER,0.5,AMU,1.,COND,1.,
     1           CP,1.,DPDZ, 1.,QW,1.)
      RHOCP=DEN*CP
C
C·· SINCE THE ZERO DEFAULT VALUES OF W(I,J) AND T(I,J) ARE SATISFACTORY,
C   THESE ARRAY ARE NOT FILLED HERE.
C
      RETURN
C*·*·*·*·*·*·*·*·*·*·*·*·*·*·*·*·*·*·*·*·*·*·*·*·*·*·*·*·*·*·*·*·*·*
      ENTRY OUTPUT
      IF(ITER.EQ.15) THEN
         KSOLVE(1)=0
         KSOLVE(2)=1
      ENDIF
      ASUM=0.
      WSUM=0.
      TSUM=0.
      DO 200 J=2,M2
      DO 200 I=2,L2
         AR=XCV(I)*YCVR(J)
         ASUM=ASUM+AR
         WSUM=WSUM+W(I,J)*AR
         TSUM=TSUM+W(I,J)*T(I,J)*AR
  200 CONTINUE
      WBAR=WSUM/ASUM
      TB=TSUM/(WSUM+SMALL)
      WP=R(M1)+0.5*PI*R(M1)
      HP=0.5*PI*R(M1)
      DH=4.*ASUM/WP
      AMUR=RK*(WBAR/DH+SMALL)**(POWER·1.)
      RE=DEN*WBAR*DH/AMUR
      FRE=-DPDZ*DH/(0.5*DEN*WBAR**2+SMALL)*RE
      DTDZ=QW*HP/(RHOCP*WSUM+SMALL)
      TWAV=0.
      DO 220 I=2,L2
         TWAV=TWAV+T(I,M1)*XCV(I)
  220 CONTINUE
      TWAV=TWAV/XL
      ANU=QW*DH/(COND*(TWAV-TB)+SMALL)
      DO 240 IUNIT=IU1,IU2
         IF(ITER.EQ.0) WRITE(IUNIT,250)
  250    FORMAT(1X,'ITER',2X,'W(3,4)',5X,'F.RE',5X,'T(5,4)',6X,'NU')
```

```
        WRITE(IUNIT,260) ITER,W(3,4),FRE,T(5,4),ANU
260     FORMAT(2X,I2,1P4E10.2)
240 CONTINUE
    IF(ITER.EQ.LAST) THEN
        DO 270 J=1,M1
        DO 270 I=1,L1
            W(I,J)=W(I,J)/WBAR
            T(I,J)=(T(I,J)-TWAV)/(TB-TWAV)
270     CONTINUE
        CALL PRINT
        CALL PLOT
    ENDIF
    RETURN
C*-*-*-*-*-*-*-*-*-*-*-*-*-*-*-*-*-*-*-*-*-*-*-*-*-*-*-*-*-*-*-*-*
    ENTRY PHI
    IF(NF.EQ.1) THEN
        REL=1.-REGAM
        DO 300 J=2,M2
        DO 300 I=2,L2
            IF(ITER.EQ.0) THEN
                GAM(I,J)=AMU
            ELSE
                WP=0.5*(W(I+1,J)+W(I,J))
                WM=0.5*(W(I,J)+W(I-1,J))
                IF(I.EQ.2) WM=W(I-1,J)
                IF(I.EQ.L2) WP=0.
                DWDX=(WP-WM)/(XCV(I)*SX(J))
                WP=0.5*(W(I,J+1)+W(I,J))
                WM=0.5*(W(I,J)+W(I,J-1))
                IF(J.EQ.2) WM=0.
                IF(J.EQ.M2) WP=0.
                DWDY=(WP-WM)/YCV(J)
                GAMT=RK*(DWDX**2+DWDY**2)**((POWER-1.)*0.5)
                IF(ITER.EQ.1) GAM(I,J)=GAMT
                GAM(I,J)=REGAM*GAMT+REL*GAM(I,J)
            ENDIF
300     CONTINUE
COME HERE TO ADJUST THE PRESSURE GRADIENT
        IF(ITER.EQ.1) THEN
            GAMAV=0.
            DO 310 J=2,M2
            DO 310 I=2,L2
```

```
                GAMAV=GAMAV+GAM(I,J)*XCV(I)*YCVR(J)
310         CONTINUE
            GAMAV=GAMAV/ASUM
            DPDZ=DPDZ*GAMAV/AMU
        ENDIF
        DO 320 J=2,M2
        DO 320 I=2,L2
            SC(I,J)=-DPDZ
320     CONTINUE
    ENDIF
    IF(NF.EQ.2) THEN
        DO 330 J=2,M2
        DO 330 I=2,L2
            GAM(I,J)=COND
            SC(I,J)=-RHOCP*W(I,J)*DTDZ
330     CONTINUE
    ENDIF
COME HERE TO SPECIFY BOUNDARY CONDITIONS
        DO 340 J=2,M2
            KBCI1(J)=2
340     CONTINUE
        DO 350 I=2,L2
            KBCJ1(I)=2
350     CONTINUE
    IF(NF.EQ.2) THEN
        DO 360 J=2,M2
            KBCL1(J)=2
360     CONTINUE
        DO 370 I=2,L2
            KBCM1(I)=2
            FLXCM1(I)=QW
370     CONTINUE
    ENDIF
    RETURN
    END
CCCCCCCCCCCCCCCCCCCCCCCCCCCCCCCCCCCCCCCCCCCCCCCCCCCCCCCCCCCCCCCCCCCCCCCCC
```

11.1-5   Results for Example 11

```
RESULTS OF CONDUCT FOR POLAR COORDINATE SYSTEM
**************************************************
```

. . . .

NON-NEWTONIAN FLOW IN A SEMICIRCULAR DUCT

. . . .

ITER	W(3,4)	F.RE	T(5,4)	NU
0	0.00E+00	0.00E+00	0.00E+00	1.22E+20
1	5.29E-02	1.24E+01	0.00E+00	1.22E+20
2	7.23E-02	2.76E+01	0.00E+00	1.22E+20
3	8.09E-02	2.64E+01	0.00E+00	1.22E+20
4	8.65E-02	2.57E+01	0.00E+00	1.22E+20
5	9.00E-02	2.52E+01	0.00E+00	1.22E+20
6	9.21E-02	2.50E+01	0.00E+00	1.22E+20
7	9.34E-02	2.48E+01	0.00E+00	1.22E+20
8	9.42E-02	2.47E+01	0.00E+00	1.22E+20
9	9.47E-02	2.47E+01	0.00E+00	1.22E+20
10	9.49E-02	2.46E+01	0.00E+00	1.22E+20
11	9.51E-02	2.46E+01	0.00E+00	1.22E+20
12	9.52E-02	2.46E+01	0.00E+00	1.22E+20
13	9.53E-02	2.46E+01	0.00E+00	1.22E+20
14	9.53E-02	2.46E+01	0.00E+00	1.22E+20
15	9.53E-02	2.46E+01	0.00E+00	1.22E+20
16	9.53E-02	2.46E+01	-5.97E-01	3.33E+00
17	9.53E-02	2.46E+01	-5.97E-01	3.33E+00
18	9.53E-02	2.46E+01	-5.97E-01	3.33E+00

I =	1	2	3	4	5	6	7
TH =	0.00E+00	7.85E-02	2.36E-01	3.93E-01	5.50E-01	7.07E-01	8.64E-01

I =	8	9	10	11	12
TH =	1.02E+00	1.18E+00	1.34E+00	1.49E+00	1.57E+00

J =	1	2	3	4	5	6	7
Y =	0.00E+00	3.13E-02	9.38E-02	1.56E-01	2.19E-01	2.81E-01	3.44E-01

J =	8	9	10	11	12	13	14
Y =	4.06E-01	4.69E-01	5.31E-01	5.94E-01	6.56E-01	7.19E-01	7.81E-01

J =	15	16	17	18

Y = 8.44E·01 9.06E·01 9.69E·01 1.00E+00

******   W/WBAR   ******

. . . .

I =	1	2	3	4	5	6	7
J							
18	0.00E+00	0.00E+00	0.00E+00	0.00E+00	0.00E+00	0.00E+00	0.00E+00
17	2.82E·01	2.82E·01	2.80E·01	2.75E-01	2.66E·01	2.54E·01	2.35E·01
16	7.72E·01	7.72E·01	7.65E·01	7.50E·01	7.26E·01	6.90E·01	6.38E·01
15	1.15E+00	1.15E+00	1.14E+00	1.11E+00	1.08E+00	1.02E+00	9.42E·01
14	1.42E+00	1.42E+00	1.40E+00	1.37E+00	1.33E+00	1.26E+00	1.16E+00
13	1.60E+00	1.59E+00	1.58E+00	1.55E+00	1.50E+00	1.42E+00	1.32E+00
12	1.70E+00	1.70E+00	1.69E+00	1.66E+00	1.61E+00	1.53E+00	1.42E+00
11	1.75E+00	1.75E+00	1.74E+00	1.71E+00	1.67E+00	1.59E+00	1.49E+00
10	1.77E+00	1.77E+00	1.76E+00	1.74E+00	1.70E+00	1.63E+00	1.53E+00
9	1.77E+00	1.77E+00	1.77E+00	1.75E+00	1.71E+00	1.64E+00	1.54E+00
8	1.77E+00	1.77E+00	1.76E+00	1.74E+00	1.70E+00	1.63E+00	1.52E+00
7	1.74E+00	1.74E+00	1.73E+00	1.70E+00	1.65E+00	1.58E+00	1.46E+00
6	1.68E+00	1.67E+00	1.66E+00	1.62E+00	1.57E+00	1.48E+00	1.35E+00
5	1.54E+00	1.53E+00	1.52E+00	1.47E+00	1.41E+00	1.32E+00	1.19E+00
4	1.29E+00	1.29E+00	1.27E+00	1.23E+00	1.17E+00	1.08E+00	9.57E·01
3	9.14E·01	9.12E·01	8.94E·01	8.59E·01	8.06E·01	7.34E·01	6.43E·01
2	3.59E·01	3.58E·01	3.50E·01	3.34E·01	3.10E·01	2.78E·01	2.40E·01
1	0.00E+00	2.83E·01	2.76E·01	2.63E·01	2.44E·01	2.18E·01	1.87E·01

I =	8	9	10	11	12
J					
18	0.00E+00	0.00E+00	0.00E+00	0.00E+00	0.00E+00
17	2.09E·01	1.72E·01	1.20E·01	4.77E·02	0.00E+00
16	5.64E·01	4.61E·01	3.18E·01	1.26E·01	0.00E+00
15	8.31E·01	6.79E·01	4.73E·01	1.92E·01	0.00E+00
14	1.03E+00	8.44E·01	5.95E·01	2.45E-01	0.00E+00
13	1.17E+00	9.68E·01	6.91E·01	2.84E·01	0.00E+00
12	1.27E+00	1.06E+00	7.61E·01	3.10E·01	0.00E+00
11	1.34E+00	1.12E+00	8.04E·01	3.23E·01	0.00E+00
10	1.37E+00	1.15E+00	8.21E·01	3.24E·01	0.00E+00
9	1.38E+00	1.15E+00	8.12E·01	3.15E·01	0.00E+00
8	1.36E+00	1.12E+00	7.77E·01	2.95E·01	0.00E+00
7	1.29E+00	1.05E+00	7.17E·01	2.67E·01	0.00E+00
6	1.18E+00	9.46E·01	6.33E·01	2.31E·01	0.00E+00
5	1.02E+00	8.03E·01	5.27E·01	1.89E·01	0.00E+00

```
4    8.07E-01 6.22E-01 3.99E-01 1.40E-01 0.00E+00
3    5.32E-01 4.02E-01 2.53E-01 8.71E-02 0.00E+00
2    1.95E-01 1.44E-01 8.89E-02 3.01E-02 0.00E+00
1    1.51E-01 1.11E-01 6.80E-02 2.29E-02 0.00E+00
```

```
******    (T-TWAV)/(TB-TWAV)    ******
          . . . . . . . . . . . . . . . . . . . . .
```

I =	1	2	3	4	5	6	7
J							
18	-5.02E-02	2.23E-01	2.06E-01	1.73E-01	1.25E-01	6.24E-02	-1.07E-02
17	3.11E-01	3.09E-01	2.93E-01	2.59E-01	2.11E-01	1.49E-01	7.55E-02
16	4.90E-01	4.88E-01	4.71E-01	4.38E-01	3.89E-01	3.27E-01	2.53E-01
15	6.68E-01	6.66E-01	6.49E-01	6.16E-01	5.67E-01	5.05E-01	4.31E-01
14	8.38E-01	8.35E-01	8.19E-01	7.86E-01	7.38E-01	6.76E-01	6.03E-01
13	9.94E-01	9.92E-01	9.76E-01	9.44E-01	8.98E-01	8.37E-01	7.66E-01
12	1.14E+00	1.13E+00	1.12E+00	1.09E+00	1.04E+00	9.85E-01	9.16E-01
11	1.26E+00	1.26E+00	1.24E+00	1.21E+00	1.17E+00	1.12E+00	1.05E+00
10	1.36E+00	1.36E+00	1.35E+00	1.32E+00	1.28E+00	1.23E+00	1.17E+00
9	1.45E+00	1.45E+00	1.43E+00	1.41E+00	1.37E+00	1.33E+00	1.28E+00
8	1.51E+00	1.51E+00	1.50E+00	1.48E+00	1.45E+00	1.41E+00	1.37E+00
7	1.56E+00	1.56E+00	1.55E+00	1.53E+00	1.51E+00	1.47E+00	1.44E+00
6	1.59E+00	1.59E+00	1.58E+00	1.57E+00	1.55E+00	1.52E+00	1.49E+00
5	1.60E+00	1.60E+00	1.59E+00	1.58E+00	1.57E+00	1.55E+00	1.53E+00
4	1.60E+00	1.60E+00	1.59E+00	1.59E+00	1.58E+00	1.57E+00	1.55E+00
3	1.59E+00	1.59E+00	1.58E+00	1.58E+00	1.58E+00	1.57E+00	1.57E+00
2	1.57E+00	1.57E+00	1.57E+00	1.57E+00	1.57E+00	1.57E+00	1.57E+00
1	-5.02E-02	1.57E+00	1.57E+00	1.57E+00	1.57E+00	1.57E+00	1.57E+00

I =	8	9	10	11	12
J					
18	-9.02E-02	-1.69E-01	-2.38E-01	-2.81E-01	-5.02E-02
17	-4.15E-03	-8.34E-02	-1.52E-01	-1.96E-01	-2.01E-01
16	1.73E-01	9.28E-02	2.27E-02	-2.26E-02	-2.83E-02
15	3.50E-01	2.69E-01	1.98E-01	1.51E-01	1.45E-01
14	5.23E-01	4.42E-01	3.71E-01	3.23E-01	3.17E-01
13	6.88E-01	6.09E-01	5.38E-01	4.91E-01	4.85E-01
12	8.41E-01	7.66E-01	6.98E-01	6.53E-01	6.47E-01
11	9.82E-01	9.11E-01	8.49E-01	8.07E-01	8.01E-01
10	1.11E+00	1.04E+00	9.88E-01	9.50E-01	9.45E-01
9	1.22E+00	1.16E+00	1.11E+00	1.08E+00	1.08E+00
8	1.32E+00	1.27E+00	1.23E+00	1.20E+00	1.20E+00

```
7    1.40E+00 1.36E+00 1.32E+00 1.30E+00 1.30E+00
6    1.46E+00 1.43E+00 1.41E+00 1.39E+00 1.39E+00
5    1.51E+00 1.49E+00 1.47E+00 1.46E+00 1.46E+00
4    1.54E+00 1.53E+00 1.52E+00 1.52E+00 1.52E+00
3    1.56E+00 1.56E+00 1.55E+00 1.55E+00 1.55E+00
2    1.57E+00 1.57E+00 1.57E+00 1.57E+00 1.57E+00
1    1.57E+00 1.57E+00 1.57E+00 1.57E+00 5.02E 02
```

11.1-6 Discussion of Results

The iterative scheme employed here seems to converge satisfactorily. No major swings in the values of w are noticeable. The temperature equation converges in one iteration as expected. Our use of the underrelaxation factor REGAM does not particularly help here; if fact, the solution would converge slightly faster if we used REGAM = 1. However, REGAM was introduced to illustrate the underrelaxation practice, which is very useful in more difficult problems such as Example 13 in Section 11.3.

To appreciate the field output of the dimensionless velocity and temperature, we should compare it with a solution for a corresponding Newtonian flow. With the power-law index n equal to 0.5, the viscosity is small near the walls and large in the middle of the duct. Therefore, compared to a Newtonian flow, we expect the present solution to exhibit steeper gradients near the duct walls and a flat region in the middle. From the field output of w/\O(w,\S\up6(—)), we see a large area where the value $w/\overline{w}$ is in the range 1.70–1.77. For a more detailed study of the nonNewtonian behavior, you are encouraged to compare the present solution with a Newtonian solution, which you can obtain simply by calling PRINT after the first iteration.

In the printout of the dimensionless temperature, some values are negative. This behavior was explained in Section 10.1 in connection with Example 7. It results from the use of the average wall temperature TWAV for defining the dimensionless temperature.

11.1-7 Final Remarks

In this example, we have seen the treatment of a nonlinear equation for the axial velocity. Since our aim was to obtain a final dimensionless solution, we could take the liberty of adjusting the pressure gradient. If we were given the dimensional pressure gradient and the actual value of K for a real power-law fluid, it would seem difficult to start the solution because no guidance is available about the magnitude of the axial velocity. In such a case, it would still be a good idea to obtain the dimensionless solution as we did and then convert it to fit the given dimensional pressure gradient.

Quite interestingly, the treatment of the nonNewtonian flow has many similarities with the treatment of turbulent flow, which will be discussed in Section 11.3. Before that, we consider a duct flow with a variable viscosity caused by temperature variation.

11.2 Duct Flow with Temperature-Dependent Viscosity (Example 12)

11.2-1 Problem Description

For many fluids, the viscosity depends on the temperature. If the variation of the viscosity is substantial, it will have a significant effect on the velocity distribution in a duct. Here, to examine the effect of the temperature-dependent viscosity, we shall choose a problem of the annular sector duct (Example 9) discussed earlier in Section 10.3. We shall use the same grid and introduce the temperature-dependent viscosity. This will allow a direct comparison of the results of the present example with the constant-viscosity results of Example 9.

Of course, we should choose a problem that leads to a fully developed flow. If any of the thermal boundary conditions used in the examples of Chapter 10 are adopted, the velocity field with a temperature-dependent viscosity will not become fully developed. The reason is that these boundary conditions lead to a continuous increase or decrease of the fluid temperature in the z direction. As a result, the viscosity field will vary in the z direction. Therefore, it is not possible to get a fully developed velocity field (i.e., a velocity distribution that does not change in the z direction) under such thermal boundary conditions.

What we need is a temperature variation in the duct cross section, but no temperature change in the z direction. This is the "uninteresting" type of fully developed heat transfer that was mentioned and pushed aside in Section 9.5. Although such temperature fields are uninteresting for convective heat transfer, they are suitable candidates for studying the effect of the temperature-dependent viscosity.

With this background, we choose the geometry of the annular sector duct used in Example 9 and shown here as Fig. 11.2. The thermal boundary conditions are given by a constant temperature T_1 on the inner boundary and the side walls of the duct, and a constant temperature T_2 on the outer boundary. The fluid viscosity μ is a linear function of the temperature T. Thus

$$\mu = a + b\,T \qquad\qquad (11.4)$$

Here we shall assume the following numerical values

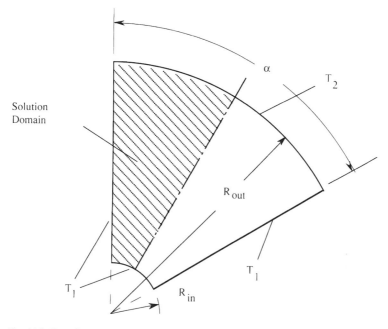

Fig. 11.2 Duct flow with temperature-dependent viscosity

$$T_1 = 100, \qquad T_2 = 0 \tag{11.5}$$

$$a = 1.0, \qquad b = 0.04 \tag{11.6}$$

11.2-2   Design of ADAPT

The subroutine ADAPT for this problem is kept very similar to the one for Example 9. Therefore, we shall discuss only those features that are different.

BEGIN. The constants in Eqs. (11.5) and (11.6) are specified and the $T(I, J)$ array is filled so as to provide correct boundary values at the boundary points.

In this problem, the temperature T does not vary in the z direction (i.e., $\partial T / \partial z = 0$). Therefore, the temperature equation has no source term. This makes the temperature field independent of the velocity field. (The temperature distribution will be same irrespective of whether there exists a flow through the duct.) On the other hand, the velocity field now depends on the temperature field since the viscosity is influenced by the temperature.

Because of this reversal of influences, we shall solve the temperature equation first, get a converged solution, and then begin the solution of the velocity equation. Both temperature and velocity equations are still linear; so we allow the usual three iterations for each equation.

OUTPUT. Here we do not calculate the temperature-related quantities such as T_b, the Nusselt number, etc. For the calculation of the Reynolds number, the viscosity is taken as the constant a in Eq. (11.4). In other words, the viscosity at $T = 0$ is used as the reference viscosity in the Reynolds number.

PHI. The only new feature is the calculation of GAM(I,J) for the velocity w according to Eq. (11.4).

11.2-3   Additional Fortran Names

ALPHA	angle between adjacent fins
AMU	constant a in the expression for μ, Eq. (11.4)
AR	area of control volume, dA
ASUM	cross-sectional area
BMU	constant b in the exression for μ, Eq. (11.4)
COND	thermal conductivity k
CP	specific heat c_p
DEN	density ρ
DH	hydraulic diameter D_h
DPDZ	axial pressure gradient dp/dz
FRE	the product fRe
PI	the constant π
RE	Reynolds number
RHOCP	heat capacity ρc_p
RIN	inner radius R_{in}
ROUT	outer radius R_{out}
T(I,J)	temperature T
T1, T2	wall temperatures T_1 and T_2
W(I,J)	axial velocity w
WBAR	cross-sectional mean velocity $\overline{w}$
WP	wetted perimeter
WSUM	$\int w\,dA$

11.2-4   Listing of ADAPT for Example 12

```
cccccccccccccccccccccccccccccccccccccccccccccccccccccccccccccccccccc
        SUBROUTINE ADAPT
C·····························································
C·····EXAMPLE 12 ·· DUCT FLOW WITH TEMPERATURE·DEPENDENT VISCOSITY
```

```
C..................................................................................
$INCLUDE:'COMMON'
C*********************************************************************************
      DIMENSION W(NI,NJ),T(NI,NJ)
      EQUIVALENCE (F(1,1,1),W(1,1)),(F(1,1,2),T(1,1))
C*.*.*.*.*.*.*.*.* * * *.*.*.*.*.* * * *.*.* * *.* *.*.*.*.*.*.*.*.*.*-*.*
      ENTRY GRID
      HEADER='DUCT FLOW WITH TEMPERATURE-DEPENDENT VISCOSITY'
      PRINTF='PRINT12'
      PLOTF='PLOT12'
      CALL DATA4(ROUT,1.,RIN,0.2,ALPHA,60.,PI,3.14159)
      XL=0.5*ALPHA*PI/180.
      CALL DATA2(R(1),RIN,YL,ROUT RIN)
      CALL INTA3(NCVLX,5,NCVLY,10,MODE,3)
      CALL EZGRID
      RETURN
C*.*.*.* * * * * *.* * * * *.*.*.* * *.*.* *.*.* * *.* * *.* *.*.*.*.*.*.* *.*
      ENTRY BEGIN
      TITLE(1)='     W/WBAR     '
      TITLE(2)='  (T T2)/(T1 T2)'
      CALL INTA6(KSOLVE(2),1,KPRINT(1),1,KPRINT(2),1,
     1           KPLOT(1),1,KPLOT(2),1,LAST,6)
      CALL DATA8(AMU,1.,BMU,0.04,COND,1.,CP,1.,DEN,1.,DPDZ, 1.,T1,100.,
     1           T2,0.)
      RHOCP=DEN*CP
      DO 100 J=1,M1
      DO 100 I=1,L1
         W(I,J)=0.
         T(I,J)=T1
         T(I,M1)=T2
  100 CONTINUE
      RETURN
C*.*.*.*.*.*.*.*.*.*.*.* * * *.*.*.* * * *.*.*.*.*.*.*.* *.*.*.*.*.*.* *.*
      ENTRY OUTPUT
      IF(ITER.EQ.3) THEN
         KSOLVE(1)=1
         KSOLVE(2)=0
      ENDIF
      ASUM=0.
      WSUM=0.
      DO 200 J=2,M2
      DO 200 I=2,L2
```

```
          AR=XCV(I)*YCVR(J)
          ASUM=ASUM+AR
          WSUM=WSUM+W(I,J)*AR
  200 CONTINUE
      WBAR=WSUM/ASUM
      WP=XL*(ROUT+RIN)+YL
      DH=4.*ASUM/WP
      RE=DH*WBAR*DEN/AMU
      FRE=-2.*DPDZ*DH/(DEN*WBAR**2+SMALL)*RE
      DO 210 IUNIT=IU1,IU2
          IF(ITER.EQ.0) WRITE(IUNIT,220)
  220     FORMAT(/,1X,'ITER',2X,'W(6,8)',5X,'W(4,11)',4X,'T(6,8)'
      1     ,5X,'T(4,11)',6X,'FRE')
          WRITE(IUNIT,230) ITER,W(6,8),W(4,11),T(6,8),T(4,11),FRE
  230     FORMAT(2X,I2,1P6E11.3)
  210 CONTINUE
      IF(ITER.EQ.LAST) THEN
          DO 240 J=1,M1
          DO 240 I=1,L1
              W(I,J)=W(I,J)/WBAR
              T(I,J)=(T(I,J)-T2)/(T1-T2)
  240     CONTINUE
          CALL PRINT
          CALL PLOT
      ENDIF
      RETURN
C* * * * * * * * * * * * * * * * * * * * * * * * * * * * * * * * * * * * * * *
      ENTRY PHI
      IF(NF.EQ.1) THEN
          DO 300 J=2,M2
          DO 300 I=2,L2
              GAM(I,J)=AMU+BMU*T(I,J)
              SC(I,J)=-DPDZ
  300     CONTINUE
      ENDIF
      IF(NF.EQ.2) THEN
          DO 310 J=2,M2
          DO 310 I=2,L2
              GAM(I,J)=COND
  310     CONTINUE
      ENDIF
COME HERE TO SPECIFY BOUNDARY CONDITIONS
```

```
      DO 320 J=2,M2
        KBCL1(J)=2
  320 CONTINUE
      RETURN
      END
CCCCCCCCCCCCCCCCCCCCCCCCCCCCCCCCCCCCCCCCCCCCCCCCCCCCCCCCCCCCCCCCCCCCCCCC
```

11.2-5   Results for Example 12

```
RESULTS OF CONDUCT FOR POLAR COORDINATE SYSTEM
**************************************************



DUCT FLOW WITH TEMPERATURE DEPENDENT VISCOSITY



ITER   W(6,8)      W(4,11)     T(6,8)      T(4,11)       FRE
  0   0.000E+00   0.000E+00   1.000E+02   1.000E+02   0.000E+00
  1   0.000E+00   0.000E+00   5.513E+01   1.108E+01   0.000E+00
  2   0.000E+00   0.000E+00   5.513E+01   1.108E+01   0.000E+00
  3   0.000E+00   0.000E+00   5.513E+01   1.108E+01   0.000E+00
  4   1.106E 02   3.614E 03   5.513E+01   1.108E+01   1.850E+02
  5   1.106E 02   3.614E 03   5.513E+01   1.108E+01   1.850E+02
  6   1.106E 02   3.614E 03   5.513E+01   1.108E+01   1.850E+02


  I =    1        2        3        4        5        6        7
 TH = 0.00E+00 5.24E 02 1.57E 01 2.62E 01 3.67E 01 4.71E 01 5.24E 01


  J =    1        2        3        4        5        6        7
  Y = 0.00E+00 4.00E 02 1.20E 01 2.00E 01 2.80E 01 3.60E 01 4.40E 01

  J =    8        9       10       11       12
  Y = 5.20E 01 6.00E 01 6.80E 01 7.60E 01 8.00E 01

******          W/WBAR          ******
     . . . . . . . . . . . . . . . . . . . . .
```

```
I =     1        2        3        4        5        6        7
J
12   0.00E+00 0.00E+00 0.00E+00 0.00E+00 0.00E+00 0.00E+00 0.00E+00
11   0.00E+00 1.93E-01 4.74E-01 6.75E-01 7.96E-01 8.53E-01 8.61E-01
10   0.00E+00 3.43E-01 9.14E-01 1.37E+00 1.66E+00 1.81E+00 1.83E+00
 9   0.00E+00 3.80E-01 1.03E+00 1.55E+00 1.91E+00 2.09E+00 2.12E+00
 8   0.00E+00 3.74E-01 1.01E+00 1.52E+00 1.88E+00 2.07E+00 2.09E+00
 7   0.00E+00 3.42E-01 9.28E-01 1.39E+00 1.71E+00 1.88E+00 1.90E+00
 6   0.00E+00 2.95E-01 8.00E-01 1.20E+00 1.47E+00 1.61E+00 1.63E+00
 5   0.00E+00 2.38E-01 6.48E-01 9.66E-01 1.18E+00 1.30E+00 1.31E+00
 4   0.00E+00 1.77E-01 4.83E-01 7.20E-01 8.81E-01 9.63E-01 9.73E-01
 3   0.00E+00 1.15E-01 3.14E-01 4.66E-01 5.69E-01 6.22E-01 6.28E-01
 2   0.00E+00 4.91E-02 1.31E-01 1.92E-01 2.32E-01 2.53E-01 2.55E-01
 1   0.00E+00 0.00E+00 0.00E+00 0.00E+00 0.00E+00 0.00E+00 0.00E+00
```

****** (T-T2)/(T1-T2) ******

```
I =     1        2        3        4        5        6        7
J
12   0.00E+00 0.00E+00 0.00E+00 0.00E+00 0.00E+00 0.00E+00 0.00E+00
11   1.00E+00 4.24E-01 1.75E-01 1.11E-01 8.77E-02 7.91E-02 7.80E-02
10   1.00E+00 7.45E-01 4.57E-01 3.27E-01 2.67E-01 2.43E-01 2.40E-01
 9   1.00E+00 8.61E-01 6.41E-01 5.08E-01 4.36E-01 4.04E-01 4.00E-01
 8   1.00E+00 9.14E-01 7.60E-01 6.50E-01 5.83E-01 5.51E-01 5.47E-01
 7   1.00E+00 9.44E-01 8.40E-01 7.59E-01 7.05E-01 6.78E-01 6.75E-01
 6   1.00E+00 9.64E-01 8.96E-01 8.40E-01 8.02E-01 7.82E-01 7.79E-01
 5   1.00E+00 9.78E-01 9.36E-01 9.00E-01 8.75E-01 8.62E-01 8.60E-01
 4   1.00E+00 9.87E-01 9.63E-01 9.43E-01 9.28E-01 9.20E-01 9.20E-01
 3   1.00E+00 9.94E-01 9.82E-01 9.72E-01 9.65E-01 9.61E-01 9.60E-01
 2   1.00E+00 9.98E-01 9.94E-01 9.91E-01 9.89E-01 9.88E-01 9.88E-01
 1   1.00E+00 1.00E+00 1.00E+00 1.00E+00 1.00E+00 1.00E+00 1.00E+00
```

11.2-6 Discussion of Results

With the linear equations for T and w, the solution converges in one iteration. The computed value of fRe is 185.0, which is much greater than the value of 58.0 computed in Example 9.

The main reason for this disparity is that, whereas the viscosity varies in the domain from 1 to 5, we used the value of 1 in defining the Reynolds number. If we had used an average value of 3 in the Reynolds number, the fRe would be 61.7, which is fairly close to the result of Example 9.

The main effect of the variable viscosity is seen in the distribution of $w/\overline{w}$. The velocities are much higher near the outer wall, where the viscosity is lower. The location of the maximum velocity has shifted towards the outer wall. By contrast, the velocities close to the inner wall have diminished.

11.2-7 Final Remarks

This example includes a number of special features. Here we have seen a case where the velocity field depends on the temperature distribution, but the temperature field is independent of the fluid velocity. Consequently, the temperature equation requires no source term. From the output of this example, we were able to study the influence of the varying viscosity on the velocity field. A variable viscosity of a different kind occurs in a turbulent flow, which we shall consider in the next example.

11.3 Turbulent Flow in a Square Duct (Example 13)

Computation of a turbulent flow is a complex subject. It is not intended here to provide a complete introduction to this subject. Only sufficient details will be given so that this adaptation of CONDUCT to the turbulent flow in a duct can be understood. For a more complete treatment of turbulent flows, you can refer to other books.

11.3-1 Problem Description

A turbulent flow is usually modeled by the same equations as for a laminar flow, except that the laminar viscosity is augmented by the so-called turbulent viscosity μ_t. The turbulent viscosity generally depends on the velocity gradients in the flow. (This is why a turbulent flow is in some respects similar to a nonNewtonian flow.)

For a fully developed turbulent flow in a duct, the product fRe is not constant but depends on the value of the Reynolds number. Therefore, the friction factor can be directly considered as a function of the Reynolds number.

Figure 11.3 shows the situation considered here. For the duct of square cross section, we wish to compute the fully developed velocity and temperature field. The thermal boundary conditions are given by a uniform heat flux on the two opposite walls, with the remaining two walls adiabatic. Because of the symmetry, our calculation will be performed on one-fourth of the duct cross section, as shown in Fig. 11.3. The objective is to obtain a dimensionless solution for the following values of the Reynolds number Re and the Prandtl number Pr.

$$Re = 1.E5, \qquad Pr = 0.7 \tag{11.7}$$

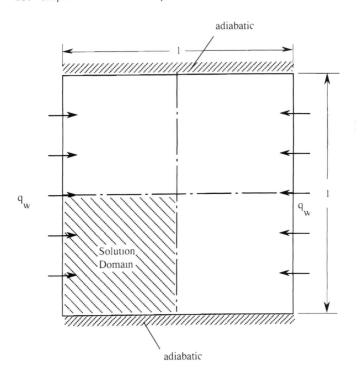

Fig. 11.3 Turbulent flow in a square duct

For this problem, we shall use the mixing-length model adapted by Patankar and Acharya (1984). The relevant equations of the model are given by

$$\mu_t = \rho L^2 \left[\left(\frac{\partial w}{\partial x} \right)^2 + \left(\frac{\partial w}{\partial y} \right)^2 \right]^{1/2} \tag{11.8}$$

where ρ is the density and L is the so-called mixing length. The mixing length is obtained from

$$L = \left[L_x^{-m} + L_y^{-m} \right]^{-1/m} \tag{11.9}$$

where m is a constant. The value m = 20 has been found satisfactory for the prediction of duct flows.

Next, the length scales L_x and L_y are obtained as products of two functions.

$$L_x = D(x^+) \wedge (S.X) \tag{11.10a}$$

$$L_y = D(y^+) \wedge (H.Y) \tag{11.10b}$$

where 2S represents the x-direction width of the duct and 2H is the y-direction width. The functions D and Λ are defined by

$$D(p) = 1 - \exp(-p/26.)$$

(11.11)

$$\Lambda(p, q) = p\left[0.14 - 0.08\left(1 - \frac{q}{p}\right)^2 - 0.06\left(1 - \frac{q}{p}\right)^4\right]$$

(11.12)

The quantities x^+ and y^+ are dimensionless distances defined by

$$x^+ = \frac{x\,(\rho\,\tau_w)^{1/2}}{\mu}$$

(11.13)

$$y^+ = \frac{y\,(\rho\,\tau_w)^{1/2}}{\mu}$$

(11.14)

Here μ is the laminar viscosity and τ_w is the wall shear stress. While calculating x^+, the shear stress τ_w is taken from the location $(0, y)$; for y^+, the value of τ_w comes from $(x, 0)$.

Similar to the turbulent viscosity μ_t, the turbulent conductivity is needed. It is given by

$$k_t = \frac{c_p\,\mu_t}{Pr_t}$$

(11.15)

where Pr_t is a turbulent Prandtl number, for which a constant value of 0.9 is used.

The turbulent viscosity μ_t is zero at the duct wall but becomes very large (in comparison with the laminar viscosity μ) in the central part of the duct. As a result, the velocity distribution has a very large gradient near the duct wall. To resolve this steep variation accurately, we shall need something like a 50 x 50 grid in the solution domain. For the purpose of this book, the output from such a grid will be inconvenient to print. Therefore, we shall use a more modest grid. It is true that the corresponding results will not be very accurate, but you will be able to run the program again with an increased number of grid points.

11.3-2   Design of ADAPT

In this problem, we define a number of two-dimensional arrays in addition to $W(I,J)$ and $T(I,J)$. They are: $AMUT(I,J)$ for μ_t, and $XPLUS(I,J)$ and $YPLUS(I,J)$ for x^+ and y^+ respectively. These three arrays are made equivalent to $F(I,J,NF)$ for $NF = 3, 4,$ and 5 respectively. As a result, we can obtain a field printout of these variables by calling PRINT. Also, the new arrays occupy the computer memory already allocated to the F array and do not consume any additional memory.

For the calculation of the functions D and L in Eqs. (11.11) and (11.12), we define the arithmetic statement functions DFACT(ARG) and CAPL(A1,A2). These will be used in the calculation of the mixing lengths.

GRID. Because of the steep variation of the velocity near the walls, it is desirable to use a very fine grid in the near-wall region. We do this by setting POWERX and POWERY equal to 3 and calling EZGRID. Generally, it is not advisable to use a value of power in excess of 2. However, for turbulent flow, the need for a fine grid near the walls is so strong that a larger value is desirable.

BEGIN. We plan to print out the field values of the turbulent viscosity μ_t. Therefore, the corresponding values of TITLE(3) and KPRINT(3) are provided.

Here, since the turbulent viscosity depends on the velocity w, the velocity equation is nonlinear. Once a converged solution for w is reached, the temperature equation can be treated as linear. We shall, therefore, perform 15 iterations for the solution of the velocity equation and 3 additional iterations for the temperature. Thus, the value of LAST is set as 18.

REGAM is the underrelaxation factor to be used for the turbulent viscosity.

OUTPUT. Nearly all the details here should be very familiar. For the turbulent flow, we do not calculate the product fRe; we simply obtain the friction factor f, which is denoted by FRIC in the program.

The solution is to be obtained for a given Reynolds number of 1.E5. Since the mean velocity $\overline{w}$ produced by the solution will be changing every iteration, we continuously adjust the laminar viscosity AMU so as to match the given Reynolds number. (In physical terms, we are looking for a fluid that will give, for the imposed pressure gradient, the desired Reynolds number.) The changing value of AMU is printed out after every iteration. As we approach the converged solution, AMU is expected to become constant.

PHI. The overall strategy here is similar to the one used in Example 11. We need some velocity gradient to obtain the turbulent viscosity from Eq. (11.8). For this purpose, the first iteration is performed as a laminar-flow computation by setting GAM(I,J) equal to the laminar viscosity AMU. Subsequently GAM(I,J) equals $(\mu + \mu_t)$, where μ_t is given by Eq. (11.8). For this calculation, an implementation of the whole set of equations (11.8)–(11.14) can be seen. The quantities FLUXI1(J,1) and FLUXJ1(I,1) are taken as the shear stresses on the duct walls.

To calculate the velocity gradient in Eq. (11.8), the values of w on the faces of each control volume are obtained by interpolation. For each direction, these interpolated values are denoted by WP and WM. Since, we have a nonuniform grid, a control-volume face does not lie *midway* between the neighboring grid points. A linear interpolation based on the actual control-volume widths is used. with appropriate adjustment for the near-boundary control volumes. A more accurate procedure for obtaining the w values at the control-volume faces would be to use an equation like (2.81): however. to include that degree of complication here did not seem necessary.

To control the change in μ_t from iteration to iteration. an underrelaxation is introduced according to Eq. (5.65) using REGAM as the underrelaxation factor.

After the first iteration. an average value of GAM(I , J) is calculated and the imposed pressure gradient is adjusted in a manner similar to the procedure described for Example 11.

For the temperature equation. GAM(I , J) is taken as $(k + k_t)$. where the turbulent conductivity is given by Eq. (11.15).

11.3-3 Additional Fortran Names

A1, A2	arguments of the function CAPL
AMU	*laminar* viscosity μ
AMUT(I,J)	*turbulent* viscosity μ_t
ANU	average Nusselt number Nu
AR	area of control volume. dA
ARG	argument of the function DFACT
ASUM	cross-sectional area
CAPL(A1,A2)	Nikuradse length scales. Λ. Eq. (11.12)
COND	thermal conductivity k
CP	specific heat c_p
DEN	density ρ
DFACT(ARG)	Van Driest damping factors D. Eq. (11.11)
DH	hydraulic diameter D_h
DPDZ	axial pressure gradient dp/dz
DTDZ	axial temperature gradient $\partial T/\partial z$
DWDX	$\partial w/\partial x$
DWDY	$\partial w/\partial y$
FM, FP	interpolation factors
FRIC	friction factor f
GAMAV	average value of *effective* viscosity. $\mu_{eff} = \mu + \mu_t$
GAMAVP	value of GAMAV from previous iteration
GAMT	temporary value of *turbulent* viscosity μ_t
HP	heated perimeter
PR	Prandtl number Pr
PRT	turbulent Prandtl number Pr_t
QW	wall heat flux q_w

RE	Reynolds number
REGAM	underrelaxation factor for *turbulent* viscosity
REL	temporary variable for 1.− REGAM
RHOCP	heat capacity ρc_p
RL	resultant mixing length L
RLX, RLY	mixing lengths L_x and L_y
T(I,J)	temperature T
TAUW	wall shear stress τ_w
TB	bulk temperature T_b
TSUM	$\int wTdA$
TWAV	average temperature of heated wall. T_w
W(I,J)	axial velocity w
WBAR	cross-sectional mean velocity $\overline{w}$
WM	interface velocity
WP	wetted perimeter or interface velocity
WSUM	$\int wdA$
XPLUS(I,J)	dimensionless distance x^+. Eq. (11.13)
YPLUS(I,J)	dimensionless distance y^+. Eq. (11.14)

11.3-4   Listing of ADAPT for Example 13

```
CCCCCCCCCCCCCCCCCCCCCCCCCCCCCCCCCCCCCCCCCCCCCCCCCCCCCCCCCCCCCCCCCCCCCCCCCC
      SUBROUTINE ADAPT
C..........................................................................
C.....EXAMPLE 13 .. TURBULENT FLOW IN A SQUARE DUCT
C..........................................................................
$INCLUDE:'COMMON'
C*************************************************************************
      DIMENSION W(NI,NJ),T(NI,NJ),AMUT(NI,NJ),XPLUS(NI,NJ),YPLUS(NI,NJ)
      EQUIVALENCE (F(1,1,1),W(1,1)),(F(1,1,2),T(1,1)),
     1            (F(1,1,3),AMUT(1,1)),(F(1,1,4),XPLUS(1,1)),
     2            (F(1,1,5),YPLUS(1,1))
C*-*-*-*-*-*-*-*-*-*-*-*-*-*-*-*-*-*-*-*-*-*-*-*-*-*-*-*-*-*-*-*-*-*-*-*-*
      DFACT(ARG)=1.-EXP(-ARG/26.)
      CAPL(A1,A2)=A1*(0.14-0.08*(1.-A2/A1)**2-0.06*(1.-A2/A1)**4)
C*-*-*-*-*-*-*-*-*-*-*-*-*-*-*-*-*-*-*-*-*-*-*-*-*-*-*-*-*-*-*-*-*-*-*-*-*
      ENTRY GRID
      HEADER='TURBULENT FLOW IN A SQUARE DUCT'
      PRINTF='PRINT13'
      PLOTF='PLOT13'
      CALL INTA2(NCVLX,10,NCVLY,10)
      CALL DATA4(XL,0.5,YL,0.5,POWERX,3.,POWERY,3.)
```

```
      CALL EZGRID
      RETURN
C*-*-*-*-*-*-*-*-*-*-*-*-*-*-*-*-*-*-*-*-*-*-*-*-*-*-*-*-*-*-*-*-*-*
      ENTRY BEGIN
      TITLE(1)='      W/WBAR      '
      TITLE(2)='(T-TWAV)/(TB-TWAV)'
      TITLE(3)=' TURB. VISCOSITY  '
      CALL INTA8(KSOLVE(1),1,KPRINT(1),1,KPLOT(1),1,KPRINT(2),1,
     1           KPLOT(2),1,KPRINT(3),1,KPLOT(3),1,LAST,18)
      CALL DATA5(AMU,1.,DEN,1.,DPDZ, 1.,RE,1.E5,REGAM,0.8)
      CALL DATA4(PR,0.7,CP,1.,PRT,0.9,QW,1.)
      COND=CP*AMU/PR
      RHOCP=DEN*CP
      GAMAVP=AMU
C
C-- SINCE THE ZERO DEFAULT VALUES OF W(I,J) AND T(I,J) ARE SATISFACTORY,
C   THESE ARRAYS ARE NOT FILLED HERE.
      RETURN
C*-*-*-*-*-*-*-*-*-*-*-*-*-*-*-*-*-*-*-*-*-*-*-*-*-*-*-*-*-*-*-*-*-*
      ENTRY OUTPUT
      IF(ITER.EQ.15)THEN
         KSOLVE(1)=0
         KSOLVE(2)=1
         COND=CP*AMU/PR
      ENDIF
      ASUM=0.
      WSUM=0.
      TSUM=0.
      DO 200 J=2,M2
      DO 200 I=2,L2
         AR=XCV(I)*YCV(J)
         ASUM=ASUM+AR
         WSUM=WSUM+W(I,J)*AR
         TSUM=TSUM+W(I,J)*T(I,J)*AR
  200 CONTINUE
      WBAR=WSUM/ASUM
      TB=TSUM/(WSUM+SMALL)
      WP=XL+YL
      DH=4.*ASUM/WP
      FRIC=-2.*DPDZ*DH/(DEN*WBAR**2+SMALL)
      HP=YL
      DTDZ=QW*HP/(RHOCP*WSUM+SMALL)
```

```
      TWAV=0.
      DO 210 J=2,M2
         TWAV=TWAV+T(1,J)*YCV(J)
  210 CONTINUE
      TWAV=TWAV/YL
      ANU=QW*DH/(COND*(TWAV-TB)+SMALL)
      IF(ITER.GE.1) AMU=DH*WBAR*DEN/RE
      DO 220 IUNIT=IU1,IU2
         IF(ITER.EQ.0) WRITE(IUNIT,230)
  230    FORMAT(1X,'ITER',3X,'W(8,8)',4X,'T(8,8)'
     1    ,3X,'  AMU',7X,'  F',9X,'NU')
         WRITE(IUNIT,240)ITER,W(8,8),T(8,8),AMU,FRIC,ANU
  240    FORMAT(1X,I3,1X,1P5E10.2)
  220 CONTINUE
      IF(ITER.EQ.LAST) THEN
         DO 250 I=1,L1
         DO 250 J=1,M1
            W(I,J)=W(I,J)/WBAR
            T(I,J)=(T(I,J)-TWAV)/(TB-TWAV)
  250    CONTINUE
         CALL PRINT
         CALL PLOT
      ENDIF
      RETURN
C*-*-*-*-*-*-*-*-*-*-*-*-*-*-*-*-*-*-*-*-*-*-*-*-*-*-*-*-*-*-*-*-*-*-*-*
      ENTRY PHI
      IF(NF.EQ.1) THEN
         REL=1.-REGAM
         DO 300 J=2,M2
         DO 300 I=2,L2
            IF(ITER.EQ.0) THEN
               GAM(I,J)=AMU
            ELSE
               TAUW=ABS(FLUXI1(J,1))
               XPLUS(I,J)=X(I)*SQRT(TAUW*DEN)/AMU
               TAUW=ABS(FLUXJ1(I,1))
               YPLUS(I,J)=Y(J)*SQRT(TAUW*DEN)/AMU
               RLX=DFACT(XPLUS(I,J))*CAPL(XL,X(I))
               RLY=DFACT(YPLUS(I,J))*CAPL(YL,Y(J))
               RL=(RLX**(-20)+RLY**(-20))**(-0.05)
               FP=1.
               IF(I.NE.L2) FP=XCV(I)/(XCV(I)+XCV(I+1))
```

```
                WP=FP*W(I+1,J)+(1.-FP)*W(I,J)
                FM=1.
                IF(I.NE.2) FM=XCV(I)/(XCV(I)+XCV(I-1))
                WM=FM*W(I-1,J)+(1.-FM)*W(I,J)
                DWDX=(WP-WM)/XCV(I)
                FP=1.
                IF(J.NE.M2) FP=YCV(J)/(YCV(J)+YCV(J+1))
                WP=FP*W(I,J+1)+(1.-FP)*W(I,J)
                FM=1.
                IF(J.NE.2) FM=YCV(J)/(YCV(J)+YCV(J-1))
                WM=FM*W(I,J-1)+(1.-FM)*W(I,J)
                DWDY=(WP-WM)/YCV(J)
                GAMT=DEN*RL**2*SQRT(DWDX**2+DWDY**2)
                IF(ITER.EQ.1) AMUT(I,J)=GAMT
                AMUT(I,J)=REGAM*GAMT+REL*AMUT(I,J)
                GAM(I,J)=AMU+AMUT(I,J)
              ENDIF
   300     CONTINUE
COME HERE TO ADJUST THE PRESSURE GRADIENT
          IF(ITER.EQ.1) THEN
              GAMAV=0.
              DO 305 J=2,M2
              DO 305 I=2,L2
                  GAMAV=GAMAV+GAM(I,J)*XCV(I)*YCV(J)
   305        CONTINUE
              GAMAV=GAMAV/ASUM
              DPDZ=DPDZ*GAMAV/GAMAVP
              GAMAVP=GAMAV
          ENDIF
          DO 307 J=2,M2
          DO 307 I=2,L2
              SC(I,J)=-DPDZ
   307     CONTINUE
        ENDIF
C
      IF(NF.EQ.2) THEN
          DO 310 J=2,M2
          DO 310 I=2,L2
              GAM(I,J)=COND+CP*AMUT(I,J)/PRT
              SC(I,J)=-RHOCP*W(I,J)*DTDZ
   310     CONTINUE
        ENDIF
```

```
COME HERE TO SPECIFY BOUNDARY CONDITIONS
      DO 320 I=2,L2
         KBCMI(I)=2
 320 CONTINUE
      DO 330 J=2,M2
         KBCLI(J)=2
 330 CONTINUE
C
      IF(NF.EQ.2) THEN
         DO 340 I=2,L2
            KBCJI(I)=2
 340     CONTINUE
         DO 350 J=2,M2
            KBCII(J)=2
            FLXCII(J)=QW
 350     CONTINUE
      ENDIF
      RETURN
      END
CCCCCCCCCCCCCCCCCCCCCCCCCCCCCCCCCCCCCCCCCCCCCCCCCCCCCCCCCCCCCCCCCCCCCC
```

11.3-5   Results for Example 13

```
RESULTS OF CONDUCT FOR CARTESIAN COORDINATE SYSTEM
**************************************************
```

```
. . . . .  . . . . . . . .    . . . . . . . . . . . . . . . . . . . . . . . . . . . . . . . . . . . . . . . .
TURBULENT FLOW IN A SQUARE DUCT
. . . . . . . . . . . . . . . . . . . . . . . . . . . . . . . . . . . . . . . . . . . . . . . . . . . . . . .
```

ITER	W(8,8)	T(8,8)	AMU	F	NU
0	0.00E+00	0.00E+00	1.00E+00	2.00E+20	1.00E+20
1	2.15E-02	0.00E+00	3.58E-07	1.56E+03	1.00E+20
2	7.32E-01	0.00E+00	7.26E-06	1.03E-03	1.00E+20
3	9.60E-02	0.00E+00	1.12E-06	4.28E-02	1.00E+20
4	1.84E-01	0.00E+00	1.95E-06	1.43E-02	1.00E+20
5	1.42E-01	0.00E+00	1.56E-06	2.22E-02	1.00E+20
6	1.54E-01	0.00E+00	1.66E-06	1.97E-02	1.00E+20
7	1.50E-01	0.00E+00	1.63E-06	2.04E-02	1.00E+20
8	1.51E-01	0.00E+00	1.63E-06	2.03E-02	1.00E+20

```
 9   1.51E-01  0.00E+00  1.63E-06  2.03E-02  1.00E+20
10   1.51E-01  0.00E+00  1.63E-06  2.03E-02  1.00E+20
11   1.51E-01  0.00E+00  1.63E-06  2.03E-02  1.00E+20
12   1.51E-01  0.00E+00  1.63E-06  2.03E-02  1.00E+20
13   1.51E-01  0.00E+00  1.63E-06  2.03E-02  1.00E+20
14   1.51E-01  0.00E+00  1.63E-06  2.03E-02  1.00E+20
15   1.51E-01  0.00E+00  1.63E-06  2.03E-02  1.00E+20
16   1.51E-01  4.91E+02  1.63E-06  2.03E-02  1.92E+02
17   1.51E-01  4.91E+02  1.63E-06  2.03E-02  1.92E+02
18   1.51E-01  4.91E+02  1.63E-06  2.03E-02  1.92E+02
```

```
I =    1        2        3        4        5        6        7
X = 0.00E+00 2.50E-04 2.25E-03 8.75E-03 2.28E-02 4.73E-02 8.53E-02

I =    8        9       10       11       12
X = 1.40E-01 2.14E-01 3.10E-01 4.32E-01 5.00E-01
```

```
J =    1        2        3        4        5        6        7
Y = 0.00E+00 2.50E-04 2.25E-03 8.75E-03 2.28E-02 4.73E-02 8.53E-02

J =    8        9       10       11       12
Y = 1.40E-01 2.14E-01 3.10E-01 4.32E-01 5.00E-01
```

```
******        W/WBAR          ******
. . . . . . . . . . . . . . . . . . . .
```

```
I =    1        2        3        4        5        6        7
J
12  0.00E+00 7.66E-02 3.97E-01 6.97E-01 8.14E-01 9.14E-01 9.96E-01
11  0.00E+00 7.60E-02 3.95E-01 6.95E-01 8.11E-01 9.11E-01 9.92E-01
10  0.00E+00 7.18E-02 3.82E-01 6.76E-01 7.89E-01 8.86E-01 9.64E-01
 9  0.00E+00 6.56E-02 3.62E-01 6.46E-01 7.55E-01 8.46E-01 9.20E-01
 8  0.00E+00 5.81E-02 3.36E-01 6.08E-01 7.11E-01 7.96E-01 8.63E-01
 7  0.00E+00 4.99E-02 3.06E-01 5.62E-01 6.58E-01 7.34E-01 7.99E-01
 6  0.00E+00 4.11E-02 2.70E-01 5.05E-01 5.93E-01 6.65E-01 7.34E-01
 5  0.00E+00 3.17E-02 2.26E-01 4.32E-01 5.15E-01 5.93E-01 6.58E-01
 4  0.00E+00 2.15E-02 1.67E-01 3.33E-01 4.32E-01 5.05E-01 5.62E-01
 3  0.00E+00 5.75E-03 4.98E-02 1.67E-01 2.26E-01 2.70E-01 3.06E-01
 2  0.00E+00 6.81E-04 5.75E-03 2.15E-02 3.17E-02 4.11E-02 4.99E-02
 1  0.00E+00 0.00E+00 0.00E+00 0.00E+00 0.00E+00 0.00E+00 0.00E+00
```

I =	8	9	10	11	12
J					
12	1.07E+00	1.13E+00	1.19E+00	1.24E+00	0.00E+00
11	1.06E+00	1.13E+00	1.18E+00	1.23E+00	1.24E+00
10	1.03E+00	1.09E+00	1.14E+00	1.18E+00	1.19E+00
9	9.82E·01	1.04E+00	1.09E+00	1.13E+00	1.13E+00
8	9.23E·01	9.82E·01	1.03E+00	1.06E+00	1.07E+00
7	8.63E·01	9.20E·01	9.64E·01	9.92E·01	9.96E·01
6	7.96E·01	8.46E·01	8.86E·01	9.11E·01	9.14E·01
5	7.11E·01	7.55E·01	7.89E·01	8.11E·01	8.14E·01
4	6.08E·01	6.46E·01	6.76E·01	6.95E·01	6.97E·01
3	3.36E·01	3.62E·01	3.82E·01	3.95E·01	3.97E·01
2	5.81E·02	6.56E·02	7.18E·02	7.60E·02	7.66E·02
1	0.00E+00	0.00E+00	0.00E+00	0.00E+00	0.00E+00

******   (T·TWAV)/(TB·TWAV)   ******
. . . .

I =	1	2	3	4	5	6	7
J							
12	1.13E+00	1.75E·01	3.97E·01	6.18E·01	7.14E·01	7.99E·01	8.73E·01
11	1.23E·01	1.71E·01	3.93E·01	6.16E·01	7.12E·01	7.98E·01	8.72E·01
10	9.32E·02	1.41E·01	3.70E·01	6.00E·01	6.99E·01	7.88E·01	8.65E·01
9	4.35E·02	9.15E·02	3.30E·01	5.73E·01	6.78E·01	7.71E·01	8.52E·01
8	·2.43E·02	2.36E·02	2.74E·01	5.35E·01	6.49E·01	7.48E·01	8.32E·01
7	1.12E·01	6.40E·02	2.03E·01	4.87E·01	6.12E·01	7.17E·01	8.05E·01
6	2.25E·01	·1.77E·01	1.09E·01	4.25E·01	5.64E·01	6.77E·01	7.85E·01
5	·3.80E·01	·3.32E·01	·2.17E·02	3.38E·01	5.03E·01	6.49E·01	7.78E·01
4	·6.13E·01	·5.65E·01	·2.22E·01	2.14E·01	4.66E·01	6.43E·01	7.78E·01
3	·1.02E+00	·9.71E·01	·6.09E·01	1.30E·01	4.59E·01	6.42E·01	7.78E·01
2	·1.04E+00	·9.91E·01	·6.28E·01	1.22E·01	4.58E·01	6.42E·01	7.78E·01
1	1.13E+00	·9.92E·01	·6.29E·01	1.21E·01	4.58E·01	6.42E·01	7.78E·01

I =	8	9	10	11	12
J					
12	9.42E·01	1.01E+00	1.07E+00	1.12E+00	1.13E+00
11	9.41E·01	1.01E+00	1.07E+00	1.12E+00	1.13E+00
10	9.35E·01	1.00E+00	1.06E+00	1.11E+00	1.12E+00
9	9.25E·01	9.93E·01	1.06E+00	1.11E+00	1.12E+00
8	9.08E·01	9.86E·01	1.06E+00	1.12E+00	1.12E+00
7	8.95E·01	9.85E·01	1.07E+00	1.12E+00	1.13E+00

```
6    8.90E-01 9.87E-01 1.07E+00 1.12E+00 1.13E+00
5    8.91E-01 9.89E-01 1.07E+00 1.13E+00 1.13E+00
4    8.91E-01 9.91E-01 1.07E+00 1.13E+00 1.14E+00
3    8.92E-01 9.91E-01 1.07E+00 1.13E+00 1.14E+00
2    8.92E-01 9.91E-01 1.07E+00 1.13E+00 1.14E+00
1    8.92E-01 9.91E-01 1.07E+00 1.13E+00 1.13E+00
```

****** 　 TURB. VISCOSITY 　 ******

```
I =    1        2        3        4        5        6        7
J
12   0.00E+00 0.00E+00 0.00E+00 0.00E+00 0.00E+00 0.00E+00 0.00E+00
11   0.00E+00 1.01E-09 1.94E-06 3.73E-05 7.43E-05 1.47E-04 2.25E-04
10   0.00E+00 9.11E-10 1.81E-06 3.58E-05 7.22E-05 1.43E-04 2.20E-04
9    0.00E+00 7.73E-10 1.61E-06 3.35E-05 6.89E-05 1.37E-04 2.16E-04
8    0.00E+00 6.22E-10 1.38E-06 3.05E-05 6.49E-05 1.31E-04 2.26E-04
7    0.00E+00 4.72E-10 1.13E-06 2.69E-05 6.08E-05 1.34E-04 2.47E-04
6    0.00E+00 3.33E-10 8.65E-07 2.28E-05 6.02E-05 1.49E-04 1.34E-04
5    0.00E+00 2.07E-10 5.94E-07 1.83E-05 6.70E-05 6.02E-05 6.08E-05
4    0.00E+00 1.00E-10 3.32E-07 1.63E-05 1.83E-05 2.28E-05 2.69E-05
3    0.00E+00 7.78E-12 5.27E-08 3.32E-07 5.94E-07 8.65E-07 1.13E-06
2    0.00E+00 1.45E-13 7.78E-12 1.00E-10 2.07E-10 3.33E-10 4.72E-10
1    0.00E+00 0.00E+00 0.00E+00 0.00E+00 0.00E+00 0.00E+00 0.00E+00


I =    8        9        10       11       12
J
12   0.00E+00 0.00E+00 0.00E+00 0.00E+00 0.00E+00
11   2.99E-04 3.46E-04 3.41E-04 2.56E-04 0.00E+00
10   2.99E-04 3.68E-04 3.77E-04 3.41E-04 0.00E+00
9    3.18E-04 4.02E-04 3.68E-04 3.46E-04 0.00E+00
8    3.45E-04 3.18E-04 2.99E-04 2.99E-04 0.00E+00
7    2.26E-04 2.16E-04 2.20E-04 2.25E-04 0.00E+00
6    1.31E-04 1.37E-04 1.43E-04 1.47E-04 0.00E+00
5    6.49E-05 6.89E-05 7.22E-05 7.43E-05 0.00E+00
4    3.05E-05 3.35E-05 3.58E-05 3.73E-05 0.00E+00
3    1.38E-06 1.61E-06 1.81E-06 1.94E-06 0.00E+00
2    6.22E-10 7.73E-10 9.11E-10 1.01E-09 0.00E+00
1    0.00E+00 0.00E+00 0.00E+00 0.00E+00 0.00E+00
```

11.3-6   Discussion of Results

The iterative procedure employed seems to converge well within about 10 iterations. Note the changing value of the laminar viscosity AMU, which eventually settles to a constant value. The use of underrelaxation through REGAM does help in this case. If we had used REGAM = 1., the convergence would have been slower.

The computed values of f and Nu are not very accurate because our grid is too coarse to resolve the steep variations near the duct walls. When the problem is run with a 50 x 50 grid, we get

$$f = 1.67E-2, \qquad Nu = 1.62E2 \tag{11.16}$$

These values agree quite well with the experimental results quoted in Patankar and Acharya (1984) for Re = 1.E5. (Incidentally, there is an error in the paper by Patankar and Acharya. Their plotted values of f are inconsistent with the definition of f. They give the same definition of f as we employ in this book. However, when they plot the experimental results for f, the f values are one-fourth of what they should be. Therefore, for Re = 1.E5, the f values from the paper should be multiplied by 4 before comparing it with the value of f in Eq. (11.16) obtained from our computations.)

The field printout of $w/\overline{w}$ shows that, although a very fine grid is used near the walls, the velocity there rises quite rapidly within a few grid points. The velocity variation is rather flat near the center of the duct. For a *laminar* flow in a square duct, the maximum value of $w/\overline{w}$ is slightly greater than 2. Here our computed maximum value for $w/\overline{w}$ is 1.24. This indicates the flatness of the turbulent velocity profile over most of the duct cross section. You can also compare our predicted profiles with the experimental results quoted in Patankar and Acharya (1984) and verify that the agreement is quite satisfactory.

The negative values in the printout of the dimensionless temperature are a result of our use of an *average* wall temperature in the defining the dimensionless temperature. This effect has already been discussed in connection with Example 7.

11.3-7   Final Remarks

Fully developed turbulent flow in noncircular ducts is known to produce a small secondary flow in the cross section of the duct. If this feature were to be included, the resulting flow will be characterized, according to the terminology used in Chapter 9, as a complex fully developed flow, which is beyond the scope of CONDUCT. Also, a much more advanced turbulence model is required to predict the secondary flow. Fortunately, the effect of the secondary flow on the overall quantities such as f and Nu is rather small. This is why we were able to get a good agreement with experimental data even though the secondary flows were ignored in our formulation.

This example shows that CONDUCT can be successfully used even with a complex mathematical model. In such situations, the computer program becomes a research tool. You can use the tool to work out the implications of a mathematical model, compare the results with experimental data, modify the model if necessary, and proceed in this manner to develop a satisfactory mathematical model for a class of physical situations.

11.4 Potential Flow over a Block (Example 14)

CONDUCT can also be used to analyze processes other than heat conduction and duct flow. In this section, we shall apply the program to the calculation of the potential flow, which is characterized by the existence of a velocity potential ϕ. The velocity components in a potential flow in two dimensions are given by

$$u = -\frac{\partial \phi}{\partial x} \qquad\qquad v = -\frac{\partial \phi}{\partial y} \qquad\qquad (11.17)$$

where u and v are the velocity components in the x and y directions respectively. For a steady, constant-density situation, the velocity components satisfy the mass conservation or the continuity equation

$$\frac{\partial u}{\partial x} + \frac{\partial v}{\partial y} = 0 \qquad\qquad (11.18)$$

A combination of Eqs. (11.17) and (11.18) gives

$$\frac{\partial}{\partial x}\left(\frac{\partial \phi}{\partial x}\right) + \frac{\partial}{\partial y}\left(\frac{\partial \phi}{\partial y}\right) = 0 \qquad\qquad (11.19)$$

This equation conforms to the general differential equation solved by CONDUCT with the choices of

$$\Gamma = 1 \quad \text{and} \quad S = 0 \qquad\qquad (11.20)$$

11.4-1 Problem Description

The flow situation of interest is shown in Fig. 11.4. A uniform flow enters a channel between two parallel plates. A solid rectangular block is mounted over the bottom plate. The objective is to calculate the potential flow in this configuration.

For a uniform undisturbed flow in the x direction, the corresponding velocity potential is

$$\phi = -Ax \qquad\qquad (11.21)$$

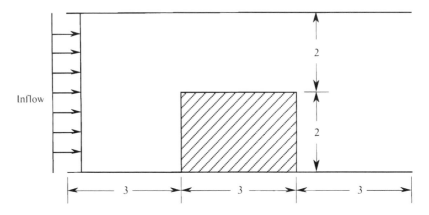

Fig. 11.4 Potential flow over a block

where A is a constant. For the situation in Fig. 11.4, it is reasonable to assume that the velocity potential at the inflow and outflow boundaries of the domain is given by Eq. (11.21). The value of the constant A can be taken as unity without any loss of generality.

11.4-2 Design of ADAPT

GRID. Here we define three new arrays $POT(I,J)$, $U(I,J)$ and $V(I,J)$ and make them equivalent to $F(I,J,NF)$ for $NF = 1$, 2, and 3, respectively. These three arrays will contain ϕ, u, and v, where the velocity potential is the main dependent variable, and the velocity components u and v will be derived from Eq. (11.17).

Since the geometry is rather simple, we use a uniform grid here and choose the number of control-volume widths in each direction such that the surfaces of the solid block coincide with the control-volume faces.

BEGIN. Here, it will be interesting to print out, in addition to the potential ϕ, the values of u and v. Therefore, $TITLE(NF)$ and $KPRINT(NF)$ are given the appropriate values for $NF = 1$, 2, and 3.

The values of $POT(I,J)$ are initialized according to Eq. (11.21) with $A = 1$. This provides the correct boundary values of ϕ at the inflow and outflow boundaries.

OUTPUT. The quantities u and v are not the dependent variables; they are to be derived from the ϕ values by using Eq. (11.17). This equation indicates that u and v are like the heat flow q in Eq. (2.31) and can be conveniently evaluated at the control-volume faces. Therefore, although we use the subscript (I , J) for U and V , the velocity U(I , J) is not located at the grid point (I , J). Rather, U(I , J) is considered to be located at the control-volume face corresponding to XU(I) and Y(J). Similarly, V(I , J) is the y-direction velocity at X(I) and YV(J). (Do you now see why we denote the locations of the control-volume faces by XU(I) and YV(J)? The quantity XU(I) is the x location for U(I , J): YV(J) is the y location for V(I , J). Incidentally, such staggered velocity locations are also used in more advanced methods for the computation of fluid flow, as described in Patankar, 1980.)

One consequence of the numbering scheme for U(I , J) and V(I , J) is that the quantities U(1 , J) and V(I , 1) are meaningless. These will be included in our printout (since it is not very easy to omit them), but no attention should be given to them.

With this background, it is now easy to interpret the calculation of U(I , J) and V(I , J) according to Eq. (11.17). The boundary velocities U(2 , J) and U(L1 , J), at the inflow and outflow locations respectively, are calculated from FLUXI1(J , 1) and FLUXL1(J , 1). This is because we use the higher-order treatment at the boundaries and its correct implications are contained in the FLUX quantities.

As we shall see later, the zero flow in the solid block will be arranged by making GAM(I , J) equal to zero in that region. In OUTPUT, we identify all control volumes in the solid block and put the velocity components at the four faces of these control volumes equal to zero.

The calculation of the velocity components U(I , J) and V(I , J) could have been deferred till the final printout. However, we would like to print out, after every iteration, a few typical values of the velocity components. Therefore, the calculation of U(I , J) and V(I , J) is performed at each iteration.

PHI. Here, we set GAM(I , J) equal to unity except for the locations within the solid block, where GAM(I , J) is made equal to zero. The top and bottom boundaries of the domain are treated as zero-flux boundaries.

11.4-3 Additional Fortran Names

POT(I , J)	velocity potential, ϕ
U(I , J)	x-direction velocity
V(I , J)	y-direction velocity

11.4-4 Listing of ADAPT for Example 14

```
CCCCCCCCCCCCCCCCCCCCCCCCCCCCCCCCCCCCCCCCCCCCCCCCCCCCCCCCCCCCCCCCCCCCCCCC
      SUBROUTINE ADAPT
```

```
C . . . . . . . . . . . . . . . . . . . . . . . . . . . . . . . . . . . . . . . . . . . . . . . . . . . . . . . . . . . . .
C · · · · · EXAMPLE 14 · · POTENTIAL FLOW OVER A BLOCK
C . . . . . . . . . . . . . . . . . . . . . . . . . . . . . . . . . . . . . . . . . . . . . . . . . . . . . . . . . . . . .
$INCLUDE:'COMMON'
C*****************************************************************************
      DIMENSION POT(NI,NJ),U(NI,NJ),V(NI,NJ)
      EQUIVALENCE (F(1,1,1),POT(1,1)),(F(1,1,2),U(1,1)),
     1     (F(1,1,3),V(1,1))
C* .* .* .* .* .* .* .* .* .* .* .* .* .* .* .* .* .* .* .* .* .* .* .* .* .* .* .* .*
      ENTRY GRID
      HEADER='POTENTIAL FLOW OVER A BLOCK'
      PRINTF='PRINT14'
      PLOTF='PLOT14'
      CALL INTA2(NCVLX,12,NCVLY,10)
      CALL DATA2(XL,9.,YL,4.)
      CALL EZGRID
      RETURN
C* .* .* .* .* .* .* .* .* .* .* .* .* .* .* .* .* .* .* .* .* .* .* .* .* .* .* .* .*
      ENTRY BEGIN
      TITLE(1)='    VEL. POT. '
      TITLE(2)='   U VELOCITY '
      TITLE(3)='   V VELOCITY '
      CALL INTA8(KSOLVE(1),1,KPRINT(1),1,KPLOT(1),1,
     1           KPRINT(2),1,KPLOT(2),1,KPRINT(3),1,KPLOT(3),1,LAST,3)
      DO 100 J=1,M1
      DO 100 I=1,L1
         POT(I,J)= X(I)
  100 CONTINUE
      RETURN
C* .* .* .* .* .* .* .* .* .* .* .* .* .* .* .* .* .* .* .* .* .* .* .* .* .* .* .* .*
      ENTRY OUTPUT
CALCULATE U AND V VELOCITIES AT THEIR STAGGERED LOCATIONS
      DO 200 J=3,M2
      DO 200 I=2,L2
         V(I,J)=·(POT(I,J)·POT(I,J·1))/(Y(J)·Y(J·1))
  200 CONTINUE
      DO 210 J=2,M2
      DO 220 I=3,L2
         U(I,J)=·(POT(I,J)·POT(I·1,J))/(X(I)·X(I·1))
  220 CONTINUE
      U(2,J)=FLUXI1(J,1)
      U(L1,J)=·FLUXL1(J,1)
```

```
  210 CONTINUE
      DO 230 J=2,M2
      DO 230 I=2,L2
         IF(X(I).GT.3..AND.X(I).LT.6..AND.Y(J).LT.2.) THEN
            U(I,J)=0.
            U(I+1,J)=0.
            V(I,J)=0.
            V(I,J+1)=0.
         ENDIF
  230 CONTINUE
      DO 240 IUNIT=IU1,IU2
         IF(ITER.EQ.0) WRITE(IUNIT,250)
  250    FORMAT(1X,'ITER',2X,'PHI(3,5)',3X,'U(3,5)',4X,
     1   'V(3,5)',3X,'PHI(5,5)',3X,'U(5,5)',4X,'V(5,5)')
         WRITE(IUNIT,260) ITER,POT(3,5),U(3,5),V(3,5),
     1                            POT(5,5),U(5,5),V(5,5)
  260    FORMAT(1X,I2,2X,1P6E10.2)
  240 CONTINUE
      IF(ITER.EQ.LAST) THEN
         CALL PRINT
C......
COME HERE TO FILL IBLOCK(I,J) BEFORE CALLING PLOT
         DO 270 J=2,M2
         DO 270 I=2,L2
            IF(X(I).GT.3..AND.X(I).LT.6..AND.Y(J).LT.2.) IBLOCK(I,J)=0.
  270    CONTINUE
C......
         CALL PLOT
      ENDIF
      RETURN
C*.*.*.*.*.* *.*.*.* *.* * *.*.* *.* * *.*.*.*.*.*.*.*.*.*.*.*.*.*.*.*
      ENTRY PHI
      DO 300 J=2,M2
      DO 300 I=2,L2
         GAM(I,J)=1.
         IF(X(I).GT.3..AND.X(I).LT.6..AND.Y(J).LT.2.) GAM(I,J)=0.
  300 CONTINUE
COME HERE TO SPECIFY BOUNDARY CONDITIONS
      DO 310 I=2,L2
         KBCJ1(I)=2
         KBCM1(I)=2
  310 CONTINUE
```

```
        RETURN
        END
CCCCCCCCCCCCCCCCCCCCCCCCCCCCCCCCCCCCCCCCCCCCCCCCCCCCCCCCCCCCCCCCCCCCCCCCC
```

11.4-5   Results for Example 14

```
RESULTS OF CONDUCT FOR CARTESIAN COORDINATE SYSTEM
*****************************************************
```

```
. . . . . . . . . . . . . . . . . . . . . . . . . . . . . . . . . . . . . . . . . . . . . . . . . . . . . .

POTENTIAL FLOW OVER A BLOCK
. . . . . . . . . . . . . . . . . . . . . . . . . . . . . . . . . . . . . . . . . . . . . . . . . . . . .
```

```
ITER  PHT(3,5)   U(3,5)    V(3,5)    PHI(5,5)   U(5,5)    V(5,5)
0    ·1.13E+00  1.00E+00  0.00E+00  ·2.63E+00  1.00E+00  0.00E+00
1    ·6.23E·01  5.51E·01  1.22E·01  ·1.28E+00  3.71E·01  4.01E·01
2    ·6.23E·01  5.51E·01  1.22E·01  ·1.28E+00  3.71E·01  4.01E·01
3    ·6.23E·01  5.51E·01  1.22E·01  ·1.28E+00  3.71E·01  4.01E·01
```

```
I =    1        2        3        4        5        6        7
X = 0.00E+00 3.75E·01 1.13E+00 1.88E+00 2.63E+00 3.38E+00 4.13E+00

I =    8        9       10       11       12       13       14
X = 4.88E+00 5.63E+00 6.38E+00 7.13E+00 7.88E+00 8.63E+00 9.00E+00
```

```
J =    1        2        3        4        5        6        7
Y = 0.00E+00 2.00E·01 6.00E·01 1.00E+00 1.40E+00 1.80E+00 2.20E+00

J =    8        9       10       11       12
Y = 2.60E+00 3.00E+00 3.40E+00 3.80E+00 4.00E+00
```

```
******        VEL. POT.        ******
              . . . . . . . . . . . . . . . . . .
```

```
I =    1        2        3        4        5        6        7
J
12    0.00E+00·2.93E·01·8.93E·01·1.55E+00·2.29E+00·3.14E+00 4.04E+00
11    0.00E+00·2.93E·01·8.91E·01·1.54E+00·2.28E+00·3.13E+00·4.04E+00
```

```
 10     0.00E+00 2.88E-01 8.76E-01 1.52E+00 2.25E+00 3.12E+00 4.04E+00
  9     0.00E+00-2.78E-01 8.46E-01 1.46E+00-2.19E+00 3.09E+00 4.03E+00
  8     0.00E+00 2.64E-01 8.01E-01 1.38E+00 2.07E+00 3.05E+00 4.02E+00
  7     0.00E+00 2.47E-01 7.45E-01 1.26E+00-1.87E+00 3.01E+00 4.01E+00
  6     0.00E+00 2.28E-01 6.82E-01 1.13E+00 1.52E+00 2.77E+00 3.93E+00
  5     0.00E+00-2.10E-01 6.23E-01 1.00E+00 1.28E+00 2.61E+00 3.88E+00
  4     0.00E+00 1.95E-01 5.74E-01 9.02E-01 1.12E+00 2.50E+00 3.84E+00
  3     0.00E+00 1.85E-01 5.40E-01 8.37E-01 1.02E+00 2.44E+00 3.82E+00
  2     0.00E+00 1.79E-01 5.23E-01 8.04E-01 9.71E-01 2.41E+00 3.81E+00
  1     0.00E+00 1.78E-01 5.20E-01 8.00E-01 9.65E-01 2.40E+00 3.80E+00


 I =      8        9       10       11       12       13       14
 J
 12    -4.96E+00 5.86E+00 6.71E+00 7.45E+00 8.11E+00 8.71E+00 9.00E+00
 11    -4.96E+00 5.87E+00 6.72E+00 7.46E+00 8.11E+00 8.71E+00 9.00E+00
 10    -4.96E+00 5.88E+00 6.75E+00 7.48E+00 8.12E+00 8.71E+00 9.00E+00
  9    -4.97E+00 5.91E+00 6.81E+00 7.54E+00 8.15E+00 8.72E+00 9.00E+00
  8    -4.98E+00 5.95E+00 6.93E+00 7.62E+00 8.20E+00 8.74E+00 9.00E+00
  7    -4.99E+00 5.99E+00 7.13E+00 7.74E+00 8.26E+00 8.75E+00 9.00E+00
  6    -5.07E+00 6.23E+00 7.48E+00 7.87E+00 8.32E+00 8.77E+00 9.00E+00
  5    -5.12E+00 6.39E+00 7.72E+00 8.00E+00 8.38E+00 8.79E+00 9.00E+00
  4    -5.16E+00 6.50E+00 7.88E+00 8.10E+00 8.43E+00 8.80E+00 9.00E+00
  3    -5.18E+00 6.56E+00 7.98E+00 8.16E+00 8.46E+00 8.82E+00 9.00E+00
  2    -5.19E+00 6.59E+00 8.03E+00 8.20E+00 8.48E+00 8.82E+00 9.00E+00
  1    -5.20E+00 6.60E+00 8.03E+00 8.20E+00 8.48E+00 8.82E+00 9.00E+00


******     U VELOCITY     ******
       . . . . . . . . . . . . . . . . . . . . .


 I =      1        2        3        4        5        6        7
 J
 12    0.00E+00 0.00E+00 0.00E+00 0.00E+00 0.00E+00 0.00E+00 0.00E+00
 11    0.00E+00 7.75E-01 7.98E-01 8.69E-01 9.90E-01 1.13E+00 1.21E+00
 10    0.00E+00 7.62E-01 7.84E-01 8.54E-01 9.83E-01 1.15E+00 1.22E+00
  9    0.00E+00 7.37E-01 7.57E-01 8.24E-01 9.65E-01 1.20E+00 1.25E+00
  8    0.00E+00 7.01E-01 7.16E-01 7.71E-01 9.20E-01 1.30E+00 1.29E+00
  7    0.00E+00 6.57E-01 6.64E-01 6.93E-01 8.07E-01 1.52E+00 1.34E+00
  6    0.00E+00 6.10E-01 6.06E-01 5.90E-01 5.24E-01 0.00E+00 0.00E+00
  5    0.00E+00 5.64E-01 5.51E-01 5.02E-01 3.71E-01 0.00E+00 0.00E+00
  4    0.00E+00 5.26E-01 5.06E-01 4.37E-01 2.87E-01 0.00E+00 0.00E+00
  3    0.00E+00 4.98E-01 4.74E-01 3.96E-01 2.42E-01 0.00E+00 0.00E+00
  2    0.00E+00 4.84E-01 4.58E-01 3.76E-01 2.23E-01 0.00E+00 0.00E+00
```

```
  1    0.00E+00 0.00E+00 0.00E+00 0.00E+00 0.00E+00 0.00E+00 0.00E+00


  I =     8        9        10       11       12       13       14
  J
  12   0.00E+00 0.00E+00 0.00E+00 0.00E+00 0.00E+00 0.00E+00 0.00E+00
  11   1.23E+00 1.21E+00 1.13E+00 9.90E-01 8.69E-01 7.98E-01 7.75E-01
  10   1.24E+00 1.22E+00 1.15E+00 9.83E-01 8.54E-01 7.84E-01 7.62E-01
   9   1.26E+00 1.25E+00 1.20E+00 9.65E-01 8.24E-01 7.57E-01 7.37E-01
   8   1.28E+00 1.29E+00 1.30E+00 9.20E-01 7.71E-01 7.16E-01 7.01E-01
   7   1.30E+00 1.34E+00 1.52E+00 8.07E-01 6.93E-01 6.64E-01 6.57E-01
   6   0.00E+00 0.00E+00 0.00E+00 5.24E-01 5.90E-01 6.06E-01 6.10E-01
   5   0.00E+00 0.00E+00 0.00E+00 3.71E-01 5.02E-01 5.51E-01 5.64E-01
   4   0.00E+00 0.00E+00 0.00E+00 2.87E-01 4.37E-01 5.06E-01 5.26E-01
   3   0.00E+00 0.00E+00 0.00E+00 2.42E-01 3.96E-01 4.74E-01 4.98E-01
   2   0.00E+00 0.00E+00 0.00E+00 2.23E-01 3.76E-01 4.58E-01 4.84E-01
   1   0.00E+00 0.00E+00 0.00E+00 0.00E+00 0.00E+00 0.00E+00 0.00E+00
```

******   V VELOCITY   ******

```
  I =     1        2        3        4        5        6        7
  J
  12   0.00E+00 0.00E+00 0.00E+00 0.00E+00 0.00E+00 0.00E+00 0.00E+00
  11   0.00E+00 1.23E-02 3.79E-02 6.46E-02 7.67E-02 3.92E-02 1.07E-02
  10   0.00E+00 2.41E-02 7.54E-02 1.33E-01 1.67E-01 7.58E-02 1.95E-02
   9   0.00E+00 3.48E-02 1.11E-01 2.09E-01 2.94E-01 1.02E-01 2.33E-02
   8   0.00E+00 4.30E-02 1.41E-01 2.88E-01 4.99E-01 9.71E-02 1.78E-02
   7   0.00E+00 4.68E-02 1.56E-01 3.49E-01 8.79E-01 0.00E+00 0.00E+00
   6   0.00E+00 4.48E-02 1.48E-01 3.13E-01 5.99E-01 0.00E+00 0.00E+00
   5   0.00E+00 3.75E-02 1.22E-01 2.44E-01 4.01E-01 0.00E+00 0.00E+00
   4   0.00E+00 2.67E-02 8.57E-02 1.64E-01 2.48E-01 0.00E+00 0.00E+00
   3   0.00E+00 1.38E-02 4.39E-02 8.17E-02 1.19E-01 0.00E+00 0.00E+00
   2   0.00E+00 0.00E+00 0.00E+00 0.00E+00 0.00E+00 0.00E+00 0.00E+00
   1   0.00E+00 0.00E+00 0.00E+00 0.00E+00 0.00E+00 0.00E+00 0.00E+00


  I =     8        9        10       11       12       13       14
  J
  12   0.00E+00  0.00E+00  0.00E+00  0.00E+00  0.00E+00  0.00E+00 0.00E+00
  11  -1.07E-02 -3.92E-02 -7.67E-02 -6.46E-02 -3.79E-02 -1.23E-02 0.00E+00
  10  -1.95E-02 -7.58E-02 -1.67E-01 -1.33E-01 -7.54E-02 -2.41E-02 0.00E+00
   9  -2.33E-02 -1.02E-01 -2.94E-01 -2.09E-01 -1.11E-01 -3.48E-02 0.00E+00
   8  -1.78E-02 -9.71E-02 -4.99E-01 -2.88E-01 -1.41E-01 -4.30E-02 0.00E+00
```

```
7    0.00E+00 0.00E+00 8.79E-01 3.49E-01 1.56E-01 4.68E-02 0.00E+00
6    0.00E+00 0.00E+00 5.99E-01 3.13E-01 1.48E-01 4.48E-02 0.00E+00
5    0.00E+00 0.00E+00-4.01E-01 2.44E-01 1.22E-01 3.75E-02 0.00E+00
4    0.00E+00 0.00E+00-2.48E-01 1.64E-01 8.57E-02 2.67E-02 0.00E+00
3    0.00E+00 0.00E+00-1.19E-01 8.17E-02 4.39E-02-1.38E-02 0.00E+00
2    0.00E+00 0.00E+00 0.00E+00 0.00E+00 0.00E+00 0.00E+00 0.00E+00
1    0.00E+00 0.00E+00 0.00E+00 0.00E+00 0.00E+00 0.00E+00 0.00E+00
```

11.4-6   Discussion of Results

Here, the problem is linear and the solution converges in one iteration. In the printout of the velocity field, the zero values can be used to identify the region occupied by the solid block. The velocity field in the left half of the domain can be seen to be exactly symmetrical with the velocity field in the right half. This is consistent with the behavior of the potential flow, which gives a symmetrical flow field over symmetrical bodies, such as cylinders and spheres. In fact, we could have confined our calculation only to the left half of the domain. The velocity field can be used to plot velocity vectors, which are useful in visualizing the flow field.

11.4-7   Final Remarks

Potential flow over bodies is an important topic in fluid mechanics. Through this example, we have seen how CONDUCT can be used for the calculation of potential flows. It is true that most bodies of interest will have shapes that do not fit nicely in our coordinate systems. However, you can use the techniques described in Section 7.7 and illustrated in Example 4 to approximate bodies with curved boundaries. Even with a coarse grid discretization, you will get surprisingly good solutions.

11.5   Water Seepage under a Dam (Example 15)

This final example deals with the flow in a porous material such as ground. Such flows are governed by Darcy's law, which states that the velocity components in a porous material are governed by

$$u = -C\frac{\partial P}{\partial x} \qquad v = -C\frac{\partial P}{\partial y} \qquad (11.22)$$

where P is the pressure and C is the Darcy constant of the material (related to its permeability).

Equation (11.22) is very similar to Eq. (11.17) for potential flow. Therefore, the use of the continuity equation (11.18) leads to

$$\frac{\partial}{\partial x}\left(C\frac{\partial P}{\partial x}\right) + \frac{\partial}{\partial y}\left(C\frac{\partial P}{\partial y}\right) = 0 \tag{11.23}$$

Thus, if we choose the pressure P as our dependent variable, then we have

$$\Gamma = C \qquad \text{and} \qquad S = 0 \tag{11.24}$$

11.5-1   Problem Description

Figure 11.5 shows the permeable ground under the foundation of a dam. The shaded region is the underground part of the dam. The top boundary of Fig. 11.5 represents the bottom of the water reservoir created by the dam. Of course, the dam extends above this line and holds a large body of water on the left side. On the right side of the dam, the water level is relatively low. Because of the difference in the water levels on the two sides of the dam, the hydrostatic pressures P_1 and P_2 act on the ground along the top boundary in Fig. 11.5, with $P_1 > P_2$. It is this pressure difference that causes seepage of water in the ground under and around the dam.

Regarding the other boundary conditions, we shall treat the three remaining boundaries as zero-flow boundaries, as shown in Fig. 11.5. The left and right boundaries can be thought of as sufficiently far from the dam so that the flow there is negligible. The same argument can be used for the bottom boundary; alternatively, there may be an impermeable rock under the bottom boundary.

The geometry of the problem is shown in Fig. 11.5. The numerical values of the relevant parameters are

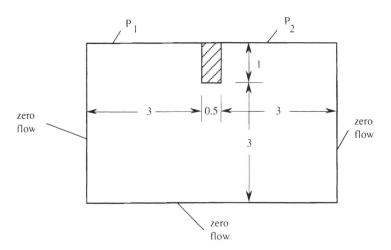

Fig. 11.5 Water seepage under a dam

$$C = 1, \qquad P_1 = 100, \qquad P_2 = 0 \qquad\qquad (11.25)$$

Our objective is to calculate the distribution of the pressure P and the associated flow field.

11.5-2 Design of ADAPT

As in Example 14, we introduce the arrays $U(I,J)$ and $V(I,J)$ for calculation of the velocity components.

GRID. A nonuniform grid is constructed by the use of ZGRID. In the x direction, the grid is made fine near the surfaces of the dam.

BEGIN. The variable DC is used for the Darcy constant C. The array $P(I,J)$ is filled with the appropriate boundary values along the top boundary.

OUTPUT. The calculation of the velocity components $U(I,J)$ and $V(I,J)$ follows the procedure described for Example 14 in Section 11.4-2.

PHI. The $GAM(I,J)$ array is made equal to the Darcy constant except for the points within the dam, where $GAM(I,J)$ is set equal to zero to simulate an impermeable material. The treatment of the boundary conditions is straightforward.

11.5-3 Additional Fortran Names

DC	Darcy contant C
P(I,J)	pressure
U(I,J)	x-direction velocity
V(I,J)	y-direction velocity

11.5-4 Listing of ADAPT for Example 15

```
CCCCCCCCCCCCCCCCCCCCCCCCCCCCCCCCCCCCCCCCCCCCCCCCCCCCCCCCCCCCCCCCCCCCCCC
      SUBROUTINE ADAPT
C-------------------------------------------------------------------
C------EXAMPLE 15 -- WATER SEEPAGE UNDER A DAM
C-------------------------------------------------------------------
$INCLUDE:'COMMON'
C*********************************************************************
      DIMENSION P(NI,NJ),U(NI,NJ),V(NI,NJ)
```

```
      EQUIVALENCE (F(1,1,1),P(1,1)),(F(1,1,2),U(1,1)),
    1    (F(1,1,3),V(1,1))
C*.*.*.*.*.*.*.*.*.*.*. *.*.*.*.*.*.* * *.*.*.*.*.*.*.*.*.*.*.*.*.*.*
      ENTRY GRID
      HEADER='WATER SEEPAGE UNDER A DAM'
      PRINTF='PRINT15'
      PLOTF='PLOT15'
      CALL INTA4(NZX,3,NCVX(1),5,NCVX(2),2,NCVX(3),5)
      CALL DATA5(XZONE(1),3.,XZONE(2),.5,XZONE(3),3.,POWRX(1), 1.2,
    1            POWRX(3),1.2)
      CALL INTA3(NZY,2,NCVY(1),6,NCVY(2),3)
      CALL DATA2(YZONE(1),3.,YZONE(2),1.)
      CALL ZGRID
      RETURN
C*.*.*.*.*.*.*.*.*.*.*.*.*.*.*-*.*.*.* * *.*.*.*.*.*.*.*.*.*.*.*.*.*
      ENTRY BEGIN
      TITLE(1)='    PRESSURE   '
      TITLE(2)='   U VELOCITY  '
      TITLE(3)='   V VELOCITY  '
      CALL INTA8(KSOLVE(1),1,KPRINT(1),1,KPLOT(1),1,
    1            KPRINT(2),1,KPLOT(2),1,KPRINT(3),1,KPLOT(3),1,LAST,3)
      DC=1.
      DO 100 J=1,M1
      DO 100 I=1,L1
         P(I,J)=0.
  100 CONTINUE
      DO 110 I=2,L2
         IF(X(I).LT.3.) P(I,M1)=100.
  110 CONTINUE
      RETURN
C*.* * *.* *.* **.* *.*.*.*.* *.* * *.*.*.* *.*.*.*.* *.*.*.*.*.*.*.*
      ENTRY OUTPUT
CALCULATE U AND V VELOCITIES AT THEIR STAGGERED LOCATIONS
      DO 200 J=3,M2
      DO 200 I=2,L2
         V(I,J)= DC*(P(I,J) P(I,J 1))/(Y(J) Y(J 1))
  200 CONTINUE
      DO 210 I=2,L2
         V(I,M1)= FLUXM1(I,1)
  210 CONTINUE
      DO 220 J=2,M2
      DO 220 I=3,L2
```

```
         U(I,J)=-DC*(P(I,J)-P(I-1,J))/(X(I)-X(I-1))
  220 CONTINUE
      DO 230 J=2,M2
      DO 230 I=2,L2
         IF(X(I).GT.3..AND.X(I).LT.3.5.AND.Y(J).GT.3.) THEN
           U(I,J)=0.
           U(I+1,J)=0.
           V(I,J)=0.
           V(I,J+1)=0.
         ENDIF
  230 CONTINUE
      DO 240 IUNIT=IU1,IU2
         IF(ITER.EQ.0) WRITE(IUNIT,250)
  250    FORMAT(1X,'ITER',2X,'P(3,5)',5X,'U(3,5)',5X,
     1   'V(3,5)',5X,'P(5,5)',5X,'U(5,5)',5X,'V(5,5)')
         WRITE(IUNIT,260) ITER,P(3,5),U(3,5),V(3,5),
     1                         P(5,5),U(5,5),V(5,5)
  260    FORMAT(1X,I2,1P6E11.2)
  240 CONTINUE
      IF(ITER.EQ.LAST) THEN
        CALL PRINT
C......
COME HERE TO FILL IBLOCK(I,J) BEFORE CALLING PLOT
        DO 270 J=2,M2
        DO 270 I=2,L2
           IF(X(I).GT.3..AND.X(I).LT.3.5.AND.Y(J).GT.3.) IBLOCK(I,J)=1
  270   CONTINUE
        CALL PLOT
C......
      ENDIF
      RETURN
C*.*.*.*.*.*.*.*.*.*.*.*.*.*.*.*.*.*.*.*.*.*.*.*.*.*.*.*.*.*.*.*.*.*
      ENTRY PHI
      DO 300 J=2,M2
      DO 300 I=2,L2
         GAM(I,J)=DC
         IF(X(I).GT.3..AND.X(I).LT.3.5.AND.Y(J).GT.3.) GAM(I,J)=0.
  300 CONTINUE
COME HERE TO SPECIFY BOUNDARY CONDITIONS
      DO 310 I=2,L2
         KBCJ1(I)=2
  310 CONTINUE
```

```
      DO 320 J=2,M2
         KBCI1(J)=2
         KBCL1(J)=2
  320 CONTINUE
      RETURN
      END
CCCCCCCCCCCCCCCCCCCCCCCCCCCCCCCCCCCCCCCCCCCCCCCCCCCCCCCCCCCCCCCCCCCCCCCC
```

11.5-5   Results for Example 15

```
RESULTS OF CONDUCT FOR CARTESIAN COORDINATE SYSTEM
****************************************************
```

```
. . . . . . . . . . . . . . . . . . . . . . . . . . . . . . . . . . . . . . . . . . . . . . . . . . . . . . . . . . . . . . .

WATER SEEPAGE UNDER A DAM

. . . . . . . . . . . . . . . . . . . . . . . . . . . . . . . . . . . . . . . . . . . . . . . . . . . . . . . . . . .
```

ITER	P(3,5)	U(3,5)	V(3,5)	P(5,5)	U(5,5)	V(5,5)
0	0.00E+00	0.00E+00	0.00E+00	0.00E+00	0.00E+00	0.00E+00
1	7.33E+01	2.92E+00	-6.14E+00	6.37E+01	9.50E+00	-4.94E+00
2	7.33E+01	2.92E+00	-6.15E+00	6.37E+01	9.49E+00	-4.95E+00
3	7.33E+01	2.92E+00	-6.15E+00	6.37E+01	9.49E+00	-4.95E+00

```
I =    1        2        3        4        5        6        7
X = 0.00E+00 3.52E-01 1.04E+00 1.69E+00 2.28E+00 2.78E+00 3.13E+00

I =    8        9       10       11       12       13       14
X = 3.38E+00 3.72E+00 4.22E+00 4.81E+00 5.46E+00 6.15E+00 6.50E+00

J =    1        2        3        4        5        6        7
Y = 0.00E+00 2.50E-01 7.50E-01 1.25E+00 1.75E+00 2.25E+00 2.75E+00

J =    8        9       10       11
Y = 3.17E+00 3.50E+00 3.83E+00 4.00E+00

******          PRESSURE        ******
       . . . . . . . . . . . . . . . . . . . . . . .
```

I =	1	2	3	4	5	6	7
J							
11	0.00E+00	1.00E+02	1.00E+02	1.00E+02	1.00E+02	1.00E+02	0.00E+00
10	9.78E+01	9.78E+01	9.76E+01	9.71E+01	9.65E+01	9.61E+01	5.56E+01
9	9.36E+01	9.35E+01	9.28E+01	9.15E+01	8.96E+01	8.80E+01	5.92E+01
8	8.95E+01	8.93E+01	8.82E+01	8.60E+01	8.27E+01	7.92E+01	5.81E+01
7	8.46E+01	8.44E+01	8.29E+01	7.97E+01	7.43E+01	6.59E+01	5.41E+01
6	7.96E+01	7.94E+01	7.75E+01	7.37E+01	6.78E+01	6.00E+01	5.27E+01
5	7.55E+01	7.53E+01	7.33E+01	6.94E+01	6.37E+01	5.72E+01	5.20E+01
4	7.25E+01	7.22E+01	7.02E+01	6.64E+01	6.12E+01	5.57E+01	5.15E+01
3	7.04E+01	7.02E+01	6.82E+01	6.46E+01	5.98E+01	5.49E+01	5.13E+01
2	6.94E+01	6.92E+01	6.72E+01	6.37E+01	5.91E+01	5.46E+01	5.12E+01
1	0.00E+00	6.90E+01	6.71E+01	6.36E+01	5.91E+01	5.45E+01	5.12E+01

I =	8	9	10	11	12	13	14
J							
11	0.00E+00	0.00E+00	0.00E+00	0.00E+00	0.00E+00	0.00E+00	0.00E+00
10	2.90E+01	3.93E+00	3.46E+00	2.85E+00	2.41E+00	2.18E+00	2.15E+00
9	3.50E+01	1.20E+01	1.04E+01	8.51E+00	7.17E+00	6.50E+00	6.41E+00
8	4.00E+01	2.08E+01	1.73E+01	1.40E+01	1.18E+01	1.07E+01	1.05E+01
7	4.59E+01	3.41E+01	2.57E+01	2.03E+01	1.71E+01	1.56E+01	1.54E+01
6	4.73E+01	4.00E+01	3.22E+01	2.63E+01	2.25E+01	2.06E+01	2.04E+01
5	4.80E+01	4.28E+01	3.63E+01	3.06E+01	2.67E+01	2.47E+01	2.45E+01
4	4.85E+01	4.43E+01	3.88E+01	3.36E+01	2.98E+01	2.78E+01	2.75E+01
3	4.87E+01	4.51E+01	4.02E+01	3.54E+01	3.18E+01	2.98E+01	2.96E+01
2	4.88E+01	4.54E+01	4.09E+01	3.63E+01	3.28E+01	3.08E+01	3.06E+01
1	4.88E+01	4.55E+01	4.09E+01	3.64E+01	3.29E+01	3.10E+01	0.00E+00

******   U VELOCITY   ******

I =	1	2	3	4	5	6	7
J							
11	0.00E+00	0.00E+00	0.00E+00	0.00E+00	0.00E+00	0.00E+00	0.00E+00
10	0.00E+00	0.00E+00	3.29E-01	6.91E-01	1.03E+00	9.40E-01	0.00E+00
9	0.00E+00	0.00E+00	9.79E-01	2.07E+00	3.17E+00	3.16E+00	0.00E+00
8	0.00E+00	0.00E+00	1.59E+00	3.43E+00	5.65E+00	6.98E+00	0.00E+00
7	0.00E+00	0.00E+00	2.25E+00	4.97E+00	9.03E+00	1.69E+01	3.43E+01
6	0.00E+00	0.00E+00	2.73E+00	5.86E+00	9.98E+00	1.56E+01	2.11E+01
5	0.00E+00	0.00E+00	2.92E+00	6.03E+00	9.49E+00	1.30E+01	1.53E+01
4	0.00E+00	0.00E+00	2.93E+00	5.85E+00	8.67E+00	1.10E+01	1.22E+01
3	0.00E+00	0.00E+00	2.88E+00	5.60E+00	8.00E+00	9.74E+00	1.05E+01

```
2    0.00E+00 0.00E+00 2.84E+00 5.45E+00 7.64E+00 9.13E+00 9.75E+00
1    0.00E+00 0.00E+00 0.00E+00 0.00E+00 0.00E+00 0.00E+00 0.00E+00


I =      8        9       10       11       12       13       14
J
11   0.00E+00 0.00E+00 0.00E+00 0.00E+00 0.00E+00 0.00E+00 0.00E+00
10   0.00E+00 0.00E+00 9.40E 01 1.03E+00 6.91E-01 3.29E 01 0.00E+00
9    0.00E+00 0.00E+00 3.16E+00 3.17E+00 2.07E+00 9.79E 01 0.00E+00
8    0.00E+00 0.00E+00 6.98E+00 5.65E+00 3.43E+00 1.59E+00 0.00E+00
7    3.29E+01 3.43E+01 1.69E+01 9.03E+00 4.97E+00 2.25E+00 0.00E+00
6    2.18E+01 2.11E+01 1.56E+01 9.98E+00 5.86E+00 2.73E+00 0.00E+00
5    1.56E+01 1.53E+01 1.30E+01 9.49E+00 6.03E+00 2.92E+00 0.00E+00
4    1.24E+01 1.22E+01 1.10E+01 8.67E+00 5.85E+00 2.93E+00 0.00E+00
3    1.06E+01 1.05E+01 9.74E+00 8.00E+00 5.60E+00 2.88E+00 0.00E+00
2    9.85E+00 9.75E+00 9.13E+00 7.64E+00 5.45E+00 2.84E+00 0.00E+00
1    0.00E+00 0.00E+00 0.00E+00 0.00E+00 0.00E+00 0.00E+00 0.00E+00
```

******   V VELOCITY   ******

```
I =      1        2        3        4        5        6        7
J
11   0.00E+00 1.31E+01 1.45E+01 1.72E+01 2.08E+01 2.34E+01 0.00E+00
10   0.00E+00 1.30E+01 1.43E+01 1.70E+01 2.08E+01 2.41E+01 0.00E+00
9    0.00E+00 1.25E+01 1.38E+01 1.64E+01 2.08E+01 2.66E+01 0.00E+00
8    0.00E+00 1.17E+01 1.28E+01 1.52E+01 2.00E+01 3.19E+01 0.00E+00
7    0.00E+00 1.01E+01 1.08E+01 1.20E+01 1.31E+01 1.18E+01 2.79E+00
6    0.00E+00 8.21E+00 8.47E+00 8.68E+00 8.10E+00 5.52E+00 1.53E+00
5    0.00E+00 6.14E+00 6.15E+00 5.92E+00 4.95E+00 2.94E+00 8.16E 01
4    0.00E+00 4.06E+00 3.98E+00 3.67E+00 2.87E+00 1.59E+00 4.38E 01
3    0.00E+00 2.01E+00 1.95E+00 1.75E+00 1.32E+00 7.12E 01 1.95E 01
2    0.00E+00 0.00E+00 0.00E+00 0.00E+00 0.00E+00 0.00E+00 0.00E+00
1    0.00E+00 0.00E+00 0.00E+00 0.00E+00 0.00E+00 0.00E+00 0.00E+00


I =      8        9       10       11       12       13       14
J
11   0.00E+00 2.34E+01 2.08E+01 1.72E+01 1.45E+01 1.31E+01 0.00E+00
10   0.00E+00 2.41E+01 2.08E+01 1.70E+01 1.43E+01 1.30E+01 0.00E+00
9    0.00E+00 2.66E+01 2.08E+01 1.64E+01 1.38E+01 1.25E+01 0.00E+00
8    0.00E+00 3.19E+01 2.00E+01 1.52E+01 1.28E+01 1.17E+01 0.00E+00
7    2.79E+00 1.18E+01 1.31E+01 1.20E+01 1.08E+01 1.01E+01 0.00E+00
6    1.53E+00 5.52E+00 8.10E+00 8.68E+00 8.47E+00 8.21E+00 0.00E+00
```

5	8.16E 01	2.94E+00	4.95E+00	5.92E+00	6.15E+00	6.14E+00	0.00E+00
4	4.38E 01	1.59E+00	2.87E+00	3.67E+00	3.98E+00	4.06E+00	0.00E+00
3	1.95E 01	7.12E 01	1.32E+00	1.75E+00	1.95E+00	2.01E+00	0.00E+00
2	0.00E+00	0.00E+00	0.00E+00	0.00E+00	0.00E+00	0.00E+00	0.00E+00
1	0.00E+00	0.00E+00	0.00E+00	0.00E+00	0.00E+00	0.00E+00	0.00E+00

11.5-6 Discussion of Results

Once again, this is a linear problem for which we obtain convergence in one iteration. The velocity field shows the expected flow pattern. The component u is positive everywhere, indicating an overall flow from left to right. The values of v are negative on the left side of the dam and positive on the right. Thus, there is a downflow on the left side and an upflow in the rest of the domain.

11.5-7 Final Remarks

The concept of the flow in porous materials based on Darcy's law has many practical applications. Using this concept, we can analyze flows in rock beds, stored grain, filters and packing materials, building insulation, walls of blood vessels, and similar situations. The analogy between heat conduction and the Darcy flow gives us not only a method of computing the flows in porous media but also enhances our understanding of the two different physical processes.

Problems

11.1 Modify Example 13 to introduce different thermal boundary conditions such as a constant wall temperature or a constant heat flux on one or more walls. Verify the experimental finding that the average Nusselt number for a turbulent flow does not depend strongly on the actual thermal boundary condition employed. (This is unlike a laminar flow; see the Nusselt number values given in Problem 10.1.)

11.2 Use CONDUCT to solve the well-known potential flow around a cylinder in cross flow. You can do this in MODE = 1 by approximating the curved boundary of the cylinder by staircase steps. Alternatively, use MODE = 3 and supply the known potential field for a uniform flow as the boundary condition on a large circle used as the outer boundary.

11.3 The buoyancy affected flow in a vertical shrouded fin array is computed by Zhang and Patankar (1984). Perform the computations for the same parameters and compare with the results in the paper.

11.4 A turbulence model for a tube with internal fins is proposed in Patankar, Ivanovic, and Sparrow (1979). Using the framework of Example 13, perform comparable computations.

11.5 For a tube with internal fins, the buoyancy affected flow is calculated in Prakash and Patankar (1981). Use CONDUCT to reproduce their results.

11.6 Consider fully developed flow in a rectangular duct with one or more walls moving in the axial direction. For different pressure gradients, you can calculate interesting flow distributions, some of which can include reverse flow.

11.7 Note that the situation in Problem 11.6 can be treated by superposition. You can first calculate the flow with stationary walls and a nonzero pressure gradient. Then calculate the flow with moving wall(s) and a zero pressure gradient. The sum of these two solutions in different proportions gives the whole family of solutions with moving walls.

11.8 You can use CONDUCT to calculate an oscillating fully developed flow in a duct. Consider that the pressure gradient is a periodic function of time. Perform a time-dependent calculation for the velocity. You will find that the velocity distributions in an oscillating flow are very different from those in a steady flow.

11.9 Use the framework of Example 14 to calculate the potential flow around different airfoil shapes. Here, you will need to approximate the curved boundaries by rectangular steps.

11.10 Modify Example 15 to introduce a large impermeable rock in the middle of the calculation domain. Observe how the presence of the rock influences the flow pattern.

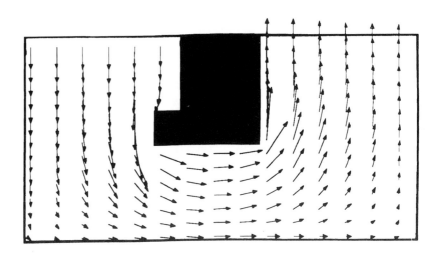

Groundwater flow under a dam

Chapter 12

Closing Remarks

In this final chapter, we shall summarize the important information and techniques presented in this book, examine some possibilities for further work, and review the strategies for appropriate use of the computer program.

12.1 The Numerical Method

The numerical method that we have employed in this book is characterized by both generality and simplicity. For the class of physical problems considered, the method can handle a wide variety of problems. For heat conduction, the problem may be steady or unsteady, the thermal conductivity may be nonuniform and temperature dependent, the heat generation may have any distribution and may be a function of the temperature, and the problem may be governed by a variety of linear or nonlinear boundary conditions. When the numerical technique is viewed as a general method for all conduction-like phenomena, we can use it for the calculation of the velocity and temperature fields in fully developed duct flows, and for other applications such as potential flow, flow in porous materials, electromagnetic fields, mass diffusion with complex chemical reactions, and so on. For the duct flow problems, we can embed solid bodies like fins or struts in the calculation domain and solve the resulting conjugate heat transfer problem if necessary. Similar interesting features can be accommodated in the other types of applications.

Despite its generality, the method is quite simple to understand and implement. This is primarily because of our use of the control-volume technique, in which the algebraic equations to be solved represent simple balances of physical quantities such heat, mass, or momentum over individual control volumes. The quantities in the algebraic equations, therefore, have a direct physical meaning, and when we obtain the final solution, we know that we have correctly made a heat balance (or mass balance, etc.) over all the little control volumes into which the whole domain has been subdivided. The method uses physical understanding in such a predominant manner, that, for the most part, we could have dispensed with calculus. After all, is it not remarkable that we can solve far more complex problems without reference to the Fourier series, the Bessel functions, the Legendre polynomials, the error function, and similar mathematical devices?

Because of its simplicity, the method not only is easy to use but also helps us to develop better insights into the relevant physical processes. As we examine the physical relationships through the algebraic equations employed and confirm them through a careful study of the computed results, we enhance and enrich our understanding of the phenomena under consideration.

12.2   The Computer Program

The description of CONDUCT in this book provides an example of the construction of a general-purpose computer program for a certain class of physical situations. We have seen that the provision of a problem-dependent subroutine gives virtually unlimited flexibility to the user in specifying the details of the problem and in designing the output. You also get an increased sense of participation in the process of solving the problem when you write the adaptation routine. Further, you still retain the option to construct, for a limited class of problems, an automatic version of the program that works by simply accepting a few data values.

A general-purpose program can be difficult to use because all its features and flexibilities require the corresponding input. In CONDUCT, we have eliminated this inconvenience by providing "default" values for many parameters. Thus, we have seen that, for a simple problem, one needs to write a very short adaptation routine. As the complexity of your problem increases, so does the length of the subroutine ADAPT.

In the framework of CONDUCT, you who have the responsibility for the problem specification. The results you get are the direct implications of what you specify in the ADAPT routine. It is true that certain quantities will remain at their default values; but you can always override them. This full control that you have over the details of a problem makes it possible to use CONDUCT in many imaginative ways.

12.3   Advanced Uses of the Program

In this book, we have applied CONDUCT to fifteen different problems. These problems were designed to illustrate various capabilities of CONDUCT and different practices incorporated in the ADAPT routine. With this background, you should now be able to use the program for a number of interesting applications.

To mention a few possibilities, you can use the program for studying different arrangements of insulation in building walls, for determining the temperature distribution in electronic circuit boards, or for predicting the periodic heat transfer in an engine cylinder. Various ducts and finned passages with a variety of thermal boundary conditions represent a vast array of possible interesting applications. You can go one step further and try to solve the buoyancy-affected fully developed flows in vertical ducts, in which the velocity and temperature fields influence each other. In Example 13, we employed the mixing-length model for a turbulent flow. You can attempt to use the more advanced k-ε model. (For more information, you may refer to Launder and Spalding, 1974.)

You can also use the program for potential flow over airfoils, moisture transport in soil, mass diffusion with chemical reactions, and electromagnetic fields.

12.4   Further Extensions of CONDUCT

Whereas you can undertake many advanced applications of CONDUCT in its present framework, you can also extend the capabilities of the program by modifying the invariant part. Of course, you should attempt to do this only after you are completely familiar with the program and confident to make the necessary changes. For those who would like to extend the program further, here are some suggestions.

Contact resistances. In CONDUCT, we can place two materials of different conductivities next to each other. In practice, since the contact surfaces are not perfectly smooth, there is a contact resistance at the interface in addition to the thermal resistance offered by the two materials themselves. In a number of cases, the values of the contact resistances are available from theoretical relationships or experimental correlations. CONDUCT can be modified to accept the user-supplied values of the contact resistances and to use them in calculating the neighbor coefficients.

Variable thickness. When we use CONDUCT for two-dimensional heat conduction in a rectangular plate, we assume that the plate has a uniform thickness everywhere. In reality, the plate could have a variable thickness, which will influence the available area for heat flow at a given location. The internal heat generation will also be affected by the variable thickness. It is possible to modify CONDUCT to account for any arbitrary variation of the plate thickness.

In a way, the treatment of variable thickness has already been illustrated in CONDUCT for a special case. In the axisymmetric geometry ($MODE = 2$), the domain is considered to have one radian extent in the circumferential direction. This means that the local depth or thickness for a point in the calculation domain equals the local radius r. As we have seen, this variation of r does affect the appropriate areas and volumes. If CONDUCT is modified to account for the arbitrary variation of the thickness, then the geometry for $MODE = 2$ will be a particular case in which the thickness varies as the radius.

Iterations within a time step. For steady nonlinear problems, we perform multiple iterations, which allow us to account for the nonlinearity by recalculating the coefficients. However, for a given time step in an unsteady problem, we do not recalculate the coefficients. We believe that the dependent variable and hence the coefficients are not likely to change very much, especially if the time step is reasonably small. This is largely true, but there may be strongly nonlinear problems for which performing multiple iterations within a time step is desirable. It is possible to modify CONDUCT to introduce the additional iteration loop.

These iterations do have their price. Currently, we do not store the old and new values of the dependent variable (ϕ_P^0 and ϕ_P separately in the computer memory. We have only one array $F(I,J,NF)$, which contains the old values ϕ_P^0 at the beginning of the time step. As soon as the new values of ϕ_P are calculated, they replace the contents of the same array $F(I,J,NF)$. Since there is no further iteration for the same time step, we have no use for the old values ϕ_P^0. If you were to introduce multiple iterations within a time step, you will have to store ϕ_P^0 and ϕ_P in separate arrays. The values of ϕ_P^0 will remain the same during these iterations, and you will be iteratively changing the values of ϕ_P until a converged solution for the time step is obtained. As you begin the next time step, you will transfer the values of ϕ_P into the array for ϕ_P^0 and proceed to obtain the new ϕ_P values appropriate to the new time step.

This additional requirement of storing ϕ_P^0 is the main reason why no provision is made in CONDUCT for performing multiple iterations within a time step.

Curvilinear orthogonal coordinates. In CONDUCT, we handle irregular and curved boundaries by approximating them with a series of rectangular steps. This procedure retains the convenience of a rectangular grid but is not suitable for a highly irregular geometry. CONDUCT can be redesigned to handle general curvilinear orthogonal coordinates so that almost any geometry can be handled by fitting an appropriate curvilinear grid within the geometry. Apart from some complications of lengths, areas, and volumes, the general framework of CONDUCT will remain the same. The modified program will be able to handle arbitrary shapes with greater elegance and accuracy.

Our use of the θr coordinates ($MODE = 3$) is, in fact, an example of curvilinear orthogonal coordinates. For this case, various lengths, areas, and volumes are simple functions of the radius r. This will have to be generalized for general orthogonal coordinates.

Curvilinear nonorthogonal coordinates. Often it is difficult to devise *orthogonal* curvilinear coordinates for a given geometry. On the other hand, *nonorthogonal* coordinates can be generated relatively easily; in principle, they can simply be sketched by hand. Then, the next step would be to prepare a version of CONDUCT for nonorthogonal curvilinear coordinates. Here, we shall have to deal with, in addition to the geometrical complication, a different form of the discretization equations. Since a control-volume face would not be normal to the line connecting the two grid points, a more complex formula is required for expressing the flux across the face. If you are interested in exploring this possibility, you may refer, among other sources, to Karki and Patankar (1988).

Three dimensions. If you work with Cartesian or cylindrical-polar coordinates, the extension of CONDUCT to three dimensions is not difficult. In our construction of the numerical method, we have used no practice that is intrinsically limited to two dimensions. In CONDUCT, we chose to implement the method in a two-dimensional framework, but a three-dimensional implementation is possible along the same lines. Of course, if you go to three dimensions, be prepared for large requirements of computer memory and computer time. But there is no fundamental limitation to producing a three-dimensional version.

12.5   Final Suggestions for Using CONDUCT

A computer program such as CONDUCT can bring you many benefits in terms of problem-solving skills and enhanced understanding. To enable you to maximize the benefits, here are some suggestions.

Develop sufficient familiarity with CONDUCT. Use the program for solving a large number of simple problems. Exercise various facilities in the program. Solve a variety of problems in addition to those of immediate interest to you.

Go from simple to complex. A common mistake in the use of a computer program such as CONDUCT is to apply it directly to a very complex problem. In such as case, an inexperienced user is likely to be unsuccessful. A better strategy is to start from a very simple idealization of your final problem, add only a few complications at a time, and through a number of successful solutions to these intermediate problems, proceed to the final complex situation.

Always modify an existing ADAPT. It is never wise to write the adaptation routine from scratch. Begin with an existing version of ADAPT, make a few changes, and test them. In this manner, gradually build the adaptation routine for the final problem to be solved.

Check the accuracy. Numerical solutions, even when they look reasonable, are not necessarily accurate unless a large number of grid points are used. If you want to rely on the numerical values predicted by your solution, you must make sure that you have a sufficiently fine grid. A further grid refinement should not make any appreciable change in the solution. Also, incorporate checks for overall heat balances, and compare (whenever possible) your numerical results with exact solutions.

Provide relevant output. In each physical situation, certain quantities are of primary interest. By arranging to print them out, you can focus attention on the main outcome of the solution. From these quantities, you can also quickly verify the reasonableness of the computed values.

Study the output. In this book, we have often discussed the use of computational solutions for enhancing physical understanding. When you adapt CONDUCT to a particular problem, you have certain expectations about the solution. You should study the output and validate your expectations. Occasionally, the results may at first look surprising; but a careful examination will enable you to modify your expectations. Alternatively, this will reveal some error that you may have made in adapting CONDUCT. In any case, a critical examination of the computed results will enrich your understanding of the underlying physical phenomena.

Discover the excitement. The simulation of complex physical phenomena through a computer program is an exciting activity. Just as you can study physical phenomena through laboratory experiments, you can often satisfy your scientific curiosity through computational simulation. A tool like CONDUCT puts at your fingertips the ability to simulate a wide range of physical phenomena. You will experience that this is a most exciting activity.

Axial velocity contours for a fully developed duct flow
with temperature-dependent viscosity
(The viscosity is much smaller near the top and right walls.)

Appendix A

Listing of the Invariant Part of CONDUCT

A.1 Include File "COMMON"

```
CCCCCCCCCCCCCCCCCCCCCCCCCCCCCCCCCCCCCCCCCCCCCCCCCCCCCCCCCCCCCCCCCCCCCCC
      PARAMETER(NI=52,NJ=52,NFMAX=5,NZMAX=10)
      CHARACTER*18 TITLE
      CHARACTER*64 HEADER,PRINTF,PLOTF
      COMMON F(NI,NJ,NFMAX),ALAM(NI,NJ),GAM(NI,NJ),CON(NI,NJ),
     1 AIP(NI,NJ),AIM(NI,NJ),AJP(NI,NJ),AJM(NI,NJ),AP(NI,NJ),
     2 IBLOCK(NI,NJ),X(NI),XU(NI),XCV(NI),ARX(NJ),Y(NJ),YV(NJ),YCV(NJ),
     3 YCVR(NJ),R(NJ),RV(NJ),SX(NJ),PTX(NI),QTX(NI),PTY(NJ),QTY(NJ),
     4 FLUXI1(NJ,NFMAX),FLUXL1(NJ,NFMAX),FLUXJ1(NI,NFMAX),
     5 FLUXM1(NI,NFMAX),KBCI1(NJ),KBCL1(NJ),KBCJ1(NI),KBCM1(NI),
     6 FLXCI1(NJ),FLXCL1(NJ),FLXCJ1(NI),FLXCM1(NI),
     7 FLXPI1(NJ),FLXPL1(NJ),FLXPJ1(NI),FLXPM1(NI)
      COMMON/GENL/NF,L1,L2,L3,M1,M2,M3,ITER,SMALL,BIG,
     1 LAST,TIME,DT,MODE,KORD,KOUT,IU1,IU2,KPGR,KSTOP
      COMMON/NFF/RELAX(NFMAX),KPRINT(NFMAX),KSOLVE(NFMAX),KBLOC(NFMAX),
     1 KPLOT(NFMAX),NTIMES(NFMAX),NTC(NFMAX),CRIT(NFMAX)
      COMMON/TTL/TITLE(NFMAX),HEADER,PRINTF,PLOTF
      COMMON/EZG/NCVLX,NCVLY,XL,YL,POWERX,POWERY
      COMMON/ZG/NZX,NZY,NCVX(NZMAX),NCVY(NZMAX),XZONE(NZMAX),
     1 YZONE(NZMAX),POWRX(NZMAX),POWRY(NZMAX)
      DIMENSION SC(NI,NJ),SP(NI,NJ)
      EQUIVALENCE (CON,SC),(AP,SP)
CCCCCCCCCCCCCCCCCCCCCCCCCCCCCCCCCCCCCCCCCCCCCCCCCCCCCCCCCCCCCCCCCCCCCCC
```

A.2 Invariant Part of CONDUCT

```
CCCCCCCCCCCCCCCCCCCCCCCCCCCCCCCCCCCCCCCCCCCCCCCCCCCCCCCCCCCCCCCCCCCCCCC
      PROGRAM CONDUCT
COMPUTER PROGRAM 'CONDUCT' BY DR. SUHAS V. PATANKAR
COPYRIGHT (C) 1991 INNOVATIVE RESEARCH, INC.
C*********************************************************************
$INCLUDE:'COMMON'
C*********************************************************************
CALCULATIONS IN THE GETTING-READY PHASE
```

```
      CALL DEFLT
      CALL GRID
      CALL READY
      CALL BEGIN
   10 CONTINUE
COME HERE TO START THE ITERATION OR TIME-STEP LOOP
      CALL OUTPUT
      IF(KSTOP.NE.0) STOP
      CALL HEART
      GO TO 10
      END
CCCCCCCCCCCCCCCCCCCCCCCCCCCCCCCCCCCCCCCCCCCCCCCCCCCCCCCCCCCCCCCCCCCCCCCCC
      SUBROUTINE DEFRD
C***********************************************************************
$INCLUDE:'COMMON'
C***********************************************************************
C
      ENTRY DEFLT
C
COME HERE TO SET THE DEFAULT VALUES
C
      HEADER='USE THE CHARACTER VARIABLE HEADER TO SPECIFY A PROBLEM TIT
     1LE'
      PRINTF='PRINT1'
      PLOTF='PLOT1'
      CALL INTA7(KSTOP,0,LAST,5,ITER,0,KORD,2,MODE,1,KPGR,1,KOUT,3)
      CALL DATA5(SMALL,1.E-20,BIG,1.E+20,TIME,0.,DT,1.E+20,R(1),0.)
      CALL DATA2(POWERX,1.,POWERY,1.)
      DO 10 NZ=1,NZMAX
      POWRX(NZ)=1.
   10 POWRY(NZ)=1.
      DO 20 N=1,NFMAX
      CRIT(N)=1.E-5
      KSOLVE(N)=0
      NTIMES(N)=10
      KBLOC(N)=1
      RELAX(N)=1.
      TITLE(N)='                      '
      KPRINT(N)=0
      KPLOT(N)=0
      DO 30 I=2,NI
      FLUXJ1(I,N)=0.
```

```
   30 FLUXM1(I,N)=0.
      DO 40 J=2,NJ
      FLUXI1(J,N)=0.
   40 FLUXL1(J,N)=0.
   20 CONTINUE
      DO 50 J=1,NJ
      DO 50 I=1,NI
      CON(I,J)=0.
      AP(I,J)=0.
      ALAM(I,J)=1.
      GAM(I,J)=1.
      IBLOCK(I,J)=0
      DO 60 N=1,NFMAX
   60 F(I,J,N)=0.
   50 CONTINUE
      DO 70 I=2,NI
      KBCJ1(I)=1
      KBCM1(I)=1
      FLXCJ1(I)=0.
      FLXCM1(I)=0.
      FLXPJ1(I)=0.
      FLXPM1(I)=0.
   70 CONTINUE
      DO 80 J=2,NJ
      KBCI1(J)=1
      KBCL1(J)=1
      FLXCI1(J)=0.
      FLXCL1(J)=0.
      FLXPI1(J)=0.
      FLXPL1(J)=0.
   80 CONTINUE
C
      RETURN
C*-*-*-*-*-*-*-*-*-*-*-*-*-*-*-*-*-*-*-*-*-*-*-*-*-*-*-*-*-*-*-*-*-*
      ENTRY READY
C
      IF(KOUT.NE.1) OPEN(UNIT=7,FILE=PRINTF)
      IU1=6
      IF(KOUT.EQ.2) IU1=7
      IU2=7
      IF(KOUT.EQ.1) IU2=6
CREATE INITIAL OUTPUT
```

```
      DO 100 IUNIT=IU1,IU2
C
      IF(MODE.EQ.1) WRITE(IUNIT,1)
    1 FORMAT(1X,'RESULTS OF CONDUCT FOR CARTESIAN COORDINATE SYSTEM'
     1/1X,50(1H*)//)
      IF(MODE.EQ.2) WRITE(IUNIT,2)
    2 FORMAT(1X,'RESULTS OF CONDUCT FOR AXISYMMETRIC COORDINATE SYSTEM'
     1/1X,53(1H*)//)
      IF(MODE.EQ.3) WRITE(IUNIT,3)
    3 FORMAT(1X,'RESULTS OF CONDUCT FOR POLAR COORDINATE SYSTEM'
     1/1X,46(1H*)//)
      WRITE(IUNIT,5) HEADER
    5 FORMAT(1X,64('·')/1X,A64/1X,64('·')//)
      IF(L1.GT.NI.OR.M1.GT.NJ.OR.L1.LT.4.OR.M1.LT.4) THEN
      WRITE(IUNIT,6)
    6 FORMAT(1X,'EXECUTION TERMINATED DUE TO ONE(OR MORE) OF THE FOLLOWI
     1NG REASON(S)'/2X,'1) L1 GREATER THAN NI'/2X,'2) M1 GREATER THAN NJ
     2'/2X,'3) L1 LESS THAN 4'/2X,'4) M1 LESS THAN 4'/)
      KSTOP=1
      ENDIF
  100 CONTINUE
      IF(KSTOP.NE.0) STOP
CALCULATE GEOMETRICAL QUANTITIES
      L2=L1-1
      L3=L2-1
      M2=M1-1
      M3=M2-1
      X(1)=XU(2)
      DO 110 I=2,L2
  110 X(I)=0.5*(XU(I+1)+XU(I))
      X(L1)=XU(L1)
      Y(1)=YV(2)
      DO 120 J=2,M2
  120 Y(J)=0.5*(YV(J+1)+YV(J))
      Y(M1)=YV(M1)
      DO 130 I=2,L2
  130 XCV(I)=XU(I+1)-XU(I)
      DO 140 J=2,M2
  140 YCV(J)=YV(J+1)-YV(J)
      IF(MODE.EQ.1) THEN
      DO 150 J=1,M1
      RV(J)=1
```

```
  150 R(J)=1
      ELSE
      RY1=R(1)·Y(1)
      DO 160 J=2,M1
  160 R(J)=Y(J)+RY1
      RV(2)=R(1)
      DO 170 J=3,M2
  170 RV(J)=RV(J·1)+YCV(J·1)
      RV(M1)=R(M1)
      ENDIF
      IF(MODE.EQ.3) THEN
      DO 180 J=1,M1
  180 SX(J)=R(J)
      ELSE
      DO 190 J=1,M1
      SX(J)=1.
  190 CONTINUE
      ENDIF
      DO 200 J=2,M2
      YCVR(J)=R(J)*YCV(J)
      IF(MODE.EQ.3) THEN
      ARX(J)=YCV(J)
      ELSE
      ARX(J)=YCVR(J)
      ENDIF
  200 CONTINUE
C
      RETURN
      END
CCCCCCCCCCCCCCCCCCCCCCCCCCCCCCCCCCCCCCCCCCCCCCCCCCCCCCCCCCCCCCCCCCCCCCCC
      SUBROUTINE HEART
C**********************************************************************
$INCLUDE:'COMMON'
C**********************************************************************
CONSTRUCT LOOP FOR ALL EQUATIONS
      DO 999 N=1,NFMAX
      NF=N
      IF(KSOLVE(NF).EQ.0) GO TO 999
C
      CALL PHI
C
CALCULATE COEFFICIENTS IN THE DISCRETIZATION EQUATION
```

```
C
      BETA=4./3.
      IF(KORD.EQ.1) BETA=1.
      RLX=(1.-RELAX(NF))/RELAX(NF)
CONSIDER VOLUMETRIC TERMS
      DO 10 J=2,M2
      DO 10 I=2,L2
      VOL=YCVR(J)*XCV(I)
      APT=ALAM(I,J)/DT
      CON(I,J)=(CON(I,J)+APT*F(I,J,NF))*VOL
      AP(I,J)=(APT-AP(I,J))*VOL
   10 CONTINUE
COEFFICIENTS FOR X-DIRECTION DIFFUSION
      DO 20 J=2,M2
      DO 20 I=2,L3
      DIFF=ARX(J)*2.*GAM(I,J)*GAM(I+1,J)/((XCV(I)*GAM(I+1,J)+
     1 XCV(I+1)*GAM(I,J)+SMALL)*SX(J))
      AIP(I,J)=DIFF+SMALL
      AIM(I+1,J)=AIP(I,J)
   20 CONTINUE
      DO 30 J=2,M2
CONSIDER I=1 BOUNDARY
      DIFF=GAM(2,J)/(0.5*XCV(2)*SX(J))+SMALL
      AIM(2,J)=BETA*DIFF
      AIP(1,J)=AIM(2,J)
      AIM(2,J)=AIM(2,J)*ARX(J)
      AIM(1,J)=(BETA-1.)*AIP(2,J)/ARX(J)
      AIP(2,J)=AIP(2,J)+AIM(1,J)*ARX(J)
      IF(KBCI1(J).EQ.1) THEN
      CON(2,J)=CON(2,J)+AIM(2,J)*F(1,J,NF)
      ELSE
      AP(1,J)=AIP(1,J)-FLXP11(J)
      CON(1,J)=FLXCI1(J)
      TEMP=AIM(2,J)/AP(1,J)
      AP(2,J)=AP(2,J)-AIP(1,J)*TEMP
      AIP(2,J)=AIP(2,J)-AIM(1,J)*TEMP
      CON(2,J)=CON(2,J)+CON(1,J)*TEMP
      ENDIF
      AP(2,J)=AP(2,J)+AIM(2,J)
      AIM(2,J)=0.
CONSIDER I=L1 BOUNDARY
      DIFF=GAM(L2,J)/(0.5*XCV(L2)*SX(J))+SMALL
```

```
      AIP(L2,J)=BETA*DIFF
      AIM(L1,J)=AIP(L2,J)
      AIP(L2,J)=AIP(L2,J)*ARX(J)
      AIP(L1,J)=(BETA-1.)*AIM(L2,J)/ARX(J)
      AIM(L2,J)=AIM(L2,J)+AIP(L1,J)*ARX(J)
      IF(KBCL1(J).EQ.1) THEN
      CON(L2,J)=CON(L2,J)+AIP(L2,J)*F(L1,J,NF)
      ELSE
      AP(L1,J)=AIM(L1,J)-FLXPL1(J)
      CON(L1,J)=FLXCL1(J)
      TEMP=AIP(L2,J)/AP(L1,J)
      AP(L2,J)=AP(L2,J)-AIM(L1,J)*TEMP
      AIM(L2,J)=AIM(L2,J)-AIP(L1,J)*TEMP
      CON(L2,J)=CON(L2,J)+CON(L1,J)*TEMP
      ENDIF
      AP(L2,J)=AP(L2,J)+AIP(L2,J)
      AIP(L2,J)=0.
   30 CONTINUE
COEFFICIENTS FOR Y-DIRECTION DIFFUSION
      DO 40 J=2,M3
      DO 40 I=2,L2
      AREA=RV(J+1)*XCV(I)
      DIFF=AREA*2.*GAM(I,J)*GAM(I,J+1)/(YCV(J)*GAM(I,J+1)+
     1 YCV(J+1)*GAM(I,J)+SMALL)
      AJP(I,J)=DIFF+SMALL
      AJM(I,J+1)=AJP(I,J)
   40 CONTINUE
      DO 50 I=2,L2
CONSIDER J=1 BOUNDARY
      AREA=RV(2)*XCV(I)
      DIFF=GAM(I,2)/(0.5*YCV(2))+SMALL
      AJM(I,2)=BETA*DIFF
      AJP(I,1)=AJM(I,2)
      AJM(I,2)=AJM(I,2)*AREA
      AJM(I,1)=(BETA-1.)*AJP(I,2)/(RV(3)*XCV(I))
      AJP(I,2)=AJP(I,2)+AJM(I,1)*AREA
      IF(KBCJ1(I).EQ.1) THEN
      CON(I,2)=CON(I,2)+AJM(I,2)*F(I,1,NF)
      ELSE
      AP(I,1)=AJP(I,1)-FLXPJ1(I)
      CON(I,1)=FLXCJ1(I)
      TEMP=AJM(I,2)/AP(I,1)
```

```
      AP(I,2)=AP(I,2)-AJP(I,1)*TEMP
      AJP(I,2)=AJP(I,2)-AJM(I,1)*TEMP
      CON(I,2)=CON(I,2)+CON(I,1)*TEMP
      ENDIF
      AP(I,2)=AP(I,2)+AJM(I,2)
      AJM(I,2)=0.
CONSIDER J=M1 BOUNDARY
      AREA=RV(M1)*XCV(I)
      DIFF=GAM(I,M2)/(0.5*YCV(M2))+SMALL
      AJP(I,M2)=BETA*DIFF
      AJM(I,M1)=AJP(I,M2)
      AJP(I,M2)=AJP(I,M2)*AREA
      AJP(I,M1)=(BETA-1.)*AJM(I,M2)/(RV(M2)*XCV(I))
      AJM(I,M2)=AJM(I,M2)+AJP(I,M1)*AREA
      IF(KBCM1(I).EQ.1) THEN
      CON(I,M2)=CON(I,M2)+AJP(I,M2)*F(I,M1,NF)
      ELSE
      AP(I,M1)=AJM(I,M1)-FLXPM1(I)
      CON(I,M1)=FLXCM1(I)
      TEMP=AJP(I,M2)/AP(I,M1)
      AP(I,M2)=AP(I,M2)-AJM(I,M1)*TEMP
      AJM(I,M2)=AJM(I,M2)-AJP(I,M1)*TEMP
      CON(I,M2)=CON(I,M2)+CON(I,M1)*TEMP
      ENDIF
      AP(I,M2)=AP(I,M2)+AJP(I,M2)
      AJP(I,M2)=0.
   50 CONTINUE
COME HERE TO INTRODUCE UNDERRELAXATION
CONSTRUCT AP(I,J) AND CON(I,J) IN THEIR FINAL FORM
      DO 60 J=2,M2
      DO 60 I=2,L2
      ANB=AIP(I,J)+AIM(I,J)+AJP(I,J)+AJM(I,J)
      AINR=ANB*RLX
      AP(I,J)=AP(I,J)+ANB+AINR
      CON(I,J)=CON(I,J)+AINR*F(I,J,NF)
   60 CONTINUE
C
CALL THE SOLVE ROUTINE TO OBTAIN THE SOLUTION OF THE DISCRETIZATION
C  EQUATIONS
C
      CALL SOLVE
  999 CONTINUE
```

```
C
      TIME=TIME+DT
      ITER=ITER+1
      IF(ITER.GE.LAST) KSTOP=1
      RETURN
      END
CCCCCCCCCCCCCCCCCCCCCCCCCCCCCCCCCCCCCCCCCCCCCCCCCCCCCCCCCCCCCCCCCCCCCCCCCC
      SUBROUTINE SOLVE
C**********************************************************************
$INCLUDE:'COMMON'
      DIMENSION RT(6)
C**********************************************************************
      BIG1=1.E+10
      SMALL1=1.0E-5
      LL2=2*L2
      LL=LL2-2
      MM2=2*M2
      MM=MM2-2
      N=NF
      NTM=NTIMES(N)
      DO 999 NT=1,NTM
      NTT=NT
      ICON=1
COME HERE TO PERFORM THE I-DIRECTION BLOCK CORRECTION
C.......................................................................
      PTX(1)=0.
      QTX(1)=0.
      DO 10 I=2,L2
      BL=SMALL
      BLP=0.
      BLM=0.
      BLC=0.
      DO 20 J=2,M2
      IF(AP(I,J).LT.BIG1) THEN
      BL=BL+AP(I,J)
      IF(AP(I,J+1).LT.BIG1) BL=BL-AJP(I,J)
      IF(AP(I,J-1).LT.BIG1) BL=BL-AJM(I,J)
      IF(AP(I+1,J).LT.BIG1) BLP=BLP+AIP(I,J)
      IF(AP(I-1,J).LT.BIG1) BLM=BLM+AIM(I,J)
CONVERGENCE CRITERION FOR THE SOLUTION ROUTINE
      RT(1)=AIP(I,J)*F(I+1,J,N)
      RT(2)=AIM(I,J)*F(I-1,J,N)
```

```
        RT(3)=AJP(I,J)*F(I,J+1,N)
        RT(4)=AJM(I,J)*F(I,J-1,N)
        RT(5)=-AP(I,J)*F(I,J,N)
        RT(6)=CON(I,J)
        RES=0
        TERM=1.0E-8
        DO 30 IRT=1,6
        RES=RES+RT(IRT)
    30 TERM=MAX(TERM,ABS(RT(IRT)))
        IF(ABS(RES/TERM).GT.CRIT(N))ICON=0
        BLC=BLC+RES
        ENDIF
    20 CONTINUE
        DENOM=BL-PTX(I-1)*BLM
        IF(ABS(DENOM/BL).LT.SMALL1) DENOM=BIG
        PTX(I)=BLP/DENOM
        QTX(I)=(BLC+BLM*QTX(I-1))/DENOM
    10 CONTINUE
        IF(NTT.NE.1.AND.ICON.EQ.1) GO TO 990
        IF(KBLOC(NF).EQ.0) GO TO 80
        BL=0.
        DO 40 I=L2,2,-1
        BL=BL*PTX(I)+QTX(I)
        DO 40 J=2,M2
        IF(AP(I,J).LT.BIG1) F(I,J,N)=F(I,J,N)+BL
    40 CONTINUE
COME HERE TO PERFORM THE J-DIRECTION BLOCK CORRECTION
C..........................................................
        PTY(1)=0.
        QTY(1)=0.
        DO 50 J=2,M2
        BL=SMALL
        BLP=0.
        BLM=0.
        BLC=0.
        DO 60 I=2,L2
        IF(AP(I,J).LT.BIG1) THEN
        BL=BL+AP(I,J)
        IF(AP(I+1,J).LT.BIG1) BL=BL-AIP(I,J)
        IF(AP(I-1,J).LT.BIG1) BL=BL-AIM(I,J)
        IF(AP(I,J+1).LT.BIG1) BLP=BLP+AJP(I,J)
        IF(AP(I,J-1).LT.BIG1) BLM=BLM+AJM(I,J)
```

```
      BLC=BLC+CON(I,J)+AIP(I,J)*F(I+1,J,N)+AIM(I,J)*F(I-1,J,N)
     1    +AJP(I,J)*F(I,J+1,N)+AJM(I,J)*F(I,J-1,N)-AP(I,J)*F(I,J,N)
      ENDIF
   60 CONTINUE
      DENOM=BL-PTY(J-1)*BLM
      IF(ABS(DENOM/BL).LT.SMALL1) DENOM=BIG
      PTY(J)=BLP/DENOM
      QTY(J)=(BLC+BLM*QTY(J-1))/DENOM
   50 CONTINUE
      BL=0.
      DO 70 J=M2,2,-1
      BL=BL*PTY(J)+QTY(J)
      DO 70 I=2,L2
      IF(AP(I,J).LT.BIG1) F(I,J,N)=F(I,J,N)+BL
   70 CONTINUE
   80 CONTINUE
CARRY OUT THE I-DIRECTION TDMA
C-----------------------------------------------------------------
      DO 90 JJ=2,MM
      J=MIN(JJ,MM2-JJ)
      PTX(1)=0.
      QTX(1)=0
      DO 100 I=2,L2
      DENOM=AP(I,J)-PTX(I-1)*AIM(I,J)
      PTX(I)=AIP(I,J)/DENOM
      TEMP=CON(I,J)+AJP(I,J)*F(I,J+1,N)+AJM(I,J)*F(I,J-1,N)
      QTX(I)=(TEMP+AIM(I,J)*QTX(I-1))/DENOM
  100 CONTINUE
      DO 110 I=L2,2,-1
  110 F(I,J,N)=F(I+1,J,N)*PTX(I)+QTX(I)
   90 CONTINUE
CARRY OUT THE J-DIRECTION TDMA
C-----------------------------------------------------------------
      DO 120 II=2,LL
      I=MIN(II,LL2-II)
      PTY(1)=0.
      QTY(1)=0
      DO 130 J=2,M2
      DENOM=AP(I,J)-PTY(J-1)*AJM(I,J)
      PTY(J)=AJP(I,J)/DENOM
      TEMP=CON(I,J)+AIP(I,J)*F(I+1,J,N)+AIM(I,J)*F(I-1,J,N)
      QTY(J)=(TEMP+AJM(I,J)*QTY(J-1))/DENOM
```

```
130 CONTINUE
    DO 140 J=M2,2,-1
140 F(I,J,N)=F(I,J+1,N)*PTY(J)+QTY(J)
120 CONTINUE
C·········································································
999 CONTINUE
    NTC(N)=NTT
    GO TO 991
990 NTC(N)=NTT-1
991 CONTINUE
CALCULATE THE UNKNOWN BOUNDARY VALUES AND FLUXES
C·········································································
    DO 160 I=2,L2
    TEMP=AJM(I,1)*(F(I,3,N)-F(I,2,N))
    IF(KBCJ1(I).EQ.2)
   1 F(I,1,N)=(AJP(I,1)*F(I,2,N)-TEMP+CON(I,1))/AP(I,1)
    FLUXJ1(I,N)=AJP(I,1)*(F(I,1,N)-F(I,2,N))+TEMP
    TEMP=AJP(I,M1)*(F(I,M3,N)-F(I,M2,N))
    IF(KBCM1(I).EQ.2)
   1 F(I,M1,N)=(AJM(I,M1)*F(I,M2,N)-TEMP+CON(I,M1))/AP(I,M1)
160 FLUXM1(I,N)=AJM(I,M1)*(F(I,M1,N)-F(I,M2,N))+TEMP
    DO 170 J=2,M2
    TEMP=AIM(1,J)*(F(3,J,N)-F(2,J,N))
    IF(KBCI1(J).EQ.2)
   1 F(1,J,N)=(AIP(1,J)*F(2,J,N)-TEMP+CON(1,J))/AP(1,J)
    FLUXI1(J,N)=AIP(1,J)*(F(1,J,N)-F(2,J,N))+TEMP
    TEMP=AIP(L1,J)*(F(L3,J,N)-F(L2,J,N))
    IF(KBCL1(J).EQ.2)
   1 F(L1,J,N)=(AIM(L1,J)*F(L2,J,N)-TEMP+CON(L1,J))/AP(L1,J)
170 FLUXL1(J,N)=AIM(L1,J)*(F(L1,J,N)-F(L2,J,N))+TEMP
C
COME HERE TO RESET CON,AP,KBC,FLXC, AND FLXP
C·········································································
    DO 180 J=2,M2
    KBCI1(J)=1
    KBCL1(J)=1
    FLXCI1(J)=0.
    FLXCL1(J)=0.
    FLXPI1(J)=0.
    FLXPL1(J)=0.
    DO 180 I=2,L2
    CON(I,J)=0.
```

```
      AP(I,J)=0.
  180 CONTINUE
      DO 190 I=2,L2
      KBCJ1(I)=1
      KBCM1(I)=1
      FLXCJ1(I)=0.
      FLXCM1(I)=0.
      FLXPJ1(I)=0.
      FLXPM1(I)=0.
  190 CONTINUE
      RETURN
      END
CCCCCCCCCCCCCCCCCCCCCCCCCCCCCCCCCCCCCCCCCCCCCCCCCCCCCCCCCCCCCCCCCCCCCC
      SUBROUTINE TOOLS
C********************************************************************
$INCLUDE:'COMMON'
C********************************************************************
      ENTRY EZGRID
C
CONSTRUCT THE X-DIRECTION GRID
      L1=NCVLX+2
      XU(2)=0.
      XU(L1)=XL
      L2=L1-1
      FCVLX=FLOAT(NCVLX)
      DO 20 I=3,L2
      DD=FLOAT(I 2)/FCVLX
      IF(POWERX.GT.0.) THEN
      XU(I)=XL*DD**POWERX
      ELSE
      XU(I)=XL*(1.-(1.-DD)**(-POWERX))
      ENDIF
   20 CONTINUE
CONSTRUCT THE Y-DIRECTION GRID
      M1=NCVLY+2
      YV(2)=0.
      YV(M1)=YL
      M2=M1-1
      FCVLY=FLOAT(NCVLY)
      DO 30 J=3,M2
      DD=FLOAT(J-2)/FCVLY
      IF(POWERY.GT.0.) THEN
```

```
      YV(J)=YL*DD**POWERY
      ELSE
      YV(J)=YL*(1.-(1.-DD)**(-POWERY))
      ENDIF
   30 CONTINUE
      RETURN
C*-*-*-*-*-*-*-*-*-*-*-*-*-*-*-*-*-*-*-*-*-*-*-*-*-*-*-*-*-*-*-*
      ENTRY ZGRID
CONSTRUCT THE GRID ZONE-BY-ZONE
C
CONSIDER THE X DIRECTION
      XU(2)=0.
      I2=2
      DO 100 NZ=1,NZX
      FCVLX=FLOAT(NCVX(NZ))
      ILAST=I2
      I1=ILAST+1
      I2=ILAST+NCVX(NZ)
      DO 100 I=I1,I2
      DD=FLOAT(I-ILAST)/FCVLX
      IF(POWRX(NZ).GT.0.) THEN
      XU(I)=XU(ILAST)+XZONE(NZ)*DD**POWRX(NZ)
      ELSE
      XU(I)=XU(ILAST)+XZONE(NZ)*(1.-(1.-DD)**(-POWRX(NZ)))
      ENDIF
  100 CONTINUE
      L1=I2
C
CONSIDER THE Y DIRECTION
      YV(2)=0.
      J2=2
      DO 110 NZ=1,NZY
      FCVLY=FLOAT(NCVY(NZ))
      JLAST=J2
      J1=JLAST+1
      J2=JLAST+NCVY(NZ)
      DO 110 J=J1,J2
      DD=FLOAT(J-JLAST)/FCVLY
      IF(POWRY(NZ).GT.0.) THEN
      YV(J)=YV(JLAST)+YZONE(NZ)*DD**POWRY(NZ)
      ELSE
      YV(J)=YV(JLAST)+YZONE(NZ)*(1.-(1.-DD)**(-POWRY(NZ)))
```

```
      ENDIF
  110 CONTINUE
      M1=J2
      RETURN
C*-*-*-*-*-*-*-*-*-*-*-*-*-*-*-*-*-*-*-*-*-*-*-*-*-*-*-*-*-*-*-*-*-*-*-*-*-*
      ENTRY PRINT
C
      DO 999 IUNIT=IU1,IU2
C
COME HERE TO ARRANGE THE PRINTOUT OF TWO-DIMENSIONAL FIELDS
      IF(KPGR.NE.0) THEN
C
CREATE PRINTOUT FOR GRID
C
      WRITE(IUNIT,1)
    1 FORMAT(' ')
      IBEG=1
      IEND=L1
      IREP=(IEND-IBEG+7)/7
      DO 200 K=1,IREP
      INCR=MIN(6,IEND-IBEG)
      ISTOP=IBEG+INCR
      WRITE(IUNIT,2) (I,I=IBEG,ISTOP)
    2 FORMAT(/2X,'I =',2X,7(I4,5X))
      IF(MODE.EQ.3) THEN
      WRITE(IUNIT,3) (X(I),I=IBEG,ISTOP)
    3 FORMAT(1X,'TH =',1P7E9.2)
      ELSE
      WRITE(IUNIT,4) (X(I),I=IBEG,ISTOP)
    4 FORMAT(2X,'X =',1P7E9.2)
      ENDIF
      IBEG=ISTOP+1
  200 CONTINUE
C
      WRITE(IUNIT,1)
      JBEG=1
      JEND=M1
      JREP=(JEND-JBEG+7)/7
      DO 210 K=1,JREP
      INCR=MIN(6,JEND-JBEG)
      JSTOP=JBEG+INCR
      WRITE(IUNIT,5) (J,J=JBEG,JSTOP)
```

```
    5 FORMAT(/2X,'J =',2X,7(I4,5X))
      WRITE(IUNIT,6) (Y(J),J=JBEG,JSTOP)
    6 FORMAT(2X,'Y =',1P7E9.2)
      JBEG=JSTOP+1
  210 CONTINUE
      ENDIF
CREATE PRINTOUT FOR THE VALUES OF DEPENDENT VARIABLES
      DO 220 N=1,NFMAX
      IF(KPRINT(N).NE.0) THEN
      WRITE(IUNIT,7) TITLE(N)
    7 FORMAT(/1X,6(1H*),3X,A18,3X,6(1H*)/9X,20(1H-))
      IBEG=1
      JBEG=1
      IEND=L1
      JEND=M1
      IREP=(IEND-IBEG+7)/7
      DO 230 K=1,IREP
      INCR=MIN(6,IEND-IBEG)
      ISTOP=IBEG+INCR
      WRITE(IUNIT,8) (I,I=IBEG,ISTOP)
    8 FORMAT(/' I =',I6,6I9)
      WRITE(IUNIT, 9)
    9 FORMAT(' J')
      DO 240 J=JEND,JBEG,-1
      WRITE(IUNIT,10) J,(F(I,J,N),I=IBEG,ISTOP)
   10 FORMAT(1X,I2,3X,1P7E9.2)
  240 CONTINUE
      IBEG=ISTOP+1
  230 CONTINUE
      ENDIF
  220 CONTINUE
  999 CONTINUE
      RETURN
C*-*-*-*-*-*-*-*-*-*-*-*-*-*-*-*-*-*-*-*-*-*-*-*-*-*-*-*-*-*-*-*-*
      ENTRY PLOT
      OPEN(UNIT=8,FILE=PLOTF)
COME HERE TO CREATE DATA FOR PLOTTING
C
      KFLOW=2
      WRITE(8,300) HEADER
  300 FORMAT(A64)
      WRITE(8,310) KFLOW,L1,M1,NFMAX,MODE,(KPLOT(I),I=1,NFMAX)
```

```
  310 FORMAT(I8I5)
      IBLOK=0
      DO 320 J=2,M2
      DO 320 I=2,L2
        IF(IBLOCK(I,J).EQ.1) THEN
            IBLOK=1
            GO TO 330
        ENDIF
  320 CONTINUE
  330 CONTINUE
      WRITE(8,310) IBLOK
      WRITE(8,340) (TITLE(N),N=1,NFMAX)
  340 FORMAT(4A18)
      WRITE(8,350) (X(I),I=1,L1),(Y(J),J=1,M1),(XU(I),I=2,L1)
     1,(YV(J),J=2,M1),(R(J),J=1,M1)
  350 FORMAT(5E12.6)
      DO 360 N=1,NFMAX
      IF(KPLOT(N).NE.0) WRITE(8,350) ((F(I,J,N),I=1,L1),J=1,M1)
  360 CONTINUE
      IF(IBLOK.EQ.1) THEN
          WRITE(8,310) ((IBLOCK(I,J),I=1,L1),J=1,M1)
      ENDIF
      CLOSE(8)
      RETURN
      END
CCCCCCCCCCCCCCCCCCCCCCCCCCCCCCCCCCCCCCCCCCCCCCCCCCCCCCCCCCCCCCCCCCCCCCCCCCC
      SUBROUTINE VALUES
C*************************************************************************
C
CREATE A FACILITY TO ASSIGN VALUES TO REAL VARIABLES
C
      ENTRY DATA9(A1,C1,A2,C2,A3,C3,A4,C4,A5,C5,A6,C6,A7,C7,A8,C8,A9,C9)
      A9=C9
      ENTRY DATA8(A1,C1,A2,C2,A3,C3,A4,C4,A5,C5,A6,C6,A7,C7,A8,C8)
      A8=C8
      ENTRY DATA7(A1,C1,A2,C2,A3,C3,A4,C4,A5,C5,A6,C6,A7,C7)
      A7=C7
      ENTRY DATA6(A1,C1,A2,C2,A3,C3,A4,C4,A5,C5,A6,C6)
      A6=C6
      ENTRY DATA5(A1,C1,A2,C2,A3,C3,A4,C4,A5,C5)
      A5=C5
      ENTRY DATA4(A1,C1,A2,C2,A3,C3,A4,C4)
```

```
      A4=C4
      ENTRY DATA3(A1,C1,A2,C2,A3,C3)
      A3=C3
      ENTRY DATA2(A1,C1,A2,C2)
      A2=C2
      ENTRY DATA1(A1,C1)
      A1=C1
      RETURN
C*-*-*-*-*-*-*-*-*-*-*-*-*-*-*-*-*-*-*-*-*-*-*-*-*-*-*-*-*-*-*-*-*-*-*-*-*
CREATE A FACILITY TO ASSIGN VALUES TO INTEGER VARIABLES
C
      ENTRY INTA9(I1,J1,I2,J2,I3,J3,I4,J4,I5,J5,I6,J6,I7,J7,I8,J8,I9,J9)
      I9=J9
      ENTRY INTA8(I1,J1,I2,J2,I3,J3,I4,J4,I5,J5,I6,J6,I7,J7,I8,J8)
      I8=J8
      ENTRY INTA7(I1,J1,I2,J2,I3,J3,I4,J4,I5,J5,I6,J6,I7,J7)
      I7=J7
      ENTRY INTA6(I1,J1,I2,J2,I3,J3,I4,J4,I5,J5,I6,J6)
      I6=J6
      ENTRY INTA5(I1,J1,I2,J2,I3,J3,I4,J4,I5,J5)
      I5=J5
      ENTRY INTA4(I1,J1,I2,J2,I3,J3,I4,J4)
      I4=J4
      ENTRY INTA3(I1,J1,I2,J2,I3,J3)
      I3=J3
      ENTRY INTA2(I1,J1,I2,J2)
      I2=J2
      ENTRY INTA1(I1,J1)
      I1=J1
      RETURN
      END
CCCCCCCCCCCCCCCCCCCCCCCCCCCCCCCCCCCCCCCCCCCCCCCCCCCCCCCCCCCCCCCCCCCCCCCCCC
```

List of Fortran Names

(This list includes only those variable names that have significant meaning. Other Fortran names that are locally used for temporary storage are not included here; their meaning can be easily inferred from the context.)

AIM(I,J)	coefficient a_w
AINR	inertia i, Eq. (5.64)
AIP(I,J)	coefficient a_E
AJM(I,J)	coefficient a_S
AJP(I,J)	coefficient a_N
ALAM(I,J)	storage capacity λ per unit volume
ANB	sum of neighbor coefficients
AP(I,J)	coefficient a_P: equivalent to S_P
APT	unsteady term coefficient a_P^0, Eq. (5.20)
AREA	area of control-volume face
ARX(J)	area of a main control-volume face normal to the x direction
BETA	boundary treatment index
BIG	a very large number such as 1.E20
BL	coefficient used in the block correction
BLC	coefficient used in the block correction
BLM	coefficient used in the block correction
BLP	coefficient used in the block correction
CON(I,J)	constant term b: equivalent to S_C
CRIT(NF)	constant for convergence criterion used in SOLVE
DIFF	diffusion conductance D
DT	time step Δt
F(I,J,NF)	various ϕ's
FLUXI1(J,NF)	diffusion flux into left boundary
FLUXJ1(I,NF)	diffusion flux into bottom boundary
FLUXL1(J,NF)	diffusion flux into right boundary
FLUXM1(I,NF)	diffusion flux into top boundary
FLXCI1(J)	boundary-flux constant f_C of left boundary
FLXCJ1(I)	boundary-flux constant f_C of bottom boundary
FLXCL1(J)	boundary-flux constant f_C of right boundary
FLXCM1(I)	boundary-flux constant f_C of top boundary
FLXPI1(J)	boundary-flux constant f_P of left boundary
FLXPJ1(I)	boundary-flux constant f_P of bottom boundary
FLXPL1(J)	boundary-flux constant f_P of right boundary
FLXPM1(I)	boundary-flux constant f_P of top boundary

GAM(I,J)   diffusion coefficient Γ

HEADER   a 64-character heading for the problem

I   index denoting the position in x

IBLOCK(I,J)   index denoting blockage

ITER   iteration counter

IU1,IU2   indices for output units

J   index denoting the position in y

KBCI1(J)   indicator for left boundary; = 1 for given ϕ_B; = 2 for given f_C and f_p in Eq. (5.39)

KBCJ1(I)   indicator for bottom boundary

KBCL1(J)   indicator for right boundary

KBCM1(I)   indicator for top boundary

KBLOC(NF)   when zero, block correction is omitted

KORD   index for boundary-condition treatment; = 1 for lower order, = 2 for higher order

KOUT   index for output: = 1 (screen), 2 (file), 3 (both)

KPGR   when nonzero, X(I) and Y(J) are printed

KPLOT(NF)   when nonzero, F(I,J,NF) is included in plotfile

KPRINT(NF)   when nonzero, F(I,J,NF) is printed

KSOLVE(NF)   when nonzero, general discretization equation is solved

KSTOP   when nonzero, computation stops

L1   value of I for the right-boundary grid line

L2   L1-1

L3   L1-2

LAST   maximum number of iterations

M1   value of J for the top-boundary grid line

M2   M1-1

M3   M1-2

MODE   indicator for coordinate system: = 1 for xy, = 2 for xr, = 3 for θr.

N   usually same as NF

NCVLX   number of x-direction control-volume widths in the domain

NCVLY   number of y-direction control-volume widths in the domain

NCVX(NZ)   number of x-direction control-volume widths in a zone

NCVY(NZ)   number of y-direction control-volume widths in a zone

NF   index denoting a particular ϕ

NFMAX   maximum value of NF

NI   maximum number of x-direction grid locations

NJ   maximum number of y-direction grid locations

NTC(NF)   actual number of repetitions in SOLVE

NTIMES(NF)   maximum number of repetitions in SOLVE

NZ   subscript for a zone

NZMAX	maximum number of zones
NZX	number of x-direction zones
NZY	number of y-direction zones
PLOTF	a 64-character variable for the name of the plotfile
POWERX	nonuniformity index for x-direction grid
POWERY	nonuniformity index for y-direction grid
POWRX(NZ)	nonuniformity index for x-direction grid in a zone
POWRY(NZ)	nonuniformity index for y-direction grid in a zone
PRINTF	a 64-character variable for the name of the printfile
PTX(I),PTY(J)	transformed coefficients in TDMA
QTX(I),QTY(J)	transformed coefficients in TDMA
R(J)	radius r for a grid point (I,J) for MODE = 2 or 3; =1 for MODE = 1
RELAX(NF)	underrelaxation factor α for F(I,J,NF)
RV(J)	value of radius r at the YV(J) interface
SC(I,J)	source term S_C. Eq. (5.13)
SMALL	a very small number such as 1.E-20
SP(I,J)	source coefficient S_P. Eq. (5.13)
SX(J)	scale factor for the x direction at grid locations Y(J)
TIME	time t for unsteady problems
TITLE(NF)	An 18-character title for F(I,J,NF)
VOL	volume of control volume
X(I)	value of x at grid location I
XCV(I)	x-direction width of the control volume
XL	x-direction length of the calculation domain
XU(I)	value of x at the control-volume face
XZONE(NZ)	x-direction length of a zone
Y(J)	value of y at grid location J
YCV(J)	y-direction width of the control volume
YCVR(J)	area rΔy for a control volume
YL	y-direction length of the calculation domain
YV(J)	value of y at the control-volume face
YZONE(NZ)	y-direction length of a zone

Appendix C

Default Values

Fortran name	Value	Explanation
ALAM(I,J)	1.	storage capacity of 1
BIG	1.E20	a big (almost infinity) value
CRIT(NF)	1.E−5	quitting criterion in TDMA
DT	1.E20	time step for steady state situations
F(I,J,N)	0.	all variables are initially zero
FLXCI1(J)	0.	$f_C = 0$ at $I=1$ boundary
FLXCJ1(I)	0.	$f_C = 0$ at $J=1$ boundary
FLXCL1(J)	0.	$f_C = 0$ at $I=L1$ boundary
FLXCM1(I)	0.	$f_C = 0$ at $J=M1$ boundary
FLXPI1(J)	0.	$f_P = 0$ at $I=1$ boundary
FLXPJ1(I)	0.	$f_P = 0$ at $J=1$ boundary
FLXPL1(J)	0.	$f_P = 0$ at $I=L1$ boundary
FLXPM1(I)	0.	$f_P = 0$ at $J=M1$ boundary
GAM(I,J)	1.	uniform Γ of 1
IBLOCK(I,J)	0	no solid blockages in the domain
ITER	0	starting iteration count
KBCI1(J)	1	specified value of ϕ at $I=1$
KBCJ1(I)	1	specified value of ϕ at $J=1$
KBCL1(J)	1	specified value of ϕ at $I=L1$
KBCM1(I)	1	specified value of ϕ at $J=M1$
KBLOC(NF)	1	perform block correction
KORD	2	higher-order treatment at boundaries
KOUT	3	output on screen and in PRINTF
KPGR	1	print grid coordinates
KPLOT(NF)	0	data not written on the plotfile
KPRINT(NF)	0	no field printout for the variable
KSOLVE(NF)	0	variable not solved
KSTOP	0	do not stop execution
LAST	5	perform 5 iterations
MODE	1	Cartesian coordinate system
NTIMES(NF)	10	perform 10 repetitions in SOLVE
PLOTF	PLOT1	name for the plotfile
POWERX	1.	uniform x spacing in the domain
POWERY	1.	uniform y spacing in the domain
POWRX(NZ)	1.	uniform x spacing in all zones

POWRY(NZ)	1.	uniform y spacing in all zones
PRINTF	PRINT1	name for the printfile
R(1)	0.	radius of the grid point $J=1$
RELAX(NF)	1.	no underrelaxation of the variable
SC(I,J)	0.	source term $S_C = 0$
SMALL	1.E–20	a small (almost zero) value
SP(I,J)	0.	source term $S_P = 0$
TIME	0.	starting time

Appendix D

User's Checklist for ADAPT

General

Provide DIMENSION and EQUIVALENCE statements to introduce convenient names for members of the F array.

GRID

Provide a HEADER and the names of a printfile and a plotfile. Then specify: MODE, L1, M1, (XU(I),I=2,L1), and (YV(J),J=2,M1). Also give R(1) if MODE ≠ 1. Note that 4 ≤ L1 ≤ NI and 4 ≤ M1 ≤ NJ. For a uniform grid, specify MODE, NCVLX, NCVLY, R(1) for MODE ≠ 1, XL, and YL, and call EZGRID. Simple nonuniformity can be introduced by using POWERX and POWERY. For more complex grids, you may use ZGRID with appropriate input.

BEGIN

Provide the values of KSOLVE(NF), KPRINT(NF), TITLE(NF), RELAX(NF), NTIMES(NF), CRIT(NF), etc. if the default values are not acceptable. Fill all the relevant F(I,J,NF) arrays with initial values, using the correct boundary values wherever they are known.

OUTPUT

Arrange any desired output. Also update the quantities that you wish to change every iteration or time step. You may call PRINT to get a field printout.

PHI

For each relevant NF, specify: ALAM(I,J), GAM(I,J), SC(I,J) and SP(I,J) for I=2,L2 and J=2,M2. Also supply the appropriate KBC values and the associated FLXC and FLXP values for the boundary points where the value of the dependent variable is not given.

Nomenclature

A	area
a	coefficient in the discretization equation
b	constant term in the discretization equation
c	specific heat
c_p	specific heat at constant pressure
D	diffusion conductance, Eq. (5.10); diameter in Chapters 9, 10, and 11
D_h	hydraulic diameter, Eq. (9.18)
f	friction factor, Eq. (9.21)
f_C	constant part of the linearized boundary flux, Eq. (5.39)
f_P	coefficient of ϕ_B in the linearized boundary flux, Eq. (5.39)
h	heat transfer coefficient
k	thermal conductivity
i	inertia term used for underrelaxation, Eq. (5.64)
J	diffusion flux, Eq. (3.10)
Nu	Nusselt number
P	TDMA coefficient, pressure
p	pressure
Q	TDMA coefficient; total heat flow rate per unit axial length
q	heat flux
r	radial coordinate
Re	Reynolds number
S	general source term, Eq. (3.6)
S_C	constant part of the linearized source term, Eq. (5.13)
S_P	coefficient of ϕ_P in the linearized source term, Eq. (5.13)
$\bar{S}$	average value of source S
T	temperature
t	time
T_b	bulk temperature, Eq. (9.30)
u	x-direction velocity, Eq. (9.1)
v	y-direction velocity, Eq. (9.2)
W	dimensionless velocity, Eq. (9.17)
w	axial velocity
$\bar{w}$	mean axial velocity, Eq. (9.19)
α	relaxation factor, Eq. (5.64); thermal diffusivity in Chapter 9
β	a factor in the boundary condition treatment, Eq. (5.30)
Δt	time step

ΔV	volume of the control volume
Δx	x-direction width of the control volume
δx	x-direction distance between two adjacent grid points
Δy	similar to Δx
ϕ	general dependent variable, Eq. (3.3)
Γ	general diffusion coefficient, Eq. (3.3)
λ	storage capacity, Eq. (3.4)
μ	viscosity
Θ	dimensionless temperature in Chapter 9
ρ	density

Subscripts

B	boundary grid point
E	neighbor in the positive x direction, i.e., on the east side
e	control volume face between P and E
N	neighbor in the positive y direction, i.e., on the north side
n	control volume face between P and N
nb	general neighbor grid point
P	central grid point under consideration
S	neighbor in the negative y direction, i.e., on the south side
s	control volume face between P and S
W	neighbor in the negative x direction, i.e., on the west side
w	control volume face between P and W; wall

Superscripts

0	old value (at time t) of the variable
*	previous-iteration value of a variable

References

Karki, K. C. and Patankar, S. V., 1988, Calculation Procedure for Viscous Incompressible Flows in Complex Geometries, *Num. Heat Transfer*, Vol. 14, p. 295.

Kays, W. M. and Crawford, M. E., 1980, *Convective Heat and Mass Transfer*, McGraw-Hill, New York.

Launder, B. E. and Spalding, D. B., 1974, The Numerical Computation of Turbulent Flow, *Comp. Methods Appl. Mech. Eng.*, Vol. 3, p. 269.

Lienhard, J. H., 1981, Heat Conduction through "Yin-Yang" Bodies, *J. Heat Transfer*, Vol. 103, p. 600.

Patankar, S. V., 1978, A Numerical Method for Conduction in Composite Materials, Flow in Irregular Geometries and Conjugate Heat Transfer, *Proc. 6th Int. Heat Transfer Conf.*, Toronto, Vol. 3, p. 297.

Patankar, S. V., 1980, *Numerical Heat Transfer and Fluid Flow*, Hemisphere Publishing Corp., Washington, D. C.

Patankar, S. V., 1981, A Calculation Procedure for Two-Dimensional Elliptic Situations, *Num. Heat Transfer*, Vol. 4, p. 409.

Patankar, S. V. and Acharya, S., 1984, Development of a Turbulence Model for Rectangular passages, *Trans. CSME*, Vol. 8, p. 146.

Patankar, S. V. and Baliga, B. R., 1978, A New Finite-Difference Scheme for Parabolic Equations, *Num. Heat Transfer*, Vol. 1, p. 27.

Patankar, S. V., Ivanovic, M., and Sparrow, E. M., 1979, Analysis of Turbulent Flow and Heat Transfer in Internally Finned Tubes and Annuli, *J. Heat Transfer*, Vol. 101, p. 29.

Prakash, C. and Patankar, S. V., 1981, Combined Free and Forced Convection in Vertical Tubes with Radial Internal Fins, *J. Heat Transfer*, Vol. 103, p. 566.

Settari, A. and Aziz, K., 1973, A Generalization of the Additive Correction Methods for the Iterative Solution of Matrix Equations, *SIAM J. Num. Analysis*, Vol. 10, p. 506.

Shah, R. K. and London, A. L., 1978, *Laminar Flow Forced Convection in Ducts*, Academic Press, New York.

Sparrow, E. M., Baliga, B. R., and Patankar, S. V., 1978, Forced Convection Heat Transfer from a Shrouded Fin Array with and without Tip Clearance, *J. Heat Transfer*, Vol. 100, p. 572.

Sparrow, E. M. and Patankar, S. V., 1977, Relationships among Boundary Conditions and Nusselt Numbers for Thermally Developed Duct Flows, *J. Heat Transfer*, Vol. 99, p. 483.

Sparrow, E. M., Patankar, S. V., and Shahrestani, H., 1978. Laminar Heat Transfer in a Pipe Subjected to a Circumferentially Varying External Heat Transfer Coefficient, *Num. Heat Transfer*, Vol. 1, p. 117.

Zhang, Z. and Patankar, S. V., 1984, Influence of Buoyancy on the Vertical Flow and Heat Transfer in a Shrouded Fin Array, *Int. J. Heat Mass Transfer*, Vol., 27, p. 137.

Index

Printed and bound by CPI Group (UK) Ltd, Croydon, CR0 4YY

23/10/2024

01778237-0005